Peterson Reference Guide to

Molt

in North American

Birds

THE PETERSON REFERENCE GUIDE SERIES

MOLT
IN NORTH AMERICAN
BIRDS

STEVE N. G. HOWELL

HOUGHTON MIFFLIN HARCOURT
BOSTON NEW YORK
2010

Sponsored by
the Roger Tory Peterson Institute
and the National Wildlife Federation

www.hmhbooks.com

Library of Congress Cataloging-in-Publication Data
Howell, Steve N. G.
Molt in North American birds / Steve N. G. Howell ; sponsored by the Roger Tory Peterson Institute
and the National Wildlife Federation.
p. cm. — (The Peterson reference guide series)
Includes bibliographical references and index.
ISBN 978-0-547-15235-6
1. Feathers—Growth. 2. Molting. 3. Birds—North America. I. Roger Tory Peterson Institute.
II. National Wildlife Federation. III. Title.
QL697.4.H69 2010
598.097—dc22
2009050643

Book design by Anne Chalmers and George Restrepo
Typefaces: Minion, Din

Printed in China
TWP 10 9 8 7 6 5 4 3 2 1

The legacy of America's great naturalist and creator of this field guide series, Roger Tory Peterson, is preserved through the programs and work of the Roger Tory Peterson Institute of Natural History (RTPI), located in his birthplace of Jamestown, New York. RTPI is a national nature education organization with a mission to continue the legacy of Roger Tory Peterson by promoting the teaching and study of nature and to thereby create knowledge of and appreciation and responsibility for the natural world. RTPI also preserves and exhibits Dr. Peterson's extraordinary collection of artwork, photography, and writing.

You can become a part of this worthy effort by joining RTPI. Simply call RTPI's membership department at 800-758-6841 ext. 226, fax 716-665-3794, or e-mail members@rtpi.org for a free one-year membership with the purchase of this book. Check out our award-winning website at www.enaturalist.org. You can link to all our programs and activities from there.

Dedicated to Phil Humphrey and the late Ken Parkes, for showing the way

And to Chris Corben, for holding the light steady

Peterson Reference Guide to

Molt

in North American

Birds

CONTENTS

PREFACE

Mark Twain is rumored to have said "Everybody talks about the weather but nobody does anything about it." Molt is the opposite: everybody does it but nobody talks about it. Yet given that feathers are critical to a bird, indeed, that feathers define a bird, it is surprising that there is little readily accessible literature pertaining to the molt strategies of North American birds. Some of this is surely because different authors have used (and continue to use) different terms to refer to molts and plumages. And the underlying physiological processes are not well known. In these respects molt is a little like migration—we may not fully understand how it works, and we may argue over terms to describe it, but regardless of these limitations we can observe it, describe it, and identify patterns.

So why should you care about molt? I assume that if you're reading this then you have at least some curiosity about the subject. And what exactly is molt? At its most basic level, molt is simply the regular, ordered growth of feathers, those lightweight structural features unique to birds. First and foremost, feathers cover and protect a bird, enabling temperature control of the body (which you and I achieve by changing clothes). The colors and patterns on feathers can signal a bird's sex, age, and breeding status; they may be bold and bright or cryptic (again, we use different clothes for different functions). And in those birds that perform the miracle of flight, the long and strong feathers of their wings and tail are rather important (we haven't yet manufactured clothes that enable us to fly).

But feathers, like clothes, are not permanent. They wear out from daily exposure to sunlight, water, vegetation, blowing sand, and various other influences, including feather-degrading bacteria.[1] Enter molt, one of the most important and fundamental processes in the life history of any bird: male or female, adult or young, breeding or nonbreeding, ostrich or hummingbird, migrant or resident, all birds have to molt regularly—or they die. Put simply, no molt = no feathers = no bird. But birds can't just go to the store and buy new feathers; they have to manufacture them, much as we do continuously with skin cells and hair.

What triggers molt? How often do birds molt? How long does a molt take? How fast do feathers grow? How long do feathers last? What is the cost of molt? When do birds molt? Where do they molt? By prompting so many questions, which I'll try to answer in this book, molt offers a fascinating window through which to appreciate how the lives of birds are built around the need to have a function-

Figure 1. The bright red epaulettes of a male Red-winged Blackbird (here of the "bicolored" subspecies *mailliardorum*) are an example of how plumage colors and patterns can be used for display. *Marin County, CA, 13 Mar. 2008. Steve N. G. Howell.*

Figure 2. Classic examples of cryptic plumage include species such as the American Bittern and nightjars, but even brightly colored plumage can be cryptic. Parrots are a well-known example—bright green birds are easily lost in green canopy—and even the complex "bold" patterns of a male Varied Thrush, shown here, can make it cryptic when feeding in the leaf-litter. *Marin County, CA, 31 Jan. 2007. Steve N. G. Howell.*

ing coat of feathers. In essence, molt relates in some way to everything a bird does—to factors such as its habitat, food, clutch size, migration distance, body size, and even its ancestry.

An understanding of molt can also enable us to establish the age of birds and find ways to separate different species simply by looking at their plumage and their patterns of molting. Knowing the age structure of a bird population, either on the breeding grounds or the nonbreeding grounds, may allow us to evaluate its health (such as, are there too many old birds, not enough young birds?) and, indirectly, the health of the environment—which we share with all other animals. Alternatively, molt timing can offer insight into challenging species identifications, such as between Dusky and Hammond's flycatchers, between Cliff and Cave swallows, and perhaps even between Fea's Petrel and the mythical Zino's Petrel (*Pterodroma madeira*), which has yet to be conclusively documented in North American waters. Hence, a knowledge of molt can have varied practical applications.

Notable contributions to the descriptive study of molt include the pioneering work of Jonathan Dwight,[2] the seminal study (in German) of Erwin and Vesta Stresemann,[3] and the recent compendium by Lukas Jenni and Raffael Winkler.[4] The thrust of Jenni and Winkler's *Moult and Ageing of European Passerines* is how to determine the age of birds caught for banding and study, but the book's introduction offers an excellent overview of literature pertaining to molt and plumage function. Although the book was written for a European audience, its introductory sections are well worth reading for anyone with a serious interest in molt.

The groundbreaking work of Phil Humphrey and Ken Parkes,[5,6] together with recent modifications to their system,[7,8] has enabled molt studies to proceed within a helpful, comparative framework. However, other than a number of careful studies by Ernest Willoughby, numerous publications by Sievert Rohwer and colleagues, and several family- or group-level revisions by Peter Pyle, relatively little critical information has been published about molt over the past 25 or so years. Molt is still an understudied and underappreciated subject. Indeed, entire family monographs have been published in recent years (including one on owls and one on wrens, dippers, and thrashers) in which molt has not even been mentioned!

Although molt may at first glance seem like an overwhelmingly varied and messy phenomenon, the molt strategies of North American birds are very orderly and are built on only four underlying plans. Recognizing this order in the universe and relating it to the birds around us is what this book is about.

REFERENCES

1. Burtt and Ichida 1999; **2.** Dwight 1900b; **3.** Stresemann and Stresemann 1966; **4.** Jenni and Winkler 1994; **5.** Humphrey and Parkes 1959; **6.** Humphrey and Parkes 1963b; **7.** Howell et al. 2003; **8.** Howell et al. 2004.

Figure 3. Feathers aren't permanent. They wear out from exposure to sun, vegetation, blowing sand, and other influences. Without molt, feathers would soon become too worn to function, and a bird would die. The feathers of this first-cycle Brant have become heavily worn and will soon need to be renewed in the bird's annual molt. *Marin County, CA, 9 July 2008. Steve N. G. Howell.*

INTRODUCTION

HOW TO USE THIS BOOK

This book explains molt strategies and relates them to the life histories of North American bird families that occur regularly north of Mexico. Some of what is presented is conjectural, aimed to stimulate discussion and perhaps even to prompt studies that could address some of the many unknowns in this fascinating field. This book is not an overview or justification of different molt terminologies, nor is it an attempt to explain the physiology of molt. And although feathers are largely responsible for the appearance of a bird, this book is not about the colors and patterns of plumage, although these are obviously related to molt and are discussed in some family accounts.

PLAN OF THE FAMILY ACCOUNTS

The family accounts start with an overview of factors such as biogeography, taxonomy, habitat, migration, and other things that might be reflected in the molting strategies of family members. Next, the molting strategy or strategies of the family are noted, as well as when and where the different molts occur. Longer discussions usually follow, often broken into one or more sections that cover the different molts and relate them to aspects of life history and ancestry. In some cases, there is additional treatment of case studies, groups of particular interest, and related subjects.

Information comes from summaries by Pyle[1,2] as well as from the Birds of North America series,[3] *Handbook of the Birds of Europe, the Middle East, and North Africa,*[4,5,6,7,8,9,10,11] and *Handbook of Australian, New Zealand, and Antarctic Birds.*[12,13,14] Other, more specific references have also often been used and are cited in the relevant family accounts. Although molt is an understudied subject, far more papers have been published on the subject than could be covered in this book. In discussing different aspects of molt and plumage, I had to pick and choose papers that treated the points being made, and I apologize to any authors whose work I may have overlooked.

The recognition of families in this book generally follows *Handbook of the Birds of the World,*[15] which differs in a few instances (for example, the upland gamebirds) from that of the American Ornithologists' Union. Following the family heading, in parentheses, are the molt strategies of the family (listed from most frequent to least frequent) and the number of species that occur regularly in North America.[16,17,18]

Figure 4. Feathers define a bird, and most species use them to fly, such as in search of food (like this Peregrine Falcon) or to avoid becoming food (like these Western Sandpipers). *Marin County, CA, 27 Aug. 2008. Steve N. G. Howell.*

As with any subject, it's helpful to start with definitions—once we agree on terminology, things go more smoothly. As a way of introducing the subject, I'll approach some of the definitions through a series of questions and answers (for recaps of the definitions, see the glossary). You may already know a lot of the information in the answers—but it may be presented here with a different spin.

What exactly is molt? What triggers it? What is a molt?

Although molt (or molting) is often thought of as feather replacement, it is really the systematic process of feather growth. For example, a bird's juvenile plumage (its first coat of "real" vaned feathers after any downy stages) is acquired by molt, even though no feathers may be replaced. The loss of feathers in later molts is usually a passive byproduct of new feathers growing in and pushing out the old ones. Feathers grow from follicles in the skin, much as our hair does, and distinct plumages are composed of distinct coats of feathers; thus, molt in birds is a cyclic process rather than a continuous renewal process as with human hair. The accidental loss of feathers, such as when a dove loses its tail to a hawk or a dog, is not considered molt. In such cases, feathers typically grow back right away rather than delaying growth until the next molt.

Molt is also a dynamic evolutionary process, and what we see today reflects millions of years of ongoing fine-tuning. In some cases, a bird's molt strategy may not make sense to us, but this may be because there has not been enough pressure to change a strategy that worked well thousands of years ago.

The processes involved in feather growth and in what triggers it are fascinating but mostly beyond the scope of this book. To anyone interested in these questions, I recommend A. A. Voitkevich's classic work *The Feathers and Plumage of Birds*.[19] Basically, though, we still don't really know what triggers molt in most species, other than a variety of factors that can include length and intensity of daylight, temperature, and the activation or suppression of certain hormones, including those relating to breeding. For example, even small changes in daylight length can stimulate molt. In one experiment, ptarmigans exposed to summerlike daylight regimes in winter molted into a colorful "breeding" plumage; when exposed to fall-like daylight regimes in spring, they molted back into a white "nonbreeding" plumage.[20] In another experiment, Dark-eyed Juncos implanted with testosterone completely failed to molt after the breeding season, presumably because artificially high testosterone levels blocked the triggers for molt.[21]

The question "What is *a* molt?" is easy to answer in theory but harder to answer in practice. By definition, a given molt involves the activation of feather follicles. For two different molts to occur, some follicles must be activated twice so that two generations of feathers are produced. Problems in distinguishing between different molts arise when birds have protracted molts and do not replace all of their feathers at one time, or when different molts overlap in timing, such as happens in large gulls[22] and certain shorebirds.[23] Even in some small songbirds, which are assumed to have discrete molting periods, the complete molt after breeding may be protracted: most of the plumage is replaced in fall, but low-level molt continues into the winter, when up to a few hundred additional small body feathers may be acquired.[24]

For now, then, it is good to accept that much remains to be learned about the physiology and genetics of molt, and that our definitions are simply attempts to put nature into boxes of convenience.

Are all feathers the same? Which feathers are which?

Most feathers can be divided into two main types, although there is effectively a continuum between the two: downy (also known as plumulaceous) and nondowny (or pennaceous). Downy feathers (such as those in some pillows and sleeping bags) are soft and weak with a poorly developed shaft and long loose barbs that are not interlocking. Nondowny feathers (such as those used for writing quills) are also known as contour feathers, because they produce the contours, or outline, of the bird. They have a stronger, well-developed shaft, and their barbs are firmly textured and interlocked to form a firm vane. The vanes of a typical body feather grade from a looser, silky or downy base out to the exposed, firmer tip; even if the pennaceous tip makes up only a small fraction of a feather, it is conventional to consider it a contour feather. In some species, such as the American Goldfinch, the body feathers themselves can be heavier and denser in winter than in summer,[25] presumably to provide extra insulation during colder conditions.

In terms of molt and molt strategies, it is the contour feathers we are talking about, which are also what we can see in the field. As in everyday birding, it's helpful when talking about molt to have an idea of the main feather groups. Among contour feathers, the main feathers of the wing are called primaries (or primary flight feathers) and secondaries (or secondary flight feathers). The primaries form the "hand" of the wing, and the secondaries form the "arm" of the wing; the innermost secondaries (those nearest the body), which act as coverts for the closed wing, are called tertials. The main feathers of the tail are known as rectrices, or simply tail feathers, and the central pair act as coverts when the tail is folded closed. The main feathers of the wings and tail are often collectively known as flight feathers. Other feather groups mentioned in the family accounts are shown in Figures 5–10 (also see the glossary).

Two other types of specialized feathers are known as filoplumes and bristles, but neither has been studied much in terms of molt. Filoplumes are hairlike feathers that expand slightly at the tip, where they have one to six short barbs. They are distributed throughout a bird's plumage (but not evenly) and have a sensory function that relates to adjusting the position of nearby feathers; motion detected at the tips of filoplumes is transmitted through the long shaft to sensory corpuscles at its base, which in turn signal the muscles that control the movement of feather tracts. Bristles consist simply of stiff hairlike shafts with a few basal barbs and, like filoplumes, have sensory corpuscles at their base. With few exceptions, bristles are confined to the heads of birds, such as the rictal bristles that surround the gape of many flycatching birds and help them detect and net their prey. The naked faces of raptors and the naked heads of vultures are actually "feathered" with bristles, which may be easier to keep clean of gore than are typical feathers.

How do feathers and plumages develop?

Although feathers cover a bird's entire body, they are not attached evenly to the skin. Rather, they grow in patches, known as tracts, from which they spread out to cover naked areas between the tracts. (In exceptional cases, such as penguins, feathers are distributed over the entire skin surface.) By growing in tracts, feathers are more easily controlled by muscles, which can raise or lower feathers for functions such as preening, display, and thermoregulation. Feathers develop within a protective wax and keratin sheath, or "pin," that projects from the follicle. The mass of these sheaths amounts to about 20 percent of the mass of the new plumage,[26] which suggests that the sheaths are remarkably important for protecting the growing feathers. As each feather grows, it emerges from the sheath and expands to its full width, a little like a butterfly emerging from its chrysalis. When the feather is fully grown, the sheath flakes off from the feather base. At this point the feather is cut off from the bird's circulatory system and is dead tissue, no longer able to grow. Further discussion of the physiology of feather growth and pigmentation, and of different feather types and structures, can be found in other references.[27,28]

Before acquiring their first coat of contour feathers, most young birds hatch from the egg either

Figure 5. First-cycle Double-crested Cormorant. Note how the scapulars cover the joint between the wings and the body. The longest and largest four scapulars are also known as the subscapulars. *Sonoma County, CA, 23 Dec. 2008. Steve N. G. Howell.*

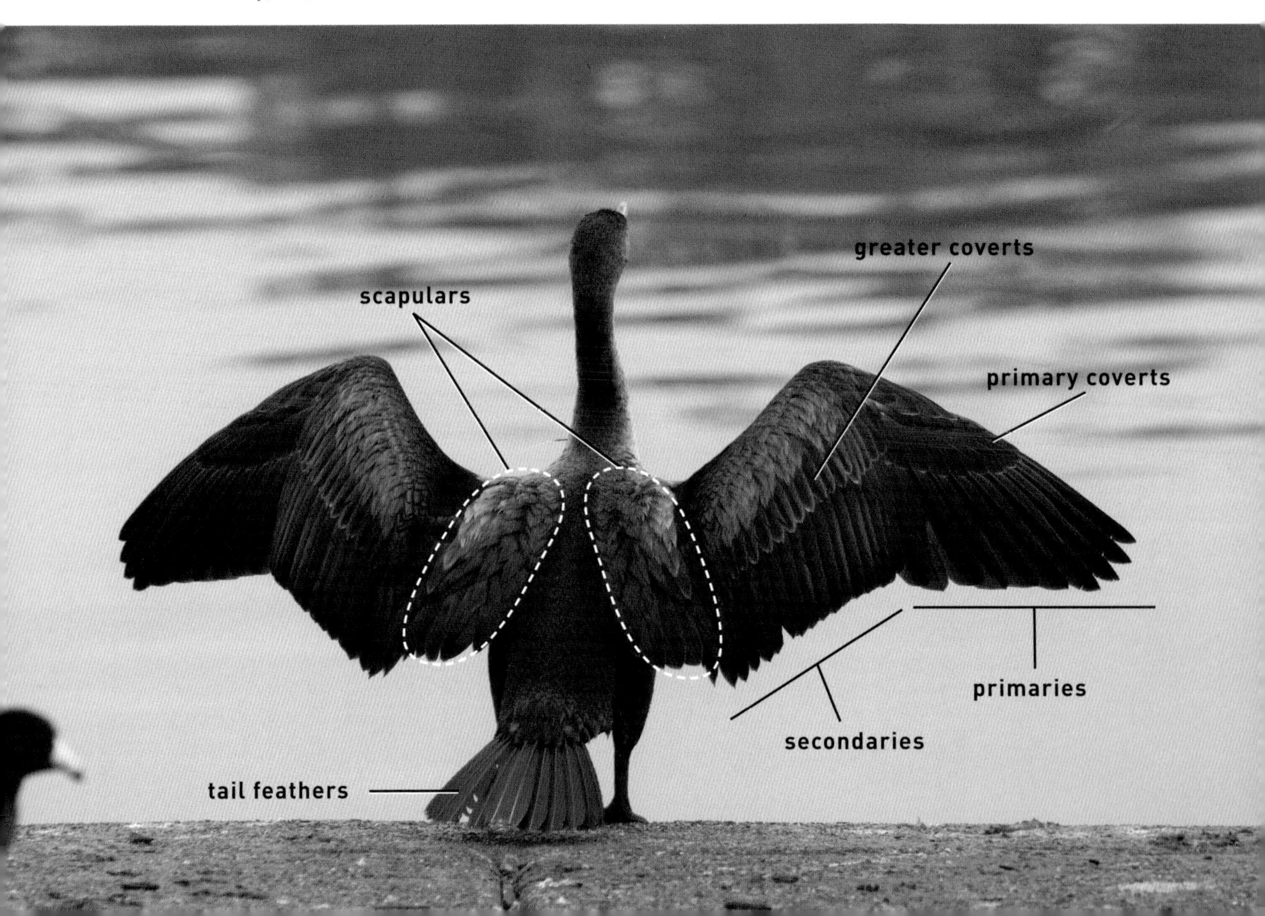

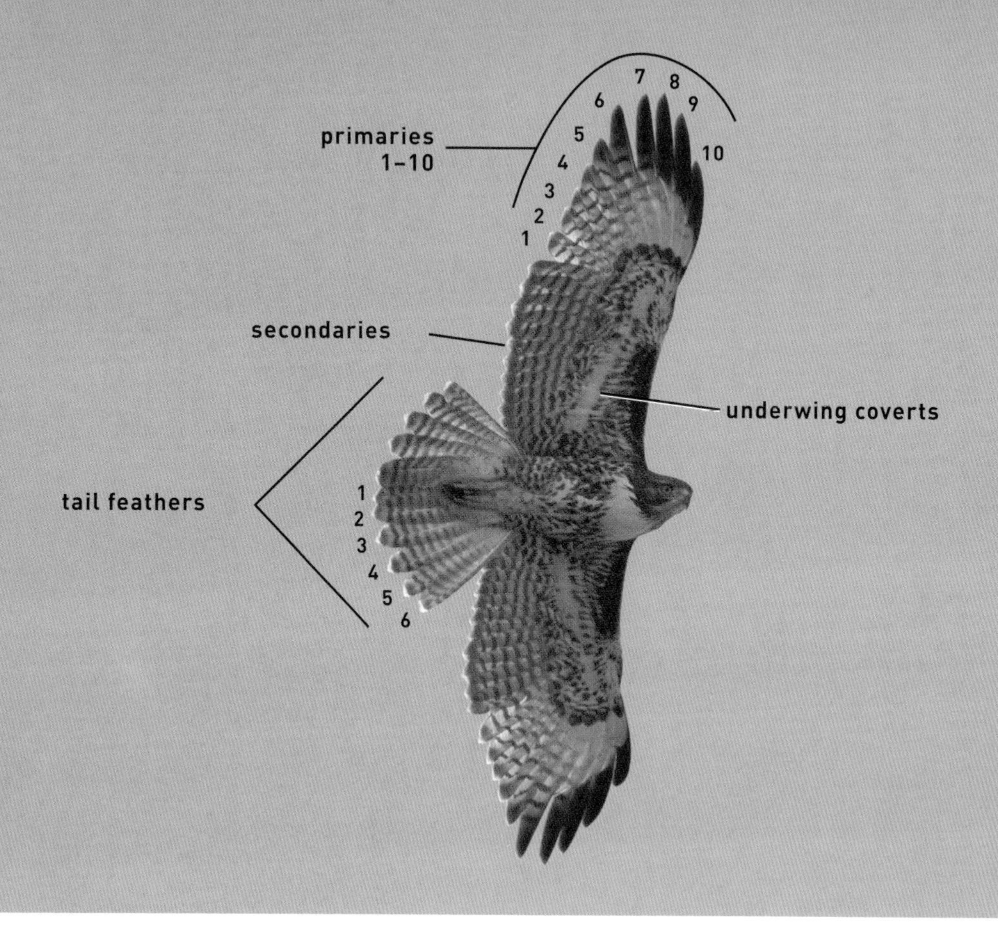

Figure 6. Juvenile Red-tailed Hawk. At this angle, the break between the primaries and secondaries is easy to see. *Marin County, CA, 6 Sept. 2008. Steve N. G. Howell.*

Figure 7. Juvenile Baird's Sandpiper. Note how the tertials (the innermost secondaries) slide under the scapulars. *Marin County, CA, 19 Aug. 2008. Steve N. G. Howell.*

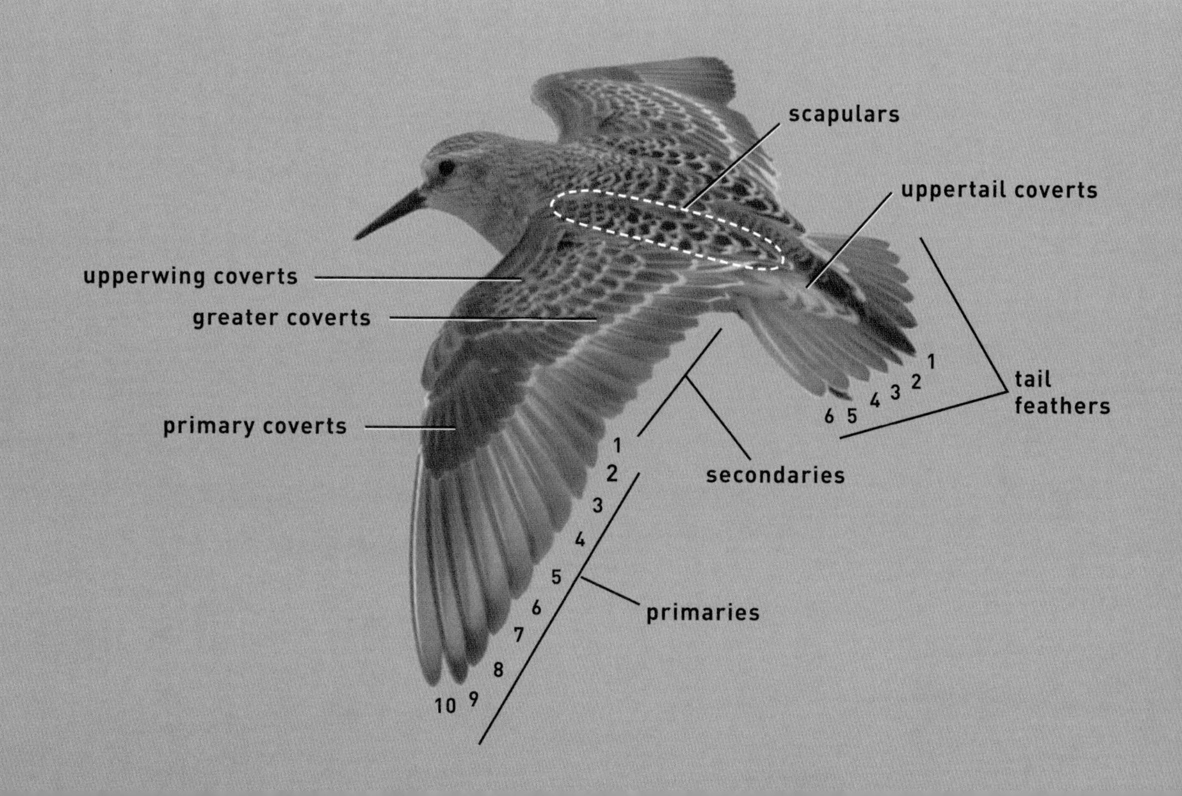

Figure 8. Juvenile Baird's Sandpiper at rest, with most of the upperwing coverts covered by the scapulars and the fluffed-out flanks. The secondaries are hidden, and only the tips of the longest primaries are exposed. *Marin County, CA, 23 Aug. 2008. Steve N. G. Howell.*

Figure 9. Hermit Thrush. Note how the tertials overlie and protect the other secondaries. Similarly, on a closed tail the central tail feathers protect the other tail feathers, which fold in like a fan and are stacked underneath. Because they are relatively exposed, the tertials and central tail feathers tend to be more frequently molted than other flight feathers. *Nayarit, Mexico, 14 Jan. 2009. Steve N. G. Howell.*

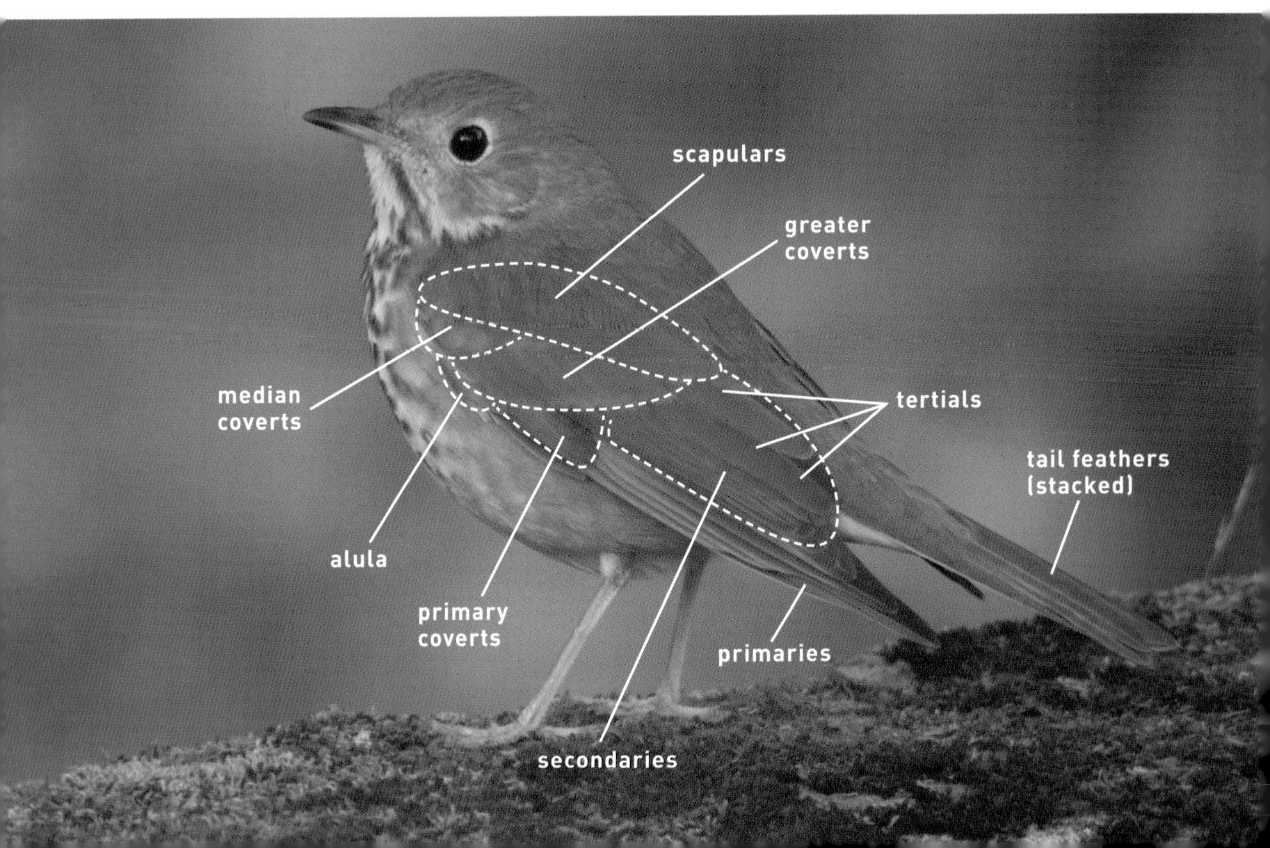

Figure 10. One nice thing about molt is that you don't need to learn the varying names of all those different head stripes. Just knowing some basic feather groups is enough. On this White-throated Sparrow, note how the tertials overlie and protect the closed secondaries and how the primaries are mostly covered by the secondaries. *Marin County, CA, 28 Jan. 2009. Steve N. G. Howell.*

Figure 11. A growing feather is initially protected in a keratin sheath, from which the feather breaks out and expands. Before the feather emerges, this white sheath is called a pin, and the feather is termed a pin feather or as being "in pin." In low-level molt, pin feathers are not easily seen on birds in the field, but on birds in heavy molt, such as this adult male Indigo Bunting, many pin feathers are apparent on the head, and there are two obvious pins in the greater coverts, most of which have been shed. *Petén, Guatemala, 5 Mar. 2008. Steve N. G. Howell.*

Figure 12. On this Anna's Hummingbird, feathers that can be seen breaking from their sheath, or pin, include several underwing coverts and p8 (p7 is mostly grown, p1–p6 are new, and p9–p10 are still to be molted). *Marin County, CA, 24 Aug. 2008. Steve N. G. Howell.*

Figure 13. North American birds hatch in various stages that range from naked and helpless, as in songbirds, to down-covered and able to run around, like this American Avocet chick. *Sonoma County, CA, 3 July 2007. Steve N. G. Howell.*

naked or covered with varying degrees of downy feathering. Hatchling birds have been classified under four main headings,[29] which effectively compose a continuum. The most "advanced" state is precocial, whereby birds hatch down-covered with their eyes open and can leave the nest within a day or two; examples include waterfowl, quail, loons, grebes, cranes, rails, and shorebirds. The kiwis of New Zealand go one step further and hatch wearing what is considered their juvenile plumage.[30] Semi-precocial young are similar but remain in or near the nest, although they are able to walk; examples include jaegers, gulls, terns, alcids, and nightjars. Semi-altricial young hatch down-covered but do not leave the nest; these include petrels, herons, ibises, hawks, falcons, and owls. Lastly, altricial young hatch with their eyes closed, have little or no down, and are unable to leave the nest; these include pelicans, cormorants, woodpeckers, kingfishers, and songbirds.

The first feathers of most young birds are downy, as typified by ducklings and nestling songbirds. Depending on rates of body growth and physiological maturity (which vary greatly among different species), at some point the follicles will produce a coat of vaned feathers, which is known as the juvenile plumage. Most species (such as gulls and sandpipers) have one downy plumage before juvenile plumage, although a few species (such as penguins[31] and flamingos[32]) have two downy plumages. Most songbirds hatch almost naked, and their wispy down is soon replaced by juvenile plumage, much of which is relatively weak and soon replaced by another molt. And in some species, such as woodpeckers and kingfishers, naked young graduate immediately to vaned feathers and molt directly into their juvenile plumage with no preceding downy phase. Thus, the juvenile plumage may represent the first, second, or third generation of feathers worn by a young bird. The plumages that follow juvenile plumage are all pennaceous, although many species have insulating down feathers (which grow from their own follicles) that underlie the protective cover of vaned feathers.

How many feathers does a bird have?

Perhaps not surprisingly, few studies have been done on this subject—imagine the time and effort involved in plucking and counting every single contour feather on even a small songbird! Even more rarely has anyone counted the number of contour feathers on multiple individuals of the same species. In the 1930s, however, Marie Siebrecht undertook the painstaking work of counting the feathers on 150 landbirds of 79 species from eastern North America.[33]

Most of these species had 1,200 to 2,700 feathers, with a male Ruby-throated Hummingbird having only 940 and a female American Robin having 2,973. The feathers usually made up 4 to 8 percent of a bird's total weight. Feather counts can be individually variable (such as two female Ruby-crowned Kinglets in mid-October having 1,119 and 1,289 feathers), and there is often a trend for birds to have more feathers in winter and fewer in summer. Seasonal differences can be most pronounced in birds that remain year-round in northern latitudes, such as in the Carolina Chickadee, with 1,704 feathers in February versus 1,256 to 1,309 in October. Presumably this trend is linked to keeping birds warm in winter. Among migrants that move south to winter in warmer climes, such as wood-warblers, seasonal differences are less pronounced.

The trend of having fewer feathers in summer may also reflect, at least in part, the formation of brood patches, whereby belly feathers are lost to allow more efficient heat transfer from adults to

Figure 14. The juvenile plumage of many shorebirds is relatively durable and can carry them through long-distance migrations, such as from the Arctic to southern South America. These Baird's Sandpipers in Chile are in somewhat worn juvenile plumage and are about to start molting. Through fading and wear, their juvenile plumage has lost the buff wash to the chest and the bold scaly patterning to the upperparts that we typically see on fresh-plumaged fall juveniles in North America (compare with Figure 8). *Tierra del Fuego, Chile, 24 Oct. 2006. Steve N. G. Howell.*

Figure 15. The juvenile plumage of many songbirds, such as this Bewick's Wren, is relatively soft and lax. These weak juvenile feathers of the head, body, and upperwing coverts soon get replaced by stronger feathers, although the bird's general appearance, or plumage aspect, doesn't change much. *Marin County, CA, 23 June 2007. Steve N. G. Howell.*

eggs and nestlings. Interestingly, the loss of feathers to create a brood patch is independent of molt and does not immediately stimulate new feather growth. In fact, brood patches are a little-studied phenomenon in terms of how feathers are shed and what triggers a brood patch to be re-feathered. Are brood patches re-feathered immediately after the young fledge, and how does this process relate to other molts? These are questions that remain to be answered fully.

Even fewer counts have been made of contour feathers on large birds. A female Mallard had 11,903 feathers,[34] and a Tundra Swan was estimated to have 25,216 feathers[35] (estimated, because only one wing was counted and the other was assumed to be the same). On the swan, 20,177 feathers were on the head and neck and only 5,039 on the rest of the bird, and the feathers made up about 10 percent of the bird's total weight.

More attention has been paid to the number of flight feathers, as these are easy to count. Thus, it is well known that most bird species have 10 primaries, which are numbered from the innermost (primary 1, or p1) out to the outermost (p10). A few families of birds have 11 primaries (such as grebes and flamingos) or 9 primaries (most notably, several songbirds). Some literature maintains that many species have 11 primaries, with the vestigial p11 being outermost; this enigmatic feather, known as the remicle, also may represent a spurious covert.[36] As a rule, variation in wing length, especially among similar species, reflects the number of secondaries. Most songbirds have 9 secondaries, but among other birds the number ranges from 6 in hummingbirds to more than 30 in some albatrosses.

The number of tail feathers, or rectrices, is also fairly consistent among bird families. Most birds have 12 rectrices, 6 on either side of the midpoint (see Figures 6 and 7), but some have 10 (such as hummingbirds), and many larger waterbirds have more, such as 14 to 24 in waterfowl and 19 to 24 in pelicans.

Do birds have to molt? How often do they molt?

Because feathers wear out, all birds have to molt. In some species the start of molt can be delayed if insufficient food is available,[37,38] but eventually birds have to molt—or they will die. In an experiment with controlled feeding to White-crowned Sparrows,[39] well-fed birds and all but the most severely malnourished individuals started to molt on a normal schedule. The drive to molt was so strong that feather renewal continued even when malnourished birds were growing deformed feathers and even up to the point that some died of malnutrition.

If there is not enough food to molt and raise young in a given year, a bird may defer breeding, but it still needs to molt. For example, some high-latitude breeders such as Brant do not nest every year if conditions are unsuitable,[40] and some seabirds in the California Current, such as Brandt's Cormorant, regularly skip breeding in years with insufficient food.[41] So at some level, molt perhaps can be more important than breeding for an individual bird.

Integral to an understanding of molt is the concept of cycles. Molt, like breeding and migration, is a cyclic activity—that is, something repeated in a usually predictable manner. Most molting and breeding cycles correspond to annual cycles; after all, the year is a reality of Earth's orbit around the sun, not an abstract concept. Winter and summer are familiar temperate-zone seasons that have a cyclic regularity, whereas tropical regions tend to have wet and dry seasons that also have an annual cycle.

As a rule, then, all birds molt once a year, during which they replace all or most of their feathers. The plumage produced by this molt common to all birds is appropriately called the basic plumage (which is produced by the prebasic molt). Some species have one or more additional molts in their annual cycle, during which they molt some feathers again. These additional molts may simply reflect the need to replace worn feathers, but in some cases they also result in the acquisition of feathers with different colors and patterns than those of the basic plumage.

Few birds have molting regimes that are not annual. Exceptions include the 2-year cycles of some southern albatrosses,[42] and the 9- to 11-month cycles of some tropical gulls and terns,[43,44] which thus could have four molt cycles in 3 years. As a rule, however, molt cycle and annual cycle are largely synonymous.

How long does a molt take? How fast do feathers grow?

Although the time taken for a complete molt typically reflects the size of the bird, other factors come into play (see Body Size and Wing Area, pages 48–49). For example, resident species of songbirds tend to have more protracted molts than do long-distance migrants, perhaps because the former have no pressure to complete their molts to accommodate a migration schedule. However, shorebirds wintering in northern latitudes tend to molt quickly before winter sets in, whereas shorebirds migrating to tropical or southern latitudes can have protracted molts. In many species, the time taken to molt the primaries is a good index of molt duration, because all or most of their other feathers tend to be replaced during the period of primary molt.

Figure 16. Most bird species breed and molt on an annual cycle. Exceptions include some southern albatrosses, which have a 2-year cycle, and some tropical terns such as this Black Noddy, which can have four cycles in 3 years. *Truk, Micronesia, 23 Apr. 2008. Steve N. G. Howell.*

Most smaller temperate-zone songbirds require only 1 to 2 months for primary molt, but the largest songbird, Northern Raven, may take 4 to 5 months for a complete molt.[45,46] Small tropical songbirds, however, may have more protracted wing molts that can last 3 to 6 months.[47] The shortest molts known among songbirds are those of two Arctic breeders, the Bluethroat and Snow Bunting, which require only 27 to 37 days for their primary molt.[48] Because these two species feed on the ground, they can afford to shed many primaries at a time and in effect become virtually flightless for a short period during the late-summer Arctic food bloom.

Among non-songbirds, most small to medium-large species require about 2 to 6 months for a complete molt, but some long-winged species may require 2 to 3 years or longer to completely renew their primaries (see Stepwise Wing Molt, pages 36–45); they still molt their head and body feathers each year, along with as many flight feathers as they can replace in the time available.

How quickly feathers grow depends on their size (such as their length, width, and weight) and structure (whether they are lighter and softer body feathers, stronger and harder wing feathers, or specialized feathers) and on the quantity and quality of food available to a molting bird. Still, there is a limit to the rate at which feathers can grow. The most easily measured growth rates apply to the long feathers of the wings and tail. In smaller birds, such as songbirds, the longest wing feathers grow about 2 to 5 millimeters per day; in larger birds, such as ducks, hawks, and cranes, they grow about 4 to 10 millimeters per day.[49]

One way of estimating feather growth rates is to examine growth bars, which are the alternating dark and light bands usually reflected on a plain feather surface turned in the light. The dark bands represent feather material grown during the day, whereas the light bands indicate nocturnal growth. Thus, one pair of bands represents a day's growth, and the total number of pairs indicates the number of days taken for a feather to grow.[50] This phenomenon, which is analagous to that of growth rings in trees, is termed "ptilochronology." A book was published in 2006 that explores how measuring differences in feather growth rates (based on growth bars) may provide an index to a bird's nutritional condition, and in turn perhaps even to environmental quality.[51]

Although it has been suggested that the primaries of individual birds grow in length at a uniform rate per day,[52,53] this is unlikely to be so for several reasons. It appears that feather mass, not length, increases at a steady rate, such that the rate of feather mass synthesis may be a limiting factor in feather growth rates.[54] Thus, shorter but heavier flight feathers will grow more slowly than longer but lighter feathers, as was found in a study of Barn Owls.[55] Feathers with a specialized structure, such as the elongated central tail feathers of a Red-tailed Tropicbird, also may grow more slowly than would a "normal" feather of the same length.[56] It has also been reported that the rate of growth can be greater in longer feathers than in shorter feathers, such as in the tail feathers of a Fork-tailed Flycatcher,[57] although how such differences in this species might be linked to feather width is unclear. And, not surprisingly, feathers grow more slowly if birds are malnourished.[58] Lastly, there also may be a limit to feather growth rates depending on how many feathers are growing at one time.

Figure 17. The Snow Bunting has one of the quickest prebasic molts of any songbird. In the late-summer food flush of the high Arctic, these birds can feed almost 24-7, and they even may become almost flightless in their rush to complete molt before cold weather sets in. The mostly dark primary coverts of this individual indicate it is a female. *Park County, CO, 9 Nov. 2008. Bill Schmoker.*

Figure 18. Specialized feathers often take longer to grow than do normal feathers. The wirelike tail streamers of a Red-tailed Tropicbird require almost 6 months to reach their full length. These feathers are molted alternately, and both tend to be full length only for a short period during courtship. The short streamers of this individual would thus indicate it either has a nest or is not yet ready for courtship and breeding. *Norfolk Island, Australia, 29 Mar. 2007. Steve N. G. Howell.*

How long do feathers last?

How long a feather lasts is a function of its structure, which often relates to its coloration, or pigmentation. The dark pigments known as melanins strengthen a feather because they comprise resistant granules deposited in the feather structure.[59] In turn, environmental factors such as diet play an important role in feather pigmentation; for example, certain amino acids need to be present in a bird's diet for it to synthesize certain pigments.

As a rule, darker areas on feathers are stronger than paler areas. This explains many of the common plumage patterns we see—the long wing feathers are usually the darkest part of a bird's plumage because they are critical for flight and need to be durable. Note how on most gulls it is only the portions of the wingtip exposed when a bird is resting that are black—this is cost-efficiency (Figure 19).

Feathers tend to be a little more durable than perhaps they need to be, but not much. It's all about balancing budgets: why pay more for something than you have to? Because molt has evolved as a cyclic, usually annual, event, the average life of most feathers is about a year. But feathers do not fade and wear at an even rate. By watching the resident birds in your yard, you can see that their feathers tend to stay in pretty good condition from fall through spring or even into early summer, but they start to deteriorate quickly in summer, shortly after which they are molted. Fading is the loss of color through exposure to sunlight, whereas wear is the physical deterioration of a feather—which is accelerated when a feather is weakened by fading.

Feather life can vary depending on how exposed a feather is. On juvenile gulls, for example, the less-exposed inner primaries are usually weaker (and paler) than the outers, and they degenerate more quickly;[60] the outer primaries, which are molted up to 4 months later than the inners, tend to be darker and stronger. In large birds that cannot replace all of their wing feathers in an annual molt, the outer primaries can last 2 or 3 years (by which time they tend to be pretty worn), or the birds may have a different molt strategy, in which the exposed outer primaries are the first to be molted, as happens with immature albatrosses.

In some cases, a careful observer might ask: why do birds molt feathers that aren't very worn? Surely they could last a little longer. The answer usually relates to fitting molt into the annual cycle. The feathers may not be too worn now, but if a bird has to migrate and breed before it can fit in another molt, then by that time the feathers might well be too worn to function efficiently.

What is the cost of molt?

The cost of molt is difficult to measure, and the two types of cost involved are not easy to separate. First is the cost of molt itself, the growth of new feathers. Second is the cost of maintaining other functions such as temperature control and flight, which are impaired when the feather coat is incomplete.

Figure 19. Franklin's Gull is one of two North American breeding species that undergo two complete molts in a year—one in late summer, which this bird is starting (the inner primaries have been shed), and one in late winter on the nonbreeding grounds in South America. Black pigment is relatively resistant to wear, and the area demarcated by the black wingtips corresponds to the exposed portion of the primaries when the bird is at rest. *Bottineau County, ND, 16 June 2008. Steve N. G. Howell.*

A further difficulty in measuring the cost of molt is that the prebasic molt corresponds to a metabolic overhaul for the bird, including widespread tissue renovation that even involves the skeleton.[61] The energy expenditure calculated for birds undergoing prebasic molt thus reflects far more than simply feather synthesis, which in and of itself may not be that costly. For example, on plucked White-crowned Sparrows in winter, birds regrowing feathers expended an amount of energy corresponding to less than 3 percent of their resting metabolic rate, compared with 25 to 54 percent expended by sparrows during their prebasic molt.[62] Whether molts other than the prebasic molt are accompanied by additional tissue renovation has not been investigated. If extra molts simply involve feather synthesis, then their cost may be relatively trivial in terms of energy expenditure.

The cost of molt may also reflect the bird's body size and the speed of its molt. A rapid molt completed in a few weeks may be more stressful than a protracted molt completed over a few months. Energy demands also vary depending on which feathers are being grown. The small body feathers can be molted with relative ease, whereas the long wing and tail feathers require more energy. And although molting is often viewed as a period of nutritional stress, this is not necessarily so if food is plentiful, as with molting Snow Geese that actually gain weight before migrating.[63] In regions with distinct seasonal peaks of food resources, molt and breeding can overlap if there is only a finite window during which both activities need to occur, whether the birds are Arctic-breeding Ivory Gulls,[64] temperate-breeding California Condors,[65] or tropical-breeding hummingbirds.[66] It's all about food, which brings us to the next questions.

When do birds molt? Where do they molt?

The answers to these intertwined questions depend on several variables, but they basically come down to food. Hence, molt regimes tend to be timed to correspond with predictable peaks in food supply, which in turn reflect bigger-picture variables such as climate. Breeding may even be constrained by the optimal molting season. In many tropical species, molt is a relatively regular phenomenon concentrated around peak periods of the most predictable food supply, whereas breeding is a more opportunistic and potentially protracted activity.[67,68]

Most food chains start with photosynthesis, and in North America food for birds is generally least plentiful in midwinter. At this season there is less and weaker sunlight, and birds need their energy to keep warm. Thus, few birds molt in midwinter in North America, and many migrate south at this season to seek food in warmer climates. Most molting in North American birds thus occurs from early spring through late fall. For birds that migrate south in winter to the tropics or to the Southern Hemisphere, however, molting can occur during the northern winter but in areas with ample food and a warmer climate.

Conventional wisdom states that molt and breeding do not overlap, because both are considered to be energy-demanding activities of a bird's life cycle (but see page 13, What is the cost of molt?). The dogma of molt–breeding separation has been derived mainly from studies of small songbirds in temperate regions with well-defined seasons. Such species can usually wait until their young have fledged and still have time to replace all of their relatively small feathers before migration or before winter sets in, although even some species of sparrows can overlap breeding and molting.[69]

As a rule, larger species have more and larger feathers and usually require more time to molt than do small birds. Thus, many larger birds often exhibit varying degrees of overlap between molting and breeding. They still tend to suspend or reduce molt during some phases of breeding (such as egg production and laying, or feeding of young), but otherwise molt may proceed steadily through the rest of the breeding season, often with some inner primaries being replaced during incubation periods.

The reason that molt and breeding are often separated may reflect indirect costs as much as direct costs. For example, even though a bird might be physiologically able to grow its longest wing feathers while breeding, this might compromise its ability to fly and hunt, and thus to provide food for its young.

A second activity that can conflict with molt is migration. As with breeding, the conventional wisdom is that birds typically don't molt while migrating, but whether this separation reflects physiological or aerodynamic constraints is unclear. And there are numerous exceptions to this rule (see Migration, pages 49–51, for more details).

Long-distance migrants exhibit various approaches to molting. If there's enough food and time before migration, birds can molt before heading south. If these resources are lacking, or perhaps if there's more-predictable food during migration and on the nonbreeding grounds, then birds can molt at migration staging sites and on the nonbreeding grounds. In some cases, birds may suspend molt during migration; that is, they start to molt before migration, then stop molting, migrate, and finish their molt after migration.

Why do some molts involve all feathers and others only some?

There are two main reasons for this. First, all things being equal, the annual molt shared by all birds tends to be complete (that is, all feathers are molted)—but all things aren't equal. Birds vary greatly in size, and feathers grow only so fast. Small birds, including all songbirds, generally have one complete molt each year. But some large and long-winged birds (such as eagles and albatrosses) don't have sufficient time between breeding seasons to replace all of their wing feathers, which means that their molts are incomplete (that is, they include some but not all of the long wing and tail feathers). Different birds have developed different strategies to address this problem (see Wing Molt Strategies, pages 34–45), but there are many cases when not all feathers can be replaced in an annual cycle.

Second, several species have a second molt in their annual cycle. These second molts are usually partial (that is, limited to head and body feathers), perhaps because the costs of replacing the longer wing and tail feathers are too high. Only two North American breeding species have been documented to regularly undergo two complete molts a year: Franklin's Gull and Bobolink. Both species live year-round in open sunny habitats, where their feathers can become worn, and both are long-distance migrants that breed in mid-latitudes and migrate to low-latitude regions of South America, where there is plentiful food in the northern winter. Thus they have time after breeding for a complete molt before migration (as do many species), but they also have the food for a second complete molt before northward migration. Neither is a large bird, or two complete molts couldn't fit into a year; probably as a reflection of poor food years in South America, the second molt of some individuals is not always complete.

Figure 20. What does the Bobolink have in common with Franklin's Gull? It's the only other North American breeding bird that undergoes two complete molts a year. A good trivia question for long drives! *McHenry County, ND, 16 June 2008. Steve N. G. Howell.*

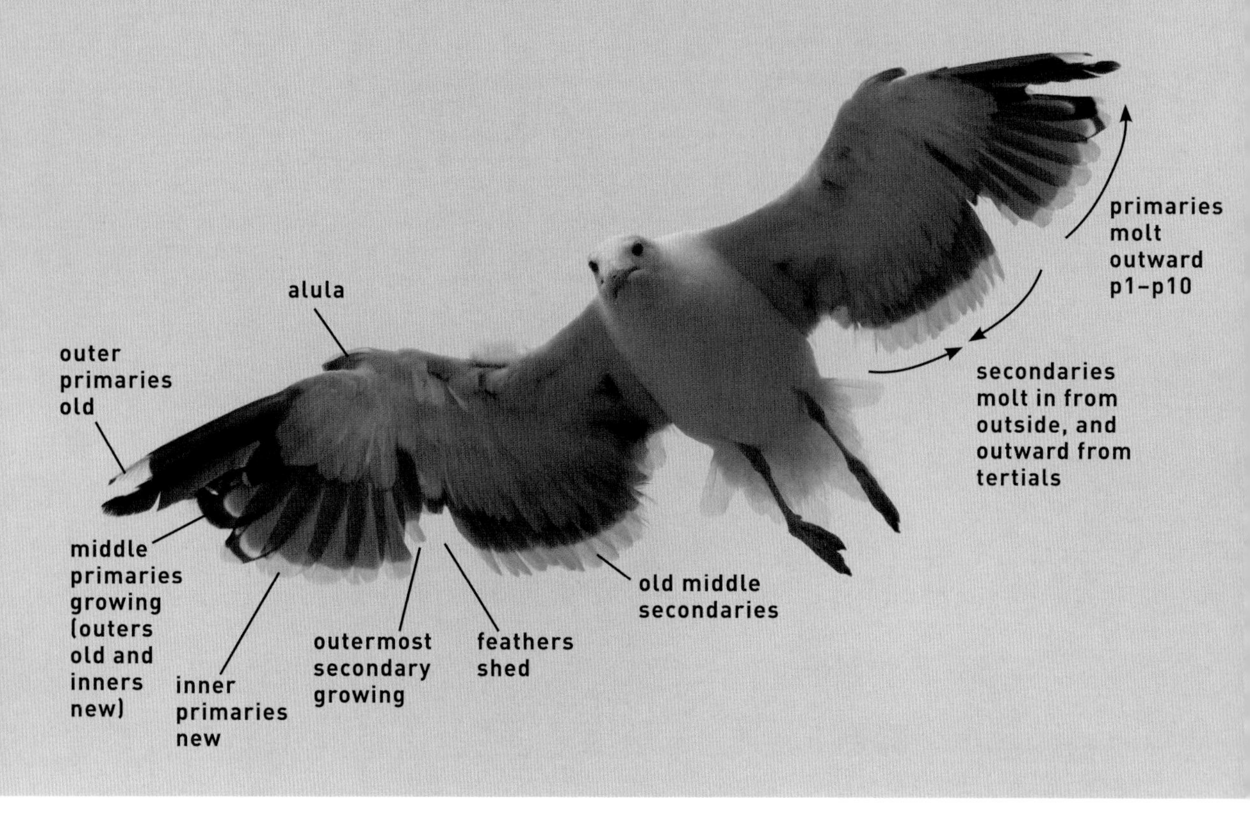

Figure 21. This adult Western Gull illustrates a sequence of wing molt common to many species of North American birds, from hawks and gulls to woodpeckers and songbirds. Molt starts at the innermost primary (p1) and moves outward sequentially to p10. At about the time when p6–p7 are shed, molt starts at the outermost secondary and moves inward toward the body, as shown here. Molt in the secondaries also proceeds outward from the tertials. *Monterey County, CA, 29 Sept. 2006. Steve N. G. Howell.*

Is there a sequence to molts—from head to tail, from tail to head?

Because it is usually easiest to observe growth of the long wing and tail feathers, some general patterns have been noted for these feathers. As a rule, the primaries (which number 10 in most birds) molt from innermost primary (p1) to outermost (p10), whereas the secondaries tend to molt toward the center from either end. In many species (from hawks and gulls to hummingbirds and songbirds), molt of the primaries usually progresses out to p6 or so before the outer secondaries start to molt. This pattern helps prevent a large gap in the wing that would hinder flight, which could occur if the inner primaries and adjacent outer secondaries molted at the same time. Among the tail feathers, there is often an overall inner-to-outer sequence, with the central feathers shed first, but many variations occur.

Information on the molt sequence of head and body feathers is more limited, and careful descriptive studies are few, such as those for the Phainopepla[70] and House Finch.[71] In general, most of the prebasic head and body molt of resident and short-distance migrant songbirds occurs within the span of primary molt, which usually starts with the innermost primary and continues to the outermost. The outermost primary and the middle secondaries, adjacent to the tertials, are usually the last feathers to become fully grown. In long-distance migrants, some molt of head and body feathers, and sometimes of inner primaries, tertials, and central tail feathers, occurs before molt is suspended for migration. Molt then completes at staging sites or on the nonbreeding grounds. One tendency is for colorful or ornamental feathers to grow in last, so they are in prime condition for display and breeding; examples include the gorget feathers of male hummingbirds and the tail streamers of terns and tropicbirds.

In songbirds, nestlings have more down on top than underneath, presumably for protection. The juvenile plumage then grows in all over the body, sometimes in different phases that may correspond to the nestling's expanding body.[72] The tail feathers tend to be the last to grow in fully,[73] which may be because nestling energy is better directed to the wing feathers, which enable escape from predators when a bird is at its most vulnerable; the tail is less important in this respect and can grow in later.

Do different sexes and ages have the same molts?

Often there is a slight difference in the timing of the prebasic molt between the sexes, which relates to different parental roles. Thus, males start to molt earlier than females in many species, but in species in which males care for the young, such as phalaropes, females are free to start molting earlier. There is also a tendency for asynchronous wing molt in breeding pairs of swans and geese, in which having at least one full-winged parent is important for brood and territory defense.[74] It is rare, though, to find cases in which the sexes of a species differ appreciably in the extent or timing of their molts. This may be because the high genetic correlations between homologous characters of males and females cause the attainment of sexual dimorphism to be very slow.[75]

When it comes to first-year versus older birds, however, there are frequent differences in molting. In fact, most species have a special molt that occurs only in their first year, producing a plumage to carry them through until they are about 1 year of age. At this point they enter into the adult cycle of molting, which is then repeated every year thereafter.

Are molts linked to plumage colors and patterns?

Although birds often change their appearance through molt, the physiological processes that govern feather pigmentation and molt timing are independent.[76] Often, however, these two phenomena are coincident. It may be that external factors, such as daylight length and temperature, trigger both molt and pigmentation hormones at the same time. Or perhaps the initiation of molt triggers pigmentation hormones, so that the two are effectively dependent. However, numerous examples demonstrate the independence of molt and pigmentation. Basically, what this means is that if a molt occurs a little late or early, or if the hormones controlling pigmentation kick in a little late or early, then colors and patterns different from the typical ones might be produced.

Age can have a bearing on the colors and patterns produced, and not until a bird reaches a certain age (and concurrent maturity with respect to the relevant hormones) can it produce some colors or even certain feather types. Plucked nestling feathers will grow back as nestling feathers up to a certain point, at which time fully formed feathers will grow, or a feather may even be nestlinglike only at the tip.[77] In the White Ibis, flight feathers plucked before a bird is 2 months old are replaced by dark feathers, at 2 to 6 months of age by lighter feathers, and after 6 months of age by white feathers.[78] In young ptarmigans, the outer two juvenile primaries grow in much later than the others (which are brown and patterned overall) and at the same time as the inner primaries are being replaced. This tim-

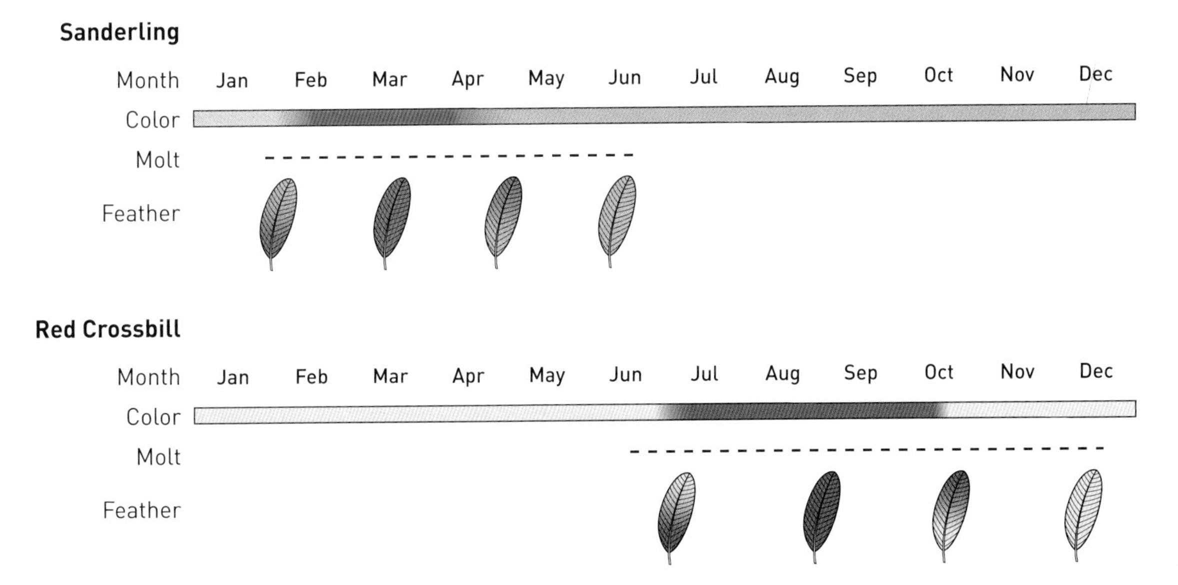

Figure 22. Examples of how colors and molts operate independently. The upper figure represents prealternate molt in a Sanderling, and the bottom figure represents prebasic molt in a Red Crossbill. The hormones that control pigment may be activated only during certain periods; if the molt does not coincide with these periods, feathers can grow in with differing colors. The change can be so quick that a feather is one color at its tip and another at its base.

Figure 23. The processes controlling molt and those controlling feather pigmentation are often independent of one another, which means that plumage coloration does not necessarily reflect our preconceptions of what plumage a bird is wearing. Many first-summer sandpipers have a gray alternate plumage that resembles the nonbreeding aspect of an adult. The right-hand Long-billed Dowitcher in this photo, however, produced a colorful basic plumage much like that associated with breeding adults in alternate plumage, and unlike the typical gray basic plumage. While it may be tempting to say that the red bird is in alternate plumage, it is in fact in basic plumage—in the Humphrey-Parkes system, the color of feathers is unimportant. I have also seen a red female Red Phalarope and a red male Bar-tailed Godwit in basic plumage, and in flocks of normal gray individuals they really stand out. *San Mateo County, CA, 11 Oct. 2008. Steve N. G. Howell.*

ing corresponds to a pigment shift between brown ("juvenile") and white ("post-juvenile"). Thus, the juvenile outer two primaries and the post-juvenile inner and middle primaries all are white, even though they represent two generations of feathers.[79]

This phenomenon helps explain the great variety in the appearance of immature gulls. In first-cycle large gulls, feathers molted in fall tend to be browner and immature-like, whereas those molted in winter and spring tend to be grayer and more adultlike, presumably because of maturing hormones that affect color and pattern. In subsequent immature stages, different colors and patterns probably reflect timing of molt combined with individual rates of hormonal development.

The first-cycle plumages of several sandpipers in summer can resemble the breeding or nonbreeding plumage of adults, or show seemingly any intermediate pattern,[80,81] presumably because of hormone levels linked to sexual maturity and breeding condition. However, the molt producing both red and gray feathers in a first-summer Sanderling, for example, is the same molt in either case; feathers can even be a mixture of red and gray, being gray-tipped and red-based, indicating that the on/off switch for color can be triggered very quickly (see Figure 22). Another example is that of first-cycle male Phainopeplas, where incoming feathers can be gray, black, or gray-tipped and black-based.[82]

Experiments with hormone injections reveal similar patterns. For example, injections of testosterone into Red-necked and Wilson's phalaropes prompted the growth of brilliant "nuptial" feathers in plucked areas of birds in juvenile and basic "nonbreeding" plumages.[83] In another experiment, Herring Gull embryos injected with testosterone resulted in very advanced-looking juvenile plumages and in an essentially adult plumage aspect being attained by second-cycle birds![84]

As well as pigmentation, it appears that feather structure can develop in intermediate stages depending on hormone levels. Examples of this can be seen in immature jaegers, in which the elongated central tail feathers are notably variable in length and shape,[85] and in first-cycle herons, which can produce head plumes of variable length and color.[86]

How is plumage colored?

Plumage pigmentation is a vast field that falls largely outside the scope of this book, but the three basic types of feather coloration are described below. A pioneering study of wood-warblers hints at the complexity of how balancing plumage coloration (for communication and camouflage) relates to factors such as feather durability and energy budgets.[87] In some cases, molt and plumage color are inextricably intertwined, and some examples are discussed in the family accounts. For those interested in pursuing this subject, a review of how the colors in birds are produced was published in 2006.[88]

Basically, there are three types of feather coloration: pigmented, structural, and cosmetic. The first is brought about by pigments deposited on the feather as it grows. The two most common types of pigments are melanins and carotenoids. Melanins are generally responsible for black, brown, and rusty coloration and can be synthesized directly by the bird. Carotenoids are responsible for many red, pink, orange, and yellow colors and are obtained from the bird's diet, as with the pinks of flamingos. Melanins are important in maintaining the strength of feathers, whereas carotenoids often produce bright colors associated with sexual selection. The production of melanins is less sensitive to nutritional stress or to parasite loads than is the production of carotenoids, such that the extent or intensity of colors with different pigment bases may convey different information about a bird's condition.[89,90]

Structural colors (such as blue and white) are produced by the interference of light rays by the fine-scale structure of the feather, and they can be either iridescent or non-iridescent. The most commonly observed "structural color" in the plumage of birds is white: keratin, the material of which feathers are made, is actually colorless, but because it scatters all incident light, feathers with no pigmentation at all appear white.[91] Iridescent colors, such as those seen on the gorgets of hummingbirds, vary with the angle of view. Non-iridescent colors, such as the blue of a male Indigo Bunting, do not change with the angle of view.

Cosmetic colors are fairly uncommon and result from pigment that is applied, or stained, onto the plumage. Examples include the buff or yellowish tones on some herons[92] and the rusty staining on Sandhill Cranes in summer. Many authors have attributed the pink flushes on gulls and terns to cosmetic coloration, but these colors are actually pigments deposited onto the growing feather.[93,94]

Figures 24–25. These two second-cycle California Gulls are unlikely to be more than a month apart in age. The differences in their appearance reflect hormone levels at the time of molt, and perhaps also the extent of their second prealternate molt. One bird produced retarded plumage patterns, suggesting a first-cycle bird, whereas the other produced more advanced, adultlike patterns. The adultlike gray scapulars and scattered upperwing coverts likely represent a mix of late-molted basic feathers and early-molted alternate feathers. Because the prebasic and prealternate molts often overlap in timing in fall, it can be impossible to know whether a given feather is basic or alternate. *Sonoma County, CA, 21 Dec. 2007. Steve N. G. Howell.*

UNDERLYING PATTERNS

Now that some questions have been answered, or at least addressed, we can look at some terminology and at molt strategies—the underlying patterns of feather growth and plumage renewal. Given that birds presumably share a common ancestor, it should not be surprising that their molts show common patterns.

THE HUMPHREY-PARKES (H-P) APPROACH

From the foregoing background questions and answers, it's clear that many factors play into molt, making it seem like a bewildering subject. However, if we ignore color and pattern, ignore when and where molts occur, and ignore all of the other confounding variables, and instead look simply at the number of molts per cycle (that is, the number of times a feather follicle is activated), we can see some fundamental patterns. The pioneers of this approach were Phil Humphrey and Ken Parkes, who proposed that molts and plumages should be given neutral terms that did not have any connotation to life history.[95,96] That is, no more "breeding plumages" or "winter plumages," which carried with them preconceptions. By applying the concept of homology (that is, shared ancestry), one could now compare the molts and plumages of any bird directly with those of any other bird. In this way Humphrey and Parkes developed what has become known as the Humphrey-Parkes (H-P) system, which helps to identify underlying patterns.

An important feature for any system of molt and plumage nomenclature is to have an unambiguous starting point. Systems of naming molts and plumages (including the H-P system) have generally focused on the adult bird, in which a pattern of molts is repeated on a cyclic schedule, with the cycle for most species being a year. Thus, all species of birds have one complete—or ostensibly complete—molt per cycle: the prebasic molt, producing basic plumage. Typically this repeating pattern of adult molts (known as the definitive cycle[97]) starts when a bird is about 1 year of age.

In the first year of life, however, there can be novel molts and plumages that lack counterparts in the adult cycle and serve to bring the young bird "up to speed" with the adult schedule. These plumages may be downy (such as chick and nestling stages) or nondowny (juvenile plumage and any subsequent novel plumages). A species may have no, one, or even two downy stages before it acquires its first coat of nondowny feathers, called its juvenile plumage. It may then have another novel molt and plumage (rarely two or more), which help the bird survive its critical first year of life.

The original H-P system takes as its starting point the highly variable molt by which juvenile plumage is replaced, whether this molt occurs at 1 week of age (and is partial) or at 1 year of age (and is complete). It has since been argued that it is more helpful to take juvenile plumage (the first set of nondowny feathers) as the starting point for studies of plumage succession,[98] so what we will use in this book is the modified H-P system. Although juvenile plumage may not be homologous across all species, it is shared by all birds and represents an unambiguous starting point for patterns of plumage succession, one that has proved helpful for understanding the relationships of molts in definitive cycles.

While some authors argue that we can never divine homologies for certain, the molts presumed to be homologous are at least comparable, and there is value in grouping them to help identify and understand patterns. We must accept, however, that even the most elegant human systems are imperfect attempts to put dynamic natural processes into boxes. There will always be exceptions. It is human nature to focus on exceptions, but to understand patterns (and exceptions) it may be more helpful to focus on features shared by the majority of birds.

DIFFERENT PLUMAGES

Before we look at the patterns revealed by the modified H-P system, we need to address a seemingly elementary question: what is plumage? This may seem like a silly question, but there are two different meanings for "plumage," and a failure to distinguish between the two can lead to confusion when we talk of molting. In the H-P sense, "plumage" refers only to a generation of feathers produced by a given molt—the color and pattern don't matter. But in everyday usage, "plumage" refers to the appearance of a bird, as in "male plumage" or "breeding plumage"—in which color and pattern do matter. In H-P lingo, the term "aspect" was coined for this second meaning of plumage, and although "aspect" has never really caught on, it is a useful term that avoids ambiguity. While neither use of "plumage" is wrong, it is important to grasp that different meanings apply to different systems. For students of molt, the term "basic plumage" is very useful, but for a field observer it is more helpful to talk of "breeding plumages" or "immature plumages."

So, how many kinds of plumages (and molts) are there per cycle, in the H-P sense? We have already met basic plumage—the plumage acquired by the complete (or near-complete) molt shared by all birds and presumed to be homologous. The molt producing basic plumage is the prebasic molt (in the H-P system, molts are named for the plumages they produce). Other plumages that need to be defined are juvenile, formative, alternate, and supplemental. Another term to be aware of is "definitive plumage" (discussed below). All of these plumages are traced here using the White-crowned Sparrow as an example (Figures 26–31).

JUVENILE PLUMAGE. Also spelled *juvenal* in many North American works, this plumage is acquired by the prejuvenile molt and is the first coat of "real" (or nondowny) feathers worn by a bird. It's a one-off plumage that provides a uniform starting point for studies of plumage succession, and as such it can be considered equivalent to the first basic plumage. Note that molts and plumages are named for the cycle in which they occur: the first cycle starts with the acquisition of juvenile (first basic) plumage, and the second cycle starts with the initiation of the second prebasic molt, which is usually when a bird is about 1 year of age. Often juvenile plumage is relatively weak and fluffy (as in many songbirds), but sometimes it is strong and durable (as in albatrosses and hawks).

One of the difficulties in establishing the pattern of molts in a bird's first cycle is the lack of critical data on the acquisition of juvenile plumage. In birds such as ducks and chickens, which grow appreciably during the period that their juvenile plumage develops, it may be difficult to distinguish the delayed activation of juvenile feather follicles (as the feather coat fills out to cover an expanding body) from the initiation of successive molts. Even among songbirds there can be a second or even third phase of juvenile feather growth, whereby additional juvenile feathers are grown after fledging;[99] these feathers grow predominantly at the edges of existing feather tracts and cover any extensive bare areas, especially on the underparts, but they also grow within other feather tracts. While the phenomenon of delayed stages of juvenile plumage development is recognized for Old World songbirds,[100] it has not been widely discussed in the New World. Thus, it may be that later stages of prejuvenile molt in North American songbirds have been described as additional molts (often called presupplemental molts in literature of the past 20 years), such as in several sparrows and buntings.[101,102,103,104] More work is needed on this subject.

FORMATIVE PLUMAGE. This is acquired by a preformative molt and is any plumage found in the first cycle but lacking a counterpart in subsequent cycles. It acts as a bridge between juvenile plumage and entry into the second cycle. Usually the preformative molt is partial, involving only the head

Figure 26. Like many sparrows, the White-crowned Sparrow has a streaky juvenile plumage (right), quite distinct from the adult plumage aspect (left). In the resident race nuttalli, of coastal California, shown here, birds pass through two more distinct plumage aspects (Figures 27–29) before attaining the adult plumage aspect with their second prebasic molt at a little over 1 year of age. *Marin County, CA, 17 Aug. 2008. Steve N. G. Howell.*

Figure 27. A rather messy-looking White-crowned Sparrow undergoing its partial preformative molt and losing the streaked juvenile plumage. *Marin County, CA, 19 Aug. 2008. Steve N. G. Howell.*

Figure 28. The fresh formative plumage of a White-crowned Sparrow looks distinct from the juvenile and adult plumage aspects. The head and body feathers are renewed in the preformative molt, as are the upperwing coverts, whereas the juvenile flight feathers are retained. Because all of these feathers are of a similar age in fall, this bird's plumage appears fairly uniform in wear. *Marin County, CA, 21 Oct. 2008. Steve N. G. Howell.*

and body feathers, but sometimes it involves flight feathers, and rarely it is complete. It can even vary considerably in extent within a species depending on time of hatching, migration distance, and wintering latitude. Frequently the formative plumage resembles the adult basic plumage in aspect, and Humphrey and Parkes themselves were thus misled into viewing formative and basic plumages as homologous; thus, in much molt literature the preformative molt has been called the first prebasic molt. In fact, the *colors and patterns* of the two plumages may be homologous, but the *molts* producing these plumages are not homologous.[105] This is a critical distinction to recognize when trying to understand molt strategies—color and pattern don't matter (see Figure 23). In some species (such as turkeys) there may be two formative plumages, but usually there is only one. If a bird has two formative plumages, the one that is additional to the formative plumage of related species is called an auxiliary formative plumage.[106]

ALTERNATE PLUMAGE. Acquired by a prealternate molt, alternate plumage is any second plumage (additional to basic plumage) found in each plumage cycle, and it alternates with the basic plumage. The alternate plumage may look identical to the basic plumage it replaces, or it may be duller or brighter—remember, color is unimportant when using the H-P system. For an alternate plumage to exist, two molts must occur. European Starlings lack an alternate plumage, but their breeding and nonbreeding aspects look quite different because of plumage wear: their sleek "breeding plumage" is produced through the abrasion of white-spotted feather tips to reveal a different pattern and is simply a worn basic or formative plumage in the H-P system.

Prealternate molts tend to be of two different types.[107] Facultative prealternate molts are those whereby feathers acquired are similar in appearance to those shed (as in tyrant-flycatchers and wrens). These molts tend to be highly variable in extent, even within a species, and reflect the replacement of feathers that become heavily worn, such as tertials, central tail feathers, and some head feathers. Obligate prealternate molts are those whereby feathers acquired are obviously different in color or pattern from those they replace, as in a Western Sandpiper or male Scarlet Tanager. While the original impetus for such molts may have been facultative, subsequent selection (through predation or sexual selection, for example) has "obliged" these species to undergo more extensive molts than they might need simply to replace a few worn feathers.

SUPPLEMENTAL PLUMAGE. Supplemental plumage is acquired by a presupplemental molt and describes the rare cases when a third plumage (additional to

Figure 29. In late winter, the nuttalli race of White-crowned Sparrow undergoes a limited prealternate molt of some head and body feathers. It acquires only partially black head stripes, unlike the solid black head stripes acquired by other races of White-crowned Sparrow. Most of the body feathers are still formative plumage, and the flight feathers are still juvenile plumage. *Marin County, CA, 25 Mar. 2008. Steve N. G. Howell.*

Figure 30. Two White-crowned Sparrows, at left a juvenile in fresh plumage and at right its parent, in heavily worn first alternate plumage. Both individuals could be in their first plumage cycle, although the fairly broad, darker brown stripe on the parent's closed wing shows that some inner primaries have been shed in the start of its second prebasic molt. The flight feathers of the parent bird, especially its tail, have been worn since fledging a year ago and are now heavily abraded; its wingbars, which are formative plumage, have all but worn away, but the more recently molted black head stripes are not as worn. With the completion of its second prebasic molt, this bird will attain the adult plumage aspect shown in Figure 31. *Marin County, CA, 19 Aug. 2008. Steve N. G. Howell.*

Figure 31. Following its second prebasic molt, the nuttalli race of White-crowned Sparrow attains its adult plumage aspect (also known as definitive plumage aspect), and it is not possible to tell if this bird is just over 1 year of age or perhaps 5 years old. In late winter it will undergo a prealternate molt of some head and body feathers, but its appearance will not change appreciably. *Marin County, CA, 21 Oct. 2008. Steve N. G. Howell.*

Figure 32. In fresh basic or formative plumage, the European Starling's glossy plumage is liberally covered with white spots (see Figure 33). *Frederick County, VA, 26 Nov. 2006. Steve N. G. Howell.*

basic and alternate) occurs in each cycle. Well-documented cases of supplemental plumages are few but include ptarmigans and some species of terns. If a bird has multiple supplemental plumages, these are termed "supplemental a," "supplemental b," and so on. Documented cases in which more than three molts occur in a cycle are rare, and they involve only limited areas of the plumage. One study of *Spizella* sparrows[108] showed that feathers on the chin and throat were molted multiple times in a cycle, perhaps as an adaptation to decrease parasitism by lice.

DEFINITIVE PLUMAGE. A definitive plumage is any plumage that does not change appreciably in appearance with the age of a bird (see Figure 31). This term is often viewed as synonymous with "adult plumage," and it can refer to basic, alternate, or supplemental plumages, such as definitive basic plumage, or definitive alternate plumage. However, formative plumages also can be definitive, as in the House Sparrow, and in some species the juvenile plumage is definitive, as in the Northern Fulmar. Most commonly, though, a bird's second basic plumage is its first definitive plumage. Thus, an American Goldfinch or Western Sandpiper in second basic plumage cannot be outwardly distinguished from one in fifth basic plumage, and the term "definitive basic plumage" may then be used. There is, however, a Catch-22 here, one that has gone largely unmentioned in molt studies: the

Figure 33. By summer, the white spots of a European Starling's fresh basic or formative plumage wear away to reveal a glossy undercoat (see Figure 32). It does not require a molt to change its seasonal appearance. *Dare County, NC, 24 May 2007. Steve N. G. Howell.*

Figure 34. This second-cycle Black Tern shows four generations of primaries, which represent four different primary molts that each started with p1. The second prebasic molt was complete out to p10, followed by a prealternate molt out to p8, a presupplemental molt out to p4, and apparently a second presupplemental molt limited to p1. It is also possible that p1 represents an early start to the third prebasic molt, which has been interrupted. Note how the new feathers are paler, and how feathers get progressively darker as they become older. The contrast between old and new feathers makes it easy to see the breaks between different molts. That p7–p8 look contrastingly (rather than gradually) paler than p5–p6 suggests there was a molt interruption at this point. *Mountrail County, ND, 18 June 2008. Steve N. G. Howell.*

only way we can distinguish a definitive plumage, or the second basic plumage of a large gull, is by its plumage *aspect*, or appearance. Thus, the H-P system proposes using one definition of plumage (= aspect) to define another definition of plumage (= feather generation). Even more confusing, perhaps, is that the H-P system proposes the term "predefinitive" for any plumages that precede definitive (often also referred to as immature plumages). Predefinitive does *not* refer to the molt by which definitive plumage is acquired. To avoid confusion, the terms "definitive plumage" and "predefinitive plumage" are not used in this book, other than being mentioned here.

FOUR FUNDAMENTAL MOLT STRATEGIES

Now that we have names for the different plumages, if we look only at the patterns of molt and ignore color and such, it is possible to identify four fundamental molt strategies in all birds.[109,110] Here I'll review these strategies, which are shown in Figures 35–36. These four strategies are useful for categorizing birds and also may convey information about their lifestyles and taxonomic relationships; for example, all songbirds collectively exhibit only two of these four strategies (Table 1). Two strategies include species that, as adults, have only basic plumages, and two strategies include species that have alternate plumages. The two simple strategies have the same number of molts in each cycle, whereas the two complex strategies have more plumages in the first cycle than in subsequent cycles.

Once the second cycle is reached, via the second prebasic molt, the molt patterns of all birds tend to be the same in subsequent cycles, which, as we learned earlier, are then termed "definitive cycles." It is only in the first cycle that novel molts occur, presumably to help young birds survive their all-important first year of life.

SIMPLE BASIC STRATEGY (SBS). At the most basic level, all bird species share only two molts, both of which produce complete (or ostensibly complete) plumages. The first is the prejuvenile (or first prebasic) molt, and the second is the prebasic molt, by which basic plumage is acquired. The simplest molt strategy involves nothing more than repeating basic plumages and is termed the "simple basic strategy." Examples of birds that exhibit the SBS include albatrosses, most petrels and storm-petrels, and barn owls. These species tend to nest relatively free from

TABLE 1

MOLT STRATEGIES OF NORTH AMERICAN BIRD FAMILIES

X denotes the predominant molt strategy or strategies (>35% of species) in a family (presumed strategies indicated in parentheses); x denotes any less-frequent molt strategy or strategies (<35% of species). A ? indicates that this strategy has been suggested or may occur.

	SBS	CBS	SAS	CAS
Swans, Geese, and Ducks		X	?	X
Plain Chachalaca		X		
Grouse and Pheasants		X		x
Wild Turkey		X		
New World Quail		?		X
Loons			X	
Grebes			?	X
Albatrosses	X			
Petrels	X	x		
Storm-Petrels	X	x		
Tropicbirds		X		
Boobies and Gannets		X		
Pelicans			X	
Cormorants			X	
Anhinga		(X)	?	
Frigatebirds	?	(X)		
Herons		X	?	
Ibises and Spoonbills	?	?	X	
Storks	(X)	?		
Flamingos		?	(X)	
New World Vultures	X	X		
Hawks	x	X		
Falcons	x	X		
Rails and Coots		X		?

Limpkin		X		
Cranes		X		
Plovers		?	?	X
Oystercatchers		?	(X)	
Stilts and Avocets				X
Northern Jacana		X		
Sandpipers		x	x	X
Gulls	x		X	X
Terns		x	x	X
Black Skimmer			?	(X)
Skuas and Jaegers		?	?	(X)
Auklets, Murres, and Puffins		?	X	X
Pigeons		X		
Parrots		X	x	
Cuckoos		X		
Barn Owl	X			
Typical Owls		X		
Nightjars		X		
Swifts	x	(X)		
Hummingbirds		X		x
Trogons		X		
Kingfishers		X		
Woodpeckers		X		

SONGBIRDS

	SBS	CBS	SAS	CAS
Tyrant-Flycatchers		X		X
Shrikes				X
Vireos		X		X

	SBS	CBS	SAS	CAS
Crows and Jays		X		
Larks		X		
Swallows		X		?
Chickadees and Titmice		X		
Verdin		X		
Bushtit		X		
Nuthatches				X
Brown Creeper		X		
Wrens		X		x
American Dipper		X		
Gnatcatchers				X
Kinglets		X		?
Old World Warblers				X
Thrushes		X		x
Wrentit		X		
Mockingbirds and Thrashers		X		?
Starlings and Allies		X		
Pipits and Wagtails				X
Waxwings		X		
Phainopepla		X		
Olive Warbler		X		?
Wood-Warblers		X		X
Tanagers		?		X
Sparrows and Allies		X		X
Cardinals and Allies		X		X
Blackbirds and Orioles		X		x
Finches		X		x
Old World Sparrows		X		

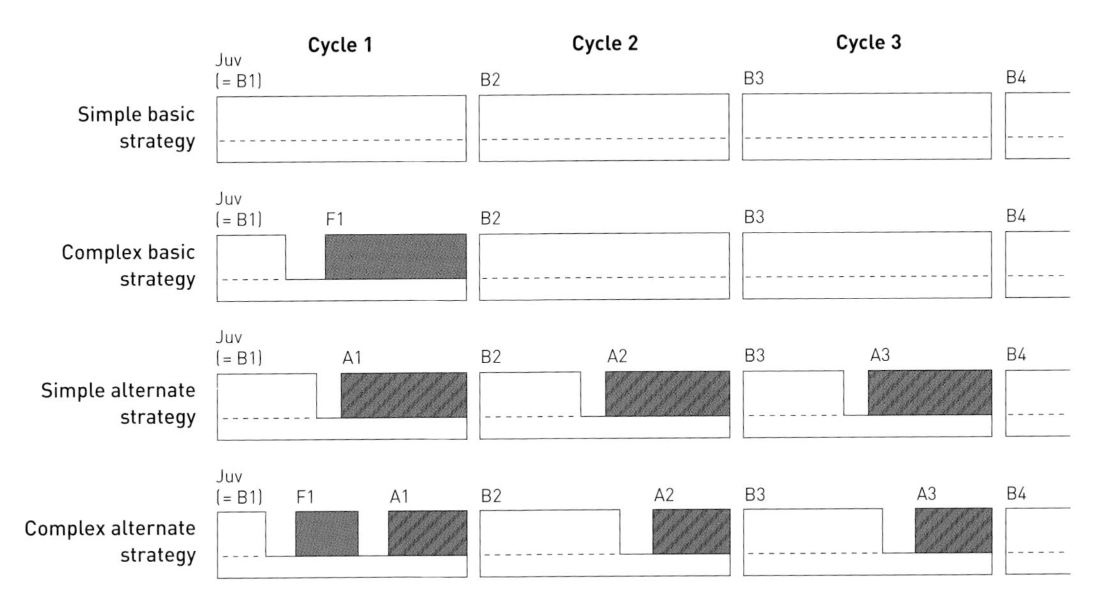

Figure 35. Diagrammatic representation of the four fundamental molt strategies, showing how they can be viewed as building on the simple basic strategy. Blocks indicate plumages (the dashed line represents a division between body feathers above the line and flight feathers below), and breaks between blocks represent molts. B1, B2, and A1, A2, etc., indicate successive basic and alternate plumages, respectively. F1 indicates formative plumage. Note how the differences between the two basic cycles and between the two alternate cycles lie in the number of molts that occur in the first cycle.

predators and have relatively long chick or nestling stages during which the young grow a strong juvenile (= first basic) plumage. Most or all individuals of these species do not breed in their first year, and their second prebasic molt typically commences earlier than the prebasic molt of breeding adults.

Although this strategy is uncommon in modern birds, it may have been the ancestral, or primitive, molt strategy. Whether or not the SBS is the most primitive strategy, it is the simplest strategy, and one way to picture the other three strategies is to view them as having additions, or modifications, to the SBS.

COMPLEX BASIC STRATEGY (CBS). The most common "modification" to the SBS is when a novel plumage (the formative plumage) is added into the first cycle following juvenile plumage, but no molt is added into subsequent cycles. This is termed the "complex basic strategy," and it is found among a wide range of species, from kestrels and kingfishers to woodpeckers and wood-warblers. One reason for the frequent occurrence of the CBS may be that juvenile plumage grows during the breeding season, whereas adult prebasic molts generally follow the breeding season. Consequently, the first cycle is usually slightly longer than adult cycles, and juvenile plumage would have to be at least as durable as adult basic plumage if it were to be retained throughout the first cycle. But juvenile plumage is generally of poorer quality than subsequent basic plumages. One possible reason for this, for example in songbirds, is that nest-bound young are often susceptible to predation and thus grow a functional, but not necessarily durable, juvenile (= first basic) plumage with which they can leave the vulnerability of the nest. They subsequently undergo a variably extensive preformative molt by which they acquire stronger feathers that protect them until the start of the second prebasic molt.

These two basic strategies, the SBS and CBS, involve only basic plumages in the adult cycle, with a formative plumage in the CBS. The other two of the four fundamental molt strategies involve the addition of alternate plumages into the basic cycles.

SIMPLE ALTERNATE STRATEGY (SAS). This strategy is defined by the addition of at most only a single plumage into the first cycle, and of a single (alternate) plumage into each subsequent cycle. Although the SAS could have developed by simply adding a single alternate plumage into each cycle of the SBS, two other pathways are possible (see The Evolution of Molt Strategies, pages 33–34). First, the SAS could develop from a CBS in which a prealternate molt developed only in the second and subsequent cycles (see the family account for ibises and spoonbills for further discussion of this idea). Second, it could develop from a complex alternate strategy (see below) by a merging of the preformative and first prealternate molts. That is, if two molts aren't needed in the first cycle, each might

Figure 36A.

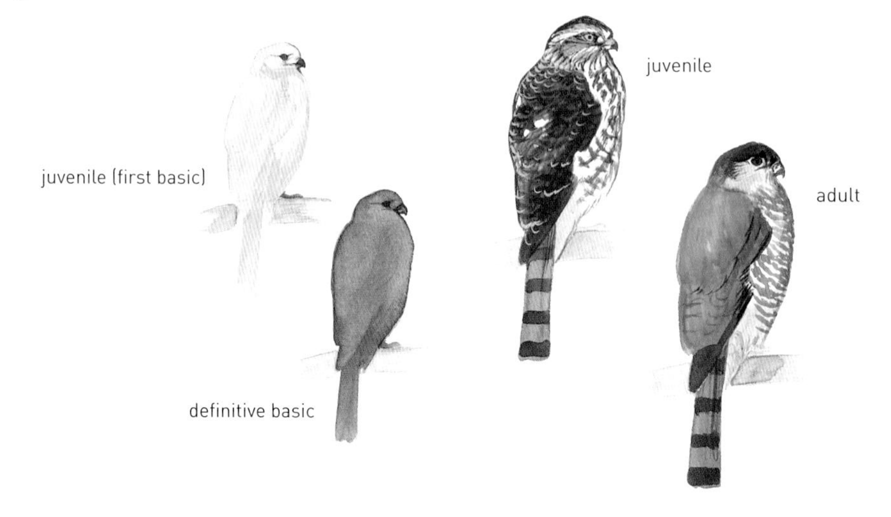

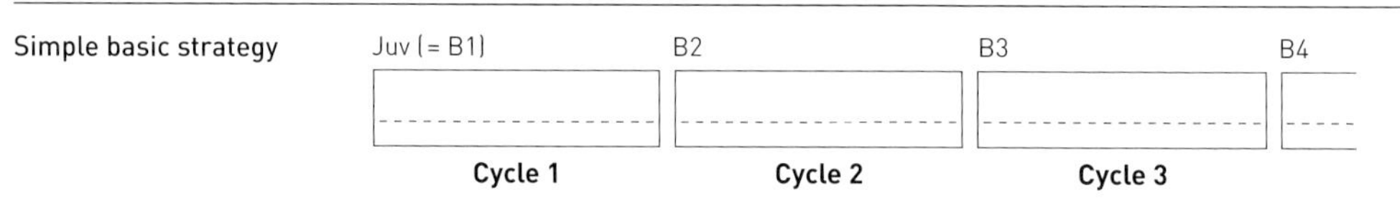

Figures 36A–36D. The four fundamental strategies are contrasted here with the life-year system, which represents what we see in the field. Since this figure was drafted, it has been suggested that the Sharp-shinned Hawk has a limited preformative molt of body feathers, which would make it a CBS species. This requires confirmation, though, and either way does not detract from the conceptual view conveyed here. Note how homologies of color often lie with basic and formative plumages (as in the Eastern Towhee and Scarlet Tanager) but how homologies of molt lie with molts (such as the complete prejuvenile and complete prebasic molts). *Illustrations used by permission of David A. Sibley, © 2002.*

Figure 36B.

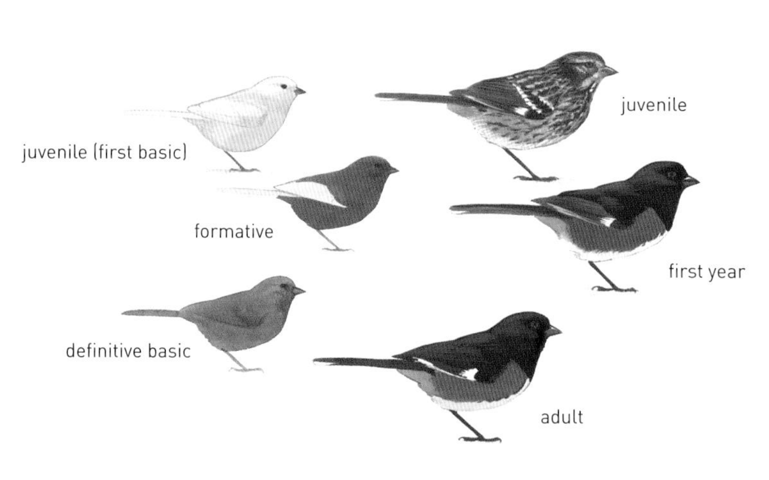

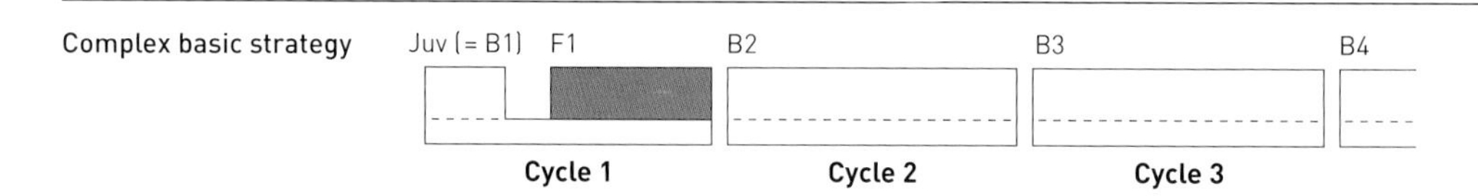

Humphrey-Parkes Terms
Life-Year Terms
Figure 36C.
juvenile
juvenile (first basic)
first alternate
first summer
adult winter/nonbreeding
definitive basic
definitive alternate
adult summer/breeding
Simple alternate strategy
Juv (= B1)
A1
B2
A2
B3
A3
B4
Cycle 1
Cycle 2
Cycle 3

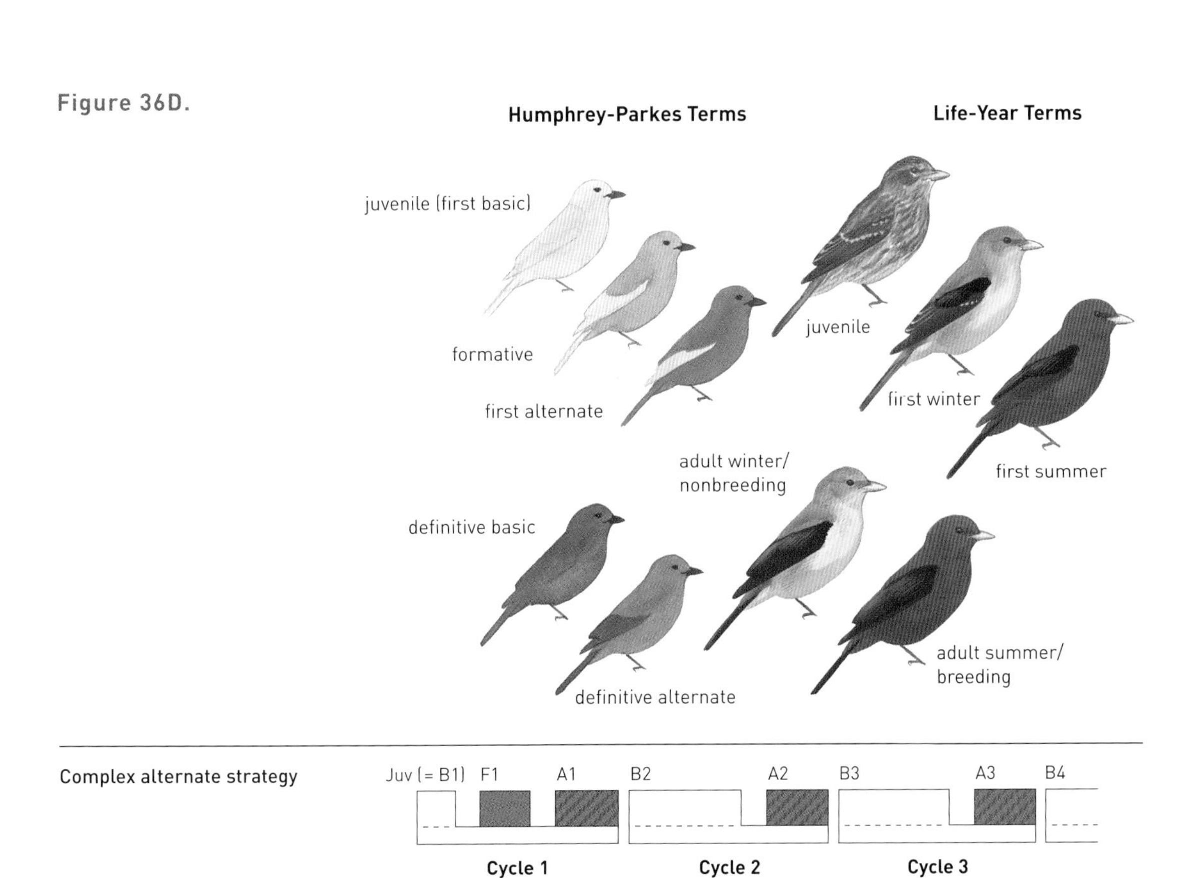
Figure 36D.
Humphrey-Parkes Terms
Life-Year Terms
juvenile (first basic)
formative
first alternate
juvenile
first winter
first summer
adult winter/
nonbreeding
definitive basic
definitive alternate
adult summer/
breeding
Complex alternate strategy
Juv (= B1)
F1
A1
B2
A2
B3
A3
B4
Cycle 1
Cycle 2
Cycle 3

Figure 37. The Northern Fulmar exhibits the simple basic molt strategy (see Figures 35–36). The juvenile plumage aspect is not distinguishable from subsequent ages unless older individuals have incomplete molts and show two generations of flight feathers. The uniformly fresh appearance of this individual suggests it is a juvenile, but it could be older. *Monterey County, CA, 15 Oct. 2006. Steve N. G. Howell.*

become reduced to the point that no feathers are replaced more than once.

The SAS is relatively rare among modern birds and is not found in songbirds. Examples of birds that exhibit the SAS are mainly relatively large and aquatic nonpasserines, namely loons, pelicans, cormorants, ibises, and large gulls. As a rule, species that follow the SAS do not breed in their first year, so a single molt, which is often protracted, is adequate to help them survive their first cycle.

COMPLEX ALTERNATE STRATEGY (CAS). This strategy involves the addition of one or more plumages (alternate, or alternate and supplemental) into each cycle of the CBS; thus there is also a formative plumage in the first cycle. Note that alternate plumages have likely developed independently in different species groups exhibiting the SAS and CAS, and thus are not necessarily homologous, unlike the juvenile and basic plumages shared by all birds. Thus, the alternate plumages of ducks are probably not homologous to those of gulls or wood-warblers. The CAS is fairly common and occurs in a diverse spectrum of species, including some ducks, many shorebirds, small gulls, and all songbirds that have alternate plumages. Shared life-history traits of many CAS species are migration, and breeding at 1 year of age. Prealternate molts also occur in some mostly resident species, such as grouse and nuthatches; extra molts in these species tend to be facultative, and life-history characteristics that may have caused their evolution include the wearing away of plumage in exposed and scrubby habitats or in tree cavities.

Figure 38. Like most finches, Lawrence's Goldfinch exhibits the complex basic molt strategy (see Figures 35–36), although it was once thought to have an alternate plumage. The juvenile, shown here, soon undergoes a preformative molt of head and body feathers and attains the adult plumage aspect. *Santa Barbara County, CA, 22 July 2007. Steve N. G. Howell.*

Figure 39. The White-faced Ibis exhibits the simple alternate molt strategy (see Figures 35–36). This flock includes several adults at different stages of prebasic molt, plus an individual (right of center) in its second basic plumage, which is typically duller than the adult basic plumage. Because immature birds are not involved in breeding, their prebasic molt occurs earlier than the adult prebasic molt, which helps explain why this bird appears to have finished its molt whereas the adults are still molting. *Imperial County, CA, 19 July 2008. Steve N. G. Howell.*

THE EVOLUTION OF MOLT STRATEGIES

As a rule, simplicity precedes complexity. For example, dugouts preceded cruise ships as a means of ferrying people around, and caves preceded apartment buildings as places for humans to live. It thus seems likely that the ancestral molt strategy was a basic strategy. For now, though, we can't say whether it was more likely to have been simple basic or complex basic, although a critical study of molt in today's living primitive birds, known as ratites (such as kiwis, ostriches, and rheas), might offer some insight. We do know that the young of all major groups of primitive birds are precocial, so this seems like a reasonable assumption for the ancestral bird. When the molt strategies of birds are well known, we may be better placed to trace the evolution of strategies. For example, when did alternate plumages develop and how many times independently? And when might formative plumages have been gained or lost in some groups? In the meantime, it is interesting to muse on how molt strategies might have developed. The following ideas are offered to promote thought and discussion.

If the ancestral molt strategy was simple basic,

Figure 40. Most terns, like this Least Tern, exhibit the complex alternate molt strategy (see Figures 35–36). Gulf Coast populations of Least Terns nest earlier than Pacific Coast birds and are usually molting out of juvenile plumage by August or September, unlike this individual, which appears to be still in full juvenile plumage. *Baja California, Mexico, 19 Sept. 2006. Steve N. G. Howell.*

it is easy to see how a complex basic molt strategy might develop with the addition of formative plumages to help increasingly altricial young get through their first year of life. It is also possible that the ancestral molt strategy was complex basic, and that the increasing size of young birds necessitated the growth of new plumages, as occurs today with turkeys (and with lizards that molt their scaly epidermis as they increase in size). From the complex basic molt strategy, a simple basic strategy could then have developed in those birds for which a formative plumage was not really needed. But how did alternate plumages develop?

At least two possibilities come to mind for the evolution of alternate plumages. The first is simply that some birds got their feathers more worn than did other birds, and thus needed to replace them before the next prebasic molt. The second is that a protracted prebasic molt could have evolved into two separate molts.

In the first scenario, species prone to getting their feathers worn (for example, through prolonged exposure to sun or water, or by migration to different habitats and climates) would be likely to develop alternate plumages. For birds starting with a simple basic molt strategy, if all ages experience similar conditions, then an alternate plumage could develop in each cycle. Thus, a simple alternate molt strategy might develop from a simple basic strategy.

For birds starting with a complex basic molt strategy, if formative plumages are adequate to protect a bird throughout its first cycle, then alternate plumages might develop only in subsequent cycles. Thus, a simple alternate molt strategy could also develop from a complex basic strategy. If the formative plumages are not adequate to protect a bird throughout its first cycle, then an alternate plumage might develop in the first cycle as well as in subsequent cycles. Thus, a complex alternate molt strategy could develop from a complex basic strategy.

Another way that alternate plumages might develop is through the interruption of a protracted prebasic molt, such as a molt begun in fall and then suspended during the colder winter conditions. If the resumption of molt in spring coincides with a flush of breeding hormones, the newly molted feathers might develop an appearance different from that of feathers grown in the fall. And if the patterns of the newly acquired feathers promote favorable social signals and interactions, or perhaps provide better camouflage, then this might lead to selection for more extensive molts in spring, with some feathers being molted twice in a cycle—and thus a prealternate molt is born. This scenario could develop in both first cycles (via a protracted preformative molt) and subsequent cycles (via protracted prebasic molts), such that a complex alternate molt strategy could develop from a complex basic strategy.

WING MOLT STRATEGIES

Wings are important for foraging, avoiding predators, migrating, and any number of other things in the life of a bird. When it comes to molting their flight feathers, birds fall into one of two basic camps: they either keep flying while molting, or they become flightless and molt all or most of their primaries at the same time. For birds that need to keep flying, various patterns of wing molt have developed to balance molting as quickly as possible with maintaining sufficient flight capability. For example, for sparrows, which hop around looking for seeds and insects, having somewhat impaired flight for a few weeks is not as problematic as it would be for aerial flycatching birds such as swallows. Thus, swallows molt their primaries more gradually and take longer to molt than do sparrows. The easiest place for us to appreciate wing molt is usually among the primaries. Most bird species have 10 primaries, which are numbered from the innermost (p1) out to the outermost (p10).

Here I divide wing molt into four categories, some of which are linked by intermediate stages. The first three categories can occur within the same family of birds (herons, for example), suggesting they are simply adaptive variations on a theme rather than fixed patterns. The fourth "category" is a grouping of exceptions to these three common patterns.

STANDARD SEQUENTIAL WING MOLT. Perhaps the most frequent sequence of primary molt among birds is the standard sequential replacement from p1 out to p10 (or to p9 in some songbirds). This is called a descendent sequence in most of the European literature,[111,112] which is a confusing and contradictory term you needn't worry about here; it reflects times when molt studies numbered the primaries from the outside in.

The first few inner primaries are relatively short and often are shed almost simultaneously, which apparently does not compromise flight too much, and then the middle and outer primaries tend to grow one or two at a time. The secondaries (s) tend to molt both outward from the tertials and inward from the outermost secondary (s1). Because a big gap in the middle of the wing could compromise efficient flight,[113] the outer secondaries do not start molting until the new inner primaries have grown in. Typically, when primary molt has reached the middle primaries (say, around when p6 is shed), the outer secondaries start to be replaced (see Figure 21). In longer-winged birds there is often another point of initiation of secondary molt at s5, whence molt proceeds toward the body. In residents and short-distance migrants, the head and body feathers typically are molted during the span of primary molt. In some long-distance migrants, however, the molt of head and body feathers can

Figure 41. A modification to the standard sequential pattern of wing molt is the so-called eccentric pattern, whereby molt starts at a primary other than p1 and moves out to the outermost primary. Eccentric wing molt normally occurs only in first-cycle birds and offers a way to acquire strong outer primaries to help see a bird through its first cycle. In this Red-necked Phalarope, the outer primaries are relatively dark and fresh compared with the faded and frayed middle primaries that are visible. But, unless primary molt was suspended (which happens sometimes with shorebirds), if all of the primaries were of one generation, the exposed outers would be more faded and worn than the relatively protected middle primaries. Thus we can infer that this bird had an eccentric preformative molt of its outer three primaries (p10 is covered here, so only p9 and p8 are visible), which is a frequent occurrence among long-distance migrant sandpipers. Looking carefully, we can also see where the tip of the juvenile p6 has been worn away to reveal a darker area on the now-exposed base of p7, indicating the point that p6 used to cover. Some of the gray back feathers appear to be new basic feathers, which means we can see four generations of feathers on this bird: juvenile middle primaries, formative outer primaries (and perhaps some tertial coverts), the brightly colored first alternate head and neck feathers, and a few freshly molted second basic scapulars! *Marin County, CA, 2 Sept. 2008. Steve N. G. Howell.*

be started before or even during migration, with the energetically more costly molt of flight feathers delayed until the birds reach their nonbreeding grounds. Birds exhibiting standard sequential wing molt include petrels, storm-petrels, grouse, quail, smaller hawks, plovers, sandpipers, gulls, terns, nightjars, swifts, woodpeckers, and songbirds.

A variation on the standard sequential wing molt is the so-called eccentric pattern, which is an incomplete wing molt characteristic of first-cycle birds. In eccentric patterns, the primary molt starts at a primary other than p1 (often at p4 or p5) and continues out to the outermost primary. Thus, the protected juvenile inner primaries are retained until the second prebasic molt, while fuller and stronger outer primaries are acquired to help the bird through its first year. As well as simply providing stronger outer primaries, eccentric molt patterns may in some cases actually increase the wing area.[114] Birds with eccentric first-cycle primary molts include some sandpipers, tyrant-flycatchers, vireos, and buntings.

Figure 42. Loons are among several heavy-bodied birds that undergo synchronous wing molt and become temporarily flightless. The adult Red-throated Loon, shown here, typically undergoes its wing molt in October–November, whereas adults of the other loons usually molt their wings in February–March. *Sonoma County, CA, 9 Oct. 2006. Steve N. G. Howell.*

SYNCHRONOUS WING MOLT. Sometimes termed simultaneous wing molt, synchronous wing molt occurs when a bird sheds all of its flight feathers at one time and becomes flightless, usually for a period of 3 to 5 weeks. This strategy occurs among some heavy-bodied birds with relatively small wings, such as ducks, for which it may be better to risk a relatively short period of flightlessness rather than undergo a prolonged period of impaired flight. In some cases the flightless period can be shortened by a reduction in body mass, such that birds can regain their powers of flight when they have only about 75 percent of their full wing area.[115] Synchronous wing molt is also a means of simply molting quickly for birds that do not need to fly much, such as rails and the Least Bittern.

It is unlikely that every primary drops at exactly the same time, and it may be that many effectively synchronous wing molts follow a predictable sequence (see the family account for auklets, murres, and puffins). When all primaries are shed within a day or two, however, it can be difficult to appreciate patterns. Probably because of the need to regain flight ability as quickly as possible, a common pattern is for the molting of wing and body feathers to be offset in timing. Thus, all molting energy is invested into the wings before involving head and body feathers.

Synchronous wing molts are typical mostly of aquatic birds such as loons, grebes, anhingas, waterfowl, rails, murres, and puffins. These species either can escape by diving or find remote and impenetrable areas safe from predators. A few other species or groups also qualify as partial, or part-time, followers of this strategy. Cranes and flamingos in remote areas relatively free from predators can have synchronous wing molts, whereas populations in other places molt their flight feathers gradually and maintain the ability to fly.

American Dippers shed their inner five or six primaries synchronously, which renders them effectively flightless for a short period; after this the outer primaries grow sequentially, one at a time, as in a typical wing molt.[116] Some Bristle-thighed Curlews on remote, predator-free islands in the tropical Pacific can shed their primaries in blocks and become flightless, whereas individuals on larger islands with predators may undergo a typical sequential wing molt.[117]

STEPWISE WING MOLT. Many large, long-winged birds simply do not have enough time between breeding seasons to undergo a standard sequential primary molt, because feathers can grow only so fast. If these birds need to fly to forage, becoming flightless is not an option. A strategy that helps them overcome this problem is known as stepwise molt, in which multiple waves of molt can be set up in the primaries. This interesting adaptation was first recognized and described in a study of Masked Boobies;[118] it has also been called *Staffelmauser*[119] and serially descendent molt.

One advantage of stepwise molt is that when different waves of molt are progressing through the wing, two or more non-adjacent feathers can grow at the same time yet create only separate small gaps in the wing's surface area. Conversely, two or more adjacent feathers growing at the same time would create a larger gap that might impair efficient flight. As a result, all of the primaries can be replaced, without compromising flight, in a shorter period than during a standard sequential molt.

It has been calculated that for a Pelagic Cormorant to replace all of its primaries one at a time in a standard sequential wing molt would require 58 weeks, considerably more time than is available. Courtesy of stepwise molt, however, about three non-adjacent primaries can grow simultaneously, such that, theoretically, molt could complete in 21 weeks without creating big gaps in the wings.[120] Such time estimates are based on rough assumptions about feather growth rates, but they are probably not too far off the mark in general terms and illustrate well the benefits of a stepwise wing molt.

Multiple waves of primary molt also mean that there is rarely if ever a large number of old and worn primaries in the wing at any one time. Perhaps for this reason, stepwise wing molts are also fairly common among species living in areas with unpredictable breeding opportunities, such as in the vast arid regions of Australia.[121,122,123] Stepwise molts thus mean that birds can interrupt wing molt to take advantage of suitable breeding conditions without the disadvantage of impaired flight.

How do stepwise molts develop? Basically, before one wave of sequential primary molt (from p1 out to p10) has completed, a second wave starts anew at p1. In some cases, a third wave may start before the first has reached p10. Over time, successive waves moving outward through the primaries develop and in extreme cases up to five waves may be running concurrently (Figures 45–54). The period between the start of each new wave is usually about a year (but see below). After interruptions to the episodes of wing molt, each wave starts up again from the point at which it stopped. In a perfect iteration, stepwise molt allows all of the primaries to be replaced in a single cycle, most often by means of two to three concurrent waves. But particularly in

Figure 43. Most "wild" waterfowl undergo their synchronous wing molt in remote areas and near cover, where they can be safe from predators. Habituated birds in many city parks, however, become flightless in full view. On this female Mallard, the white tips to the secondaries have broken from their sheaths, as have the tips to the primaries and primary coverts. Molt in other tracts is often suspended or reduced during this period so that all available energy is directed into regrowing the flight feathers; note the heavily worn tail, which will be molted later. *Sonoma County, CA, 14 July 2008. Steve N. G. Howell.*

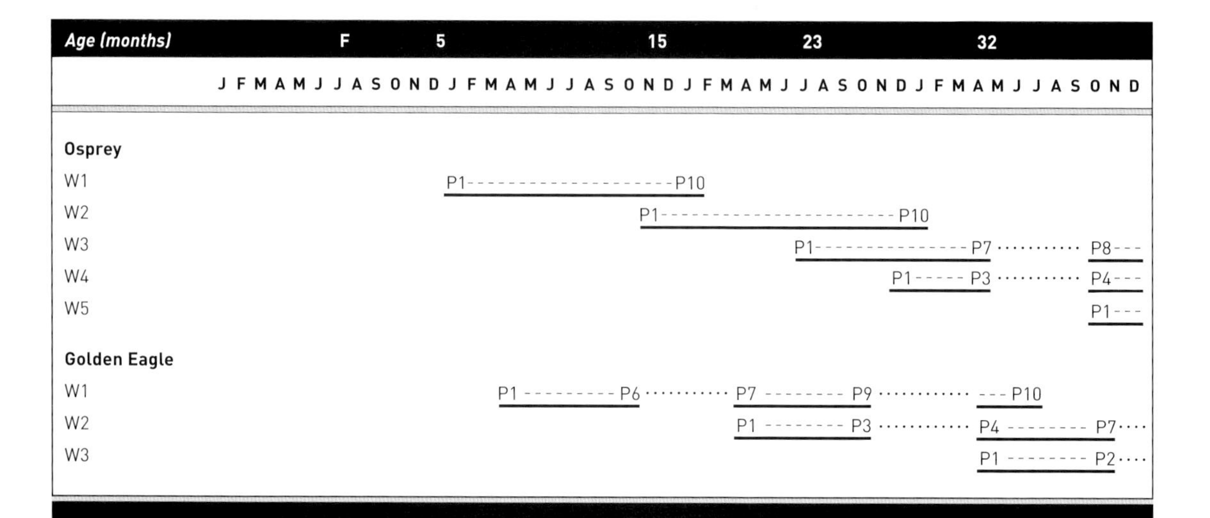

Figure 44. Diagrammatic representation of the development of successive accelerated waves (W1, W2, etc.) of stepwise primary molt in an Osprey (Prevost 1983) relative to the development of normal stepwise waves in a Golden Eagle (Bloom and Clark 2001). Note how each new stepwise wave starts before the preceding wave has ended, and that the number of waves of molt in the Osprey and eagle are not comparable. By 23 months of age the Osprey has "gained" a primary molt relative to the eagle. Interestingly, in terms of its cyclic pattern, the Osprey's third wave of primary molt appears to be the "odd one out" by starting in July, whereas the other waves start in November–January.

F represents fledging; p1----p6 and such indicate the number of primaries molted per molting period; ..... indicates an interruption of molt (as for migration, breeding, and winter).

migratory or very long-winged species, two or even three cycles may still be needed to replace all of the flight feathers.

The head and body feathers are often molted at the same time as the flight feathers, or in some cases the first wave of wing molt can start before or after the head and body feathers begin to molt. This latter strategy may be a way of ensuring that the stepwise pattern, which can be important in subsequent years, gets underway without being compromised by competing energy demands from growing head and body feathers.

Despite considerable discussion about the benefits of stepwise molt and how it develops, the principles of how it works have often been misunderstood.[124,125,126,127,128,129,130] Indeed, knowing exactly what was meant by the Stresemanns[131] when they originally described *Staffelmauser* has been debated.[132] One problem may be that there appear to be two pathways by which stepwise molts can be set up, something that has not been previously recognized. These two pathways are explained and contrasted more fully in the family account for hawks and can be called the "normal" and "accelerated" schedules.

Normal schedules develop when the first primary molt is part of the second prebasic molt, which usually starts in the bird's first summer after it has hatched and results in three to nine inner primaries being replaced before molt is suspended for the winter. In the bird's second summer, the third prebasic wing molt starts anew with p1 while the earlier wave starts up again where it left off. And so on for each prebasic molt cycle. Thus the first wave starts at about 9 to 10 months of age, the second at 21 to 22 months of age, and so on each year (see the Golden Eagle in Figure 44). This schedule is typical of species that live year-round in temperate regions (such as condors and large hawks) where molt tends to arrest in midwinter.

Accelerated schedules develop when an extra wing molt is inserted into the picture and "kick-starts" the set-up of stepwise waves, as with the Osprey.[133] Thus, the first wave often starts when a bird is only 5 to 7 months of age, the second starts at about 15 months of age, and the third at about 23 months, which is about when the second wave in a normal schedule is starting. Immature birds on accelerated schedules start the first two to three molt waves in fairly quick succession, and in some cases there is little or no appreciable pause between the first and second waves, or even between the second and third. Thus, a second-summer Osprey that has remained on the nonbreeding grounds, say in Mexico, may start its third wave of primary molt at only 2 years of age (see the Osprey in Figure 44).[134]

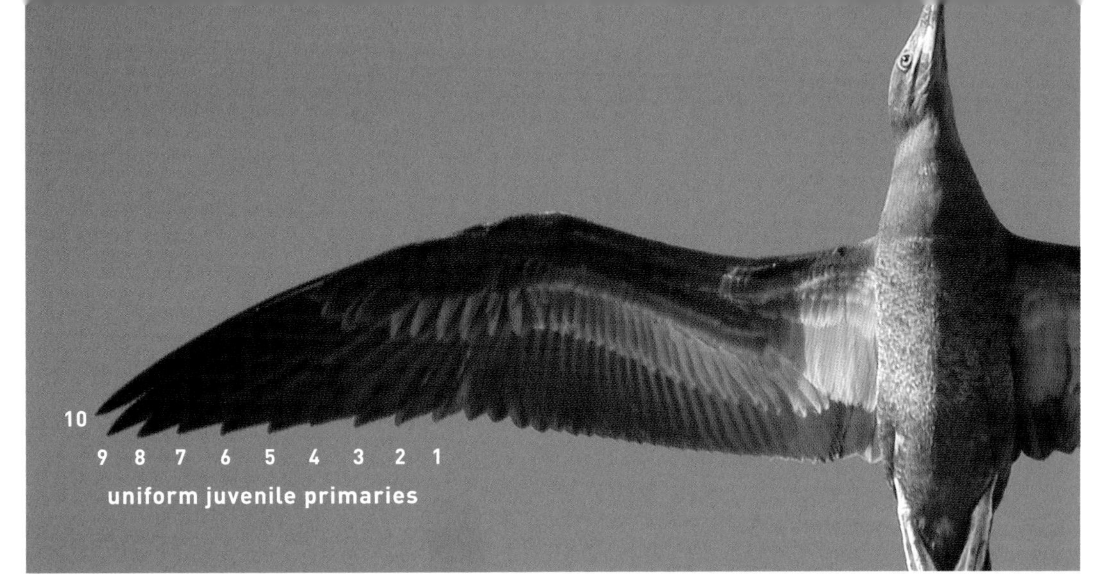

Figure 45A. First-cycle Brown Booby with uniformly fresh juvenile plumage and no wing molt. *Colima, Mexico, 15 Feb. 2008. Steve N. G. Howell.*

Figure 45B. First-cycle Brown Booby with p1 shed, indicating the start of its first wave of stepwise primary molt. *Oaxaca, Mexico, 15 Dec. 2008. Steve N. G. Howell.*

Figure 45C. First-cycle Brown Booby with p1 growing and p2 shed in its first wave of stepwise primary molt. *Oaxaca, Mexico, 8 Mar. 2007. Steve N. G. Howell.*

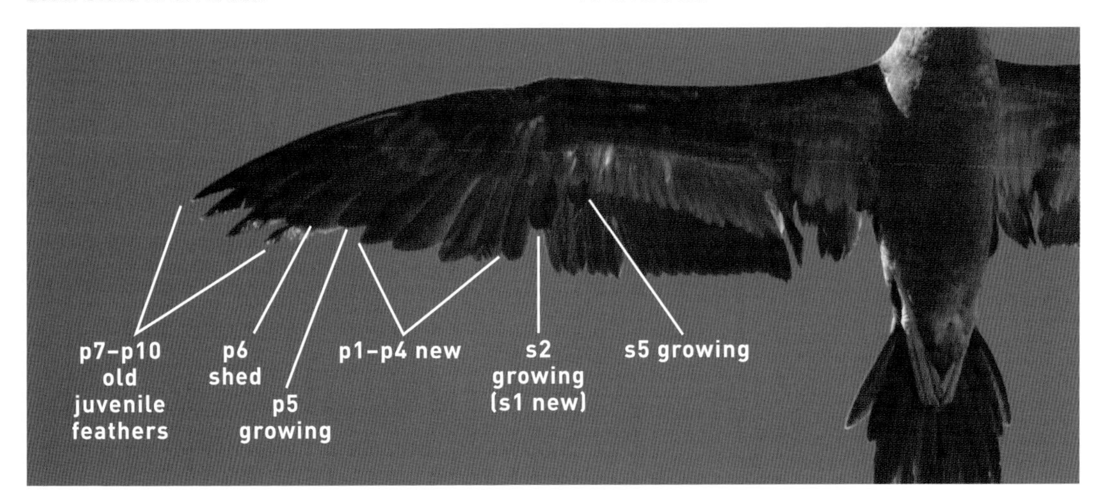

Figure 45D. First-cycle Brown Booby with p5 growing and p6 shed in its first wave of stepwise primary molt. The secondaries are also starting to molt: s1 is new, and s2 and s5 are growing. Many species of long-winged birds start secondary molt at two or more points among the secondaries, and s5 is a frequent point of molt initiation in addition to s1 (see Figures 45E and 48). *Oaxaca, Mexico, 15 Dec. 2008. Steve N. G. Howell.*

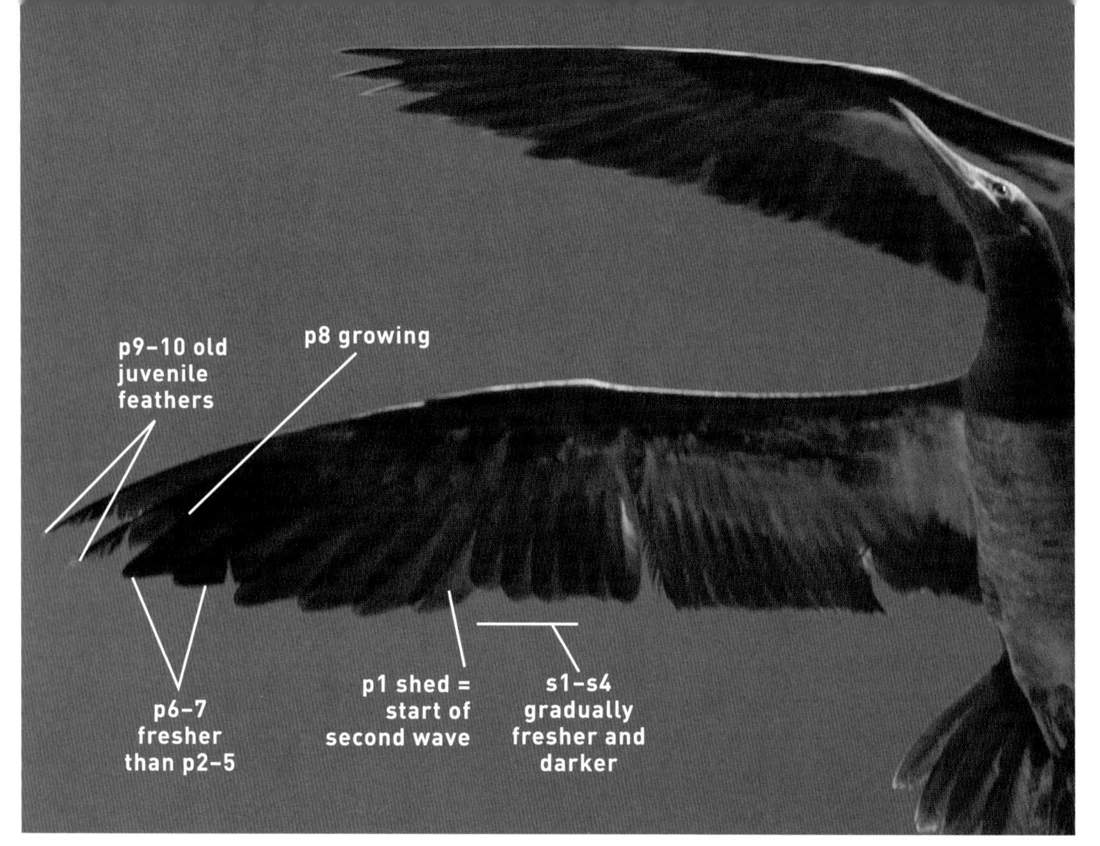

Figure 45E. Second-cycle Brown Booby with the first wave of stepwise primary molt having reached p8 growing, and the second wave having started with p1 shed. A slight contrast between the blacker p6 and browner p5 suggests that molt may have been interrupted at this point. The progressively blacker s1 to s4 indicate increasingly fresher feathers, with a contrast between s4 and the browner and older s5 (see Figure 45D). *Oaxaca, Mexico, 15 Dec. 2008. Steve N. G. Howell.*

Figure 45F. Second-cycle Brown Booby with the first wave of stepwise primary molt finishing with p10 growing, and the second wave having reached p2 growing. *Oaxaca, Mexico, 15 Dec. 2008. Steve N. G. Howell.*

Figure 45G. Second-cycle Brown Booby with the first wave of stepwise primary molt having completed, and the second wave having reached p5 being shed and p4 mostly grown. The second wave often stops at around p6 (see Figure 46). Subsequent molts include two waves, restarting at p7 and at p1. In a perfect iteration, stepwise primary molt replaces all primaries in a single episode of molt, such as one wave renewing p1–p6 and the other wave renewing p7–p10. *Oaxaca, Mexico, 15 Dec. 2008. Steve N. G. Howell.*

Figure 46. Second-cycle Brown Booby with primary molt interrupted. Note that p1–p6 and p9–p10 appear contrastingly black and fresh, whereas p7–p8 have distinctly frayed tips. This pattern results from the second wave of primary molt being interrupted at p6 (see Figure 45G). *Monterey County, CA, 15 Oct. 2006. Steve N. G. Howell.*

The accelerated schedule also occurs in seabirds such as boobies, pelicans, tropicbirds, and perhaps frigatebirds. These species all inhabit tropical regions where wing molt timing is not as constrained as it tends to be in the midwinter conditions of temperate regions. For further discussions about the evolution and benefits of accelerated stepwise schedules, see the family accounts for boobies and gannets, cormorants, and hawks.

The "rules" of stepwise molt decree that each episode of molt restarts at p1 and at the point(s) where the last wave(s) stopped. What this means is that the shorter, relatively protected inner primaries, which take less time to grow, are replaced more frequently over the life of a bird than the longer and more exposed outer primaries, which take longer to grow. In boobies, for example, the inner primaries are renewed after 6 to 8 months and the outer ones after more than 15 months.[135,136] While this may not seem optimal, the overall benefits of stepwise molt apparently outweigh the differential replacement schedules of inner and outer primaries. And this disparity in feather ages is usually only relevant in the first one or two waves of molt. After the third wave starts, all or almost all primaries can be replaced in one episode of molt such that all feathers are effectively the same age. The result of an acceler-

Figure 47. A first-cycle Brown Pelican with uniformly fresh juvenile plumage and no wing molt. When attempting to identify molt waves in the primaries, it is always helpful have a uniform generation of flight feathers for reference. *Marin County, CA, 16 July 2008. Steve N. G. Howell.*

Figure 48. In this first-cycle Brown Pelican, the first wave of primary molt has reached p7 being shed (p1–p5 are new and p6 is growing), and the outer three primaries (p8–p10) are juvenile feathers. Also note that molt has started at two centers in the outer secondaries, inward from s1 (s1 is new, s2 growing) and s5 (which is growing). Molt also starts at one or two centers among the inner secondaries, which soon leads to complicated-looking patterns of replacement among the secondaries (Pyle 2006). With the shedding of another one or two primaries, a second wave of primary molt can be expected to start (see Figure 49). *Marin County, CA, 16 July 2008. Steve N. G. Howell.*

Figure 49. This second-cycle Brown Pelican's molt is slightly advanced over that of the individual in Figure 48 and shows two waves of primary molt. The first wave (preformative) has reached p8 growing, with p9 on the right wing shed and on the left wing just growing in; p10 is still a juvenile feather. In the second wave (second prebasic), p1 is mostly grown and p2 has been shed; no juvenile feathers are left among the outer and middle secondaries (compare Figure 48). *Marin County, CA, 16 July 2008. Steve N. G. Howell.*

Figure 50. This second-cycle Brown Pelican is presumably of a similar age as the bird in Figure 49 and illustrates variation in the rate of primary molt waves. The first wave (preformative) on this individual has reached p8 growing, with p9–p10 still juvenile feathers. In the second wave (second prebasic), p1–p2 are new and p3 is growing. Note how p4–p7 become gradually fresher and darker, reflecting their sequential renewal. *Marin County, CA, 16 July 2008. Steve N. G. Howell.*

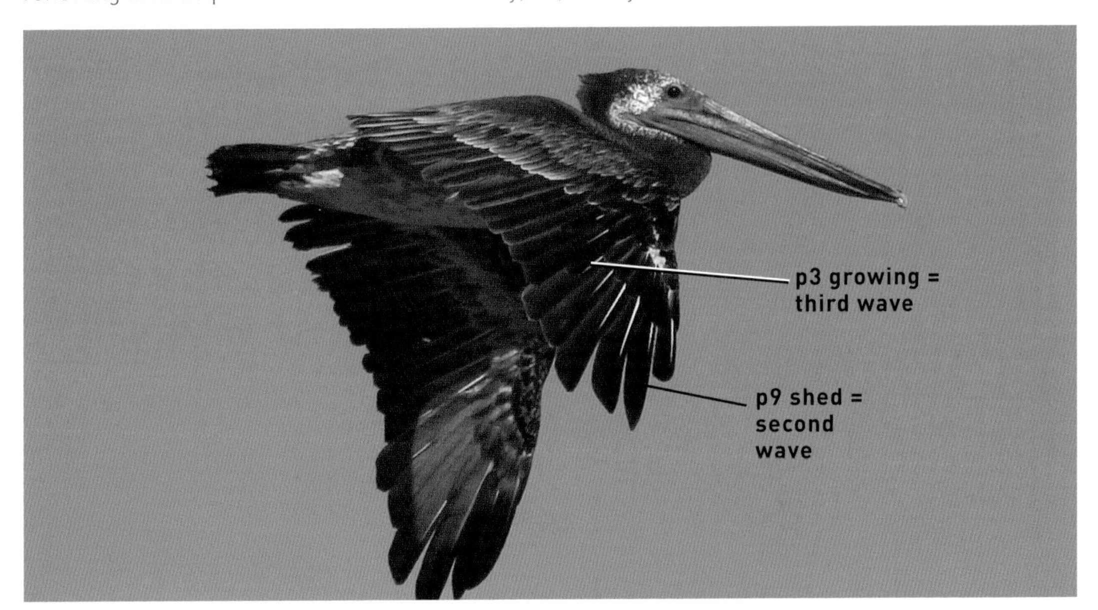

Figure 51. This presumed third-cycle Brown Pelican appears to be one year older than the bird in Figure 50: note that the retained p10 is relatively fresh and blunt-tipped, not tapered and worn as a juvenile primary tends to look by the time two waves of wing molt have developed. The advanced-looking plumage aspect of the head and neck also suggests an older age. The molt waves of this bird are at a similar stage as those of the birds in Figures 49–50. Thus, the outer wave (second prebasic) has reached p7–p8 growing, with p9 shed and p10 old, and the inner wave (third prebasic) has reached p3 growing, with p1–p2 new.

If only two waves are required to renew all of the primaries in one molting period, these wave patterns will be typical of all subsequent ages. Thus, for example, the inner wave might always replace p1–p6, and the outer wave p7–p10. When the next episode of wing molt starts, the waves resume again at p7 and at p1, and so on. *Marin County, CA, 16 July 2008. Steve N. G. Howell.*

Figure 52. This adult Brown Pelican has three waves of primary molt that have gone out of sync, with the waves on one wing being about one primary advanced relative to those on the other. On the near wing, the waves have started at p9 (growing, with p10 shed), p5 (mostly grown, with p6 growing), and p1 (mostly grown, with p2 shed). On the far wing, the waves have started at p8 (growing, with p9 shed and p10 old), p4 (mostly grown, with p5 shed), and p1 (shed, p2 still old). *Marin County, CA, 16 July 2008. Steve N. G. Howell.*

Figure 53. This first-cycle Osprey is starting its first wave of primary molt, which will develop into a stepwise pattern. Most of the flight feathers are juvenile and uniform in wear, except that p1 is growing and p2 has been shed. This first wave of primary molt usually continues out to about p8 before the second wave starts again at p1. *Nayarit, Mexico, 15 Jan. 2007. Steve N. G. Howell.*

Figure 54. This adult Osprey shows five waves of primary molt, ending at p10 (growing), continuing at p8 (growing), p5 (growing), and p3 (growing), and starting at p1 (which has been shed). The outer growing primaries are nearer to their full length than are the inner growing primaries, suggesting that the outer waves started well before p1 was shed. And the old p9 and p7 are appreciably more worn and frayed than the old p2 and p4. With long-distance migrants such as the Osprey, it is difficult to know how suspensions of molt for migration might modify the development of waves. *Nayarit, Mexico, 8 Jan. 2007. Steve N. G. Howell.*

ated stepwise schedule is often to set up two waves of molt, such as p1–p6 and p7–p10. If all feathers can be replaced with these two waves, then they are repeated each year.

Over time, it is not uncommon for stepwise waves to get out of sync between wings (Figure 52), and p1 does not necessarily molt at the start of each episode of molt.[137] Thus, although p1 still appears to be renewed in each episode, after a certain number of cycles might a wave starting at p1 even get skipped in some years? Particularly for species that can suspend their wing molts for migration, such as the Osprey and Swainson's Hawk, stepwise patterns can become quite complex.

ALTERNATIVE WING MOLTS. Most species of birds exhibit standard sequential wing molt, synchronous wing molt, or stepwise wing molt, but a few groups have developed strategies that fall outside these three main patterns. These exceptions are discussed more fully in the family accounts and are summarized here.

Albatrosses, falcons, parrots, barn owls, some typical owls, and some kingfishers have what can be termed "divergent primary molt patterns," whereby molt does not start at p1 but instead starts at a middle primary (usually between p4 and p8) and moves outward and inward from that point. The replacement of all primaries may occur in one molt cycle (as in falcons) or may require up to 3 or more cycles (as in albatrosses and some large owls).

The Limpkin appears to molt its primaries "backward" and perhaps in stepwise waves, starting at p10 and molting inward to p1.[138] Cuckoos have "scrambled" primary molts that start somewhere in the middle primaries and then skip around seemingly at random in a pattern that has been termed "transilient molt."[139] Hummingbirds simply manifest a slight variation on the standard sequence of primary molt and replace the outer two primaries in reverse order; thus p10 molts before p9.

BIRDS IN THE REAL WORLD

Now that we have an idea of theory and terminology, we can apply it to birds we see in the world around us, which brings us back to some of the subjects discussed in the background questions. In general, a bird's molt strategy reflects a balance between its ancestry and its present-day life history, in essence a variation on the age-old theme of nature versus nurture. Factors that can affect the when and where of molt, and also its extent, include a bird's environment (including its food), its body size and wing area, and its migration strategy. As well as viewing molt as part of a bird's life cycle, we can use our knowledge of molt strategies to determine the ages of birds we see and to think about how molt strategies can help with field identification.

Figure 55. In falcons, such as this Crested Caracara, the primary molt starts at p4 and spreads sequentially outward to p10 and inward to p1. Molt of the secondaries spreads in both directions from s5 and also moves outward from the tertials. *Chiapas, Mexico, 14 Mar. 2007. Steve N. G. Howell.*

ANCESTRY

Sometimes the molt patterns we observe today don't seem to make sense. In such cases we should consider whether these patterns could be ghosts of times and places past, when different conditions or selective pressures existed. Although conditions may have changed since then, there may not have been sufficient selective pressure to change the molt strategy: if it's not broken, don't fix it. For example, why would Horned Larks breeding on the tundra have complete preformative molts when the other songbirds nesting there have partial molts? It turns out that complete preformative molts are typical of larks, and this may just be an ancestral trait that has not (yet) changed. Another example is found in Baltimore and Bullocks's orioles breeding alongside each other (and even hybridizing with each other) on the Great Plains. After breeding, the Baltimore Orioles stay around and molt before migrating, whereas the Bullock's Orioles (including birds with

Figure 56. This adult Peregrine Falcon is in the late stages of its primary molt, which finishes with p10 and p1 after having started at p4 (see Figure 55). On this individual, p9 is growing and p10 is still old, and p1–p2 have been shed or are just starting to grow but are not yet visible beyond the tips of the coverts. *Sacramento County, CA, 28 Sept. 2008. Lyann A. Comrack.*

ABOVE: **Figure 57.** Wing molt in albatrosses is not well understood. In Black-footed Albatrosses, the outer three primaries are molted outward every year in a p8–p10 sequence, whereas from none to all of the inner seven primaries are replaced every year, perhaps via stepwise waves that move inward from p7 to p1. On this individual, p1–p5 are new, p6 is contrastingly old, p7–p8 are new, and p9–p10 are growing and short (especially p10). The very white head of this individual reflects strong bleaching of the feathers. *Marin County, CA, 14 Aug. 2007. Steve N. G. Howell.*

RIGHT: **Figure 58.** In large owls, three or more molt cycles may be required to replace all of the juvenile primaries. Primary molt usually starts at p7, and subsequent molts move outward and inward from this point. The second prebasic molt can include just p7, or sometimes p6–p8. On this female Snowy Owl, note the subtle contrast of p7–p8 in pattern (sparser bars), shape (slightly blunter tips), and wear (slightly fresher, with blacker bars) relative to the other primaries, indicating a bird in its second winter. *Solano County, CA, 22 Jan. 2007. Larry Sansone.*

Figure 59. This Anna's Hummingbird shows a feature of primary molt sequence unique to hummingbirds. Instead of a standard sequential p1 to p10 progression, the outer primaries are molted in the sequence p8–p10–p9. On this individual, primary molt is almost completed, with p1–p8 and p10 fresh and new, and p9 growing in (its tip visible between the tips of p7 and p8). Because p9 is the largest and heaviest primary, precision flight might be compromised if p9 were shed first, when p10 could be quite worn. Thus, replacing p10 first provides more support for the growth of p9. *Marin County, CA, 24 Aug. 2008. Steve N. G. Howell.*

which Baltimores might even have mated) migrate off to molt in the Southwest and northwestern Mexico![140] These different strategies are ingrained into each species and are an ancestral reflection of the broad patterns of late-summer food supply in the West and East.

ENVIRONMENT AND FOOD

Basically, molt is all about food, which in turn is related to climate. Where different populations live and migrate can influence both the timing and extent of molts. For example, locally breeding male Ruddy Ducks where I live in coastal central California usually molt into alternate plumage in January or February; in the same area, most wintering birds from other populations retain basic plumage through March or April, when they leave for northern and interior breeding areas where spring comes later.

Superimposed on all this is the potential for interannual variation in food supply due to environmental variables. For example, Black-legged Kittiwakes wintering in the Pacific Ocean often complete their prebasic primary molt by December or January. In years with low ocean productivity, however, this molt can be protracted until May or later, potentially taking a year to complete.[141] And even within a cycle, considerable individual variation and plasticity can be seen in the timing and extent of a molt, perhaps related to factors such as nestling food quality, fledging date, and migration distance. For example, some first-cycle Glaucous-winged Gulls wintering in central California start an extensive preformative molt in October, whereas others retain most or all of their juvenile plumage through February or later.[142]

BODY SIZE AND WING AREA

All else being equal, and because feathers can grow only so fast, larger species take longer to replace their more numerous and larger feathers than do smaller species. Frequent exceptions to the size rule occur, however, because of differences in lifestyle or in body size relative to wing area. For example, storm-petrels are relatively small pelagic birds that live in a demanding marine habitat, lay only a single egg, and have a prolonged breeding season. Their protracted prebasic molt can take 8 to 9 months, and it overlaps with breeding.[143] Conversely, much larger terrestrial birds such as jays are omnivores that tend to have ready access to food and lay clutches of 3 to 6 eggs. Their complete molt can take 8 to 9 weeks, as with Steller's Jay,[144] and it is easily completed between breeding seasons. Some heavy-bodied but not especially small birds such as ducks become flightless for a short period during molt. In this way a Mallard can replace its primaries in only 3 to 4 weeks, compared with 5 to 6 months for the much smaller Ashy Storm-Petrel.[145]

An extreme example of different lifestyles and molt strategies occurs in penguins and hummingbirds. Emperor Penguins *(Aptenodytes forsteri)*, weighing up to about 40 kilograms, have a relatively low metabolism and need seamlessly insulated plumage to swim in cold water. They stock up with food and then stand around on ice floes without eating for about 5 weeks, during which time they undergo an intensive molt of their dense coat of small feathers.[146] Conversely, the small gorgeted hummingbirds in North America, such as Ruby-throated, Anna's, and Calliope (weighing only about 3 to 4 grams), have a high metabolism and depend on their remarkable flight powers for feeding—which they must do continually. They undergo protracted molts, which, for their relatively large primaries, can take 4 to 6 months to complete.[147,148,149]

MIGRATION

In general, birds don't molt their main wing feathers during migration because any reduction in flight capability might compromise a successful migration. Whether they molt before or after migration reflects where and when they can find the most reliable food, in combination with other factors such as daylight length and ambient temperature. In general, shorter-distance migrants (such as Purple Sandpiper and Lesser Nighthawk) tend to molt on or near the breeding grounds before migrating, whereas longer-distance migrants (such as Sanderling and Common Nighthawk) molt mostly on the nonbreeding grounds.[150,151] One reason for this strategy in short-distance migrants may be a combination of ample food on the breeding grounds and time to molt before migrating the short distance to temperate nonbreeding grounds, where conditions such as climate, food supply, and competition for food in a relatively small area may be less favorable for molt. Another factor, especially for high-latitude breeders, is that feather synthesis at night may require costly modifications to the metabolism of amino acids, compared with daytime synthesis;[152] hence, the overall cost of molt may be lower in areas with relatively long days. Conversely, long-distance migrants need to migrate greater distances, but in winter they experience longer daylight hours, warmer climates, and potentially larger areas where they have sufficient food and time for molt. Warmer climates for molt would also be favored if the costs of heat loss were greater than the costs of feather synthesis.[153]

Molt isn't necessarily a "before-or-after" event relative to migration between the breeding and nonbreeding grounds. Another strategy is to molt mostly or entirely at staging sites, a phenomenon sometimes termed molt migration. Staging sites are areas between the breeding and nonbreeding grounds where birds rely on predictable food resources to fuel their molts. Examples include the

Figure 60. In their first winter, some Glaucous-winged Gulls undergo an extensive molt of head and body feathers by early winter, whereas others, like this individual, retain most of their juvenile plumage through the winter and start head and body molt in February–March. This bird is still mainly in juvenile plumage but is replacing some scapulars and some feathers on the head and sides of the chest; the new feathers look subtly grayer and fresher. *Sonoma County, CA, 12 Mar. 2008. Steve N. G. Howell.*

Figure 61. Many long-distance migrants, such as this Common Nighthawk, undergo their wing molt on the nonbreeding grounds where days are warm and food is plentiful. *Tooele County, UT, 29 June 2008. Steve N. G. Howell.*

huge fall gatherings of Eared Grebes at Mono Lake, California,[154] and the fall staging of Lazuli Buntings in the monsoon-enriched Southwest deserts.[155] In some cases, staging sites may actually lie in the opposite direction from the nonbreeding grounds, and molt can be suspended during migration, as birds track food resources. For example, adult Elegant Terns start their prebasic molt on the nesting grounds in the Gulf of California, suspend it and migrate north to the California Current where they continue (and rarely complete) molt, and then suspend molt again and migrate south to the Humboldt Current off Peru and northern Chile, where they finish their prebasic molt.

Different populations of a species can have different molt strategies relative to migration. Dunlin populations breeding from east of the Ural Mountains across Siberia into North America molt before migrating south in fall. Dunlin populations breeding from west of the Ural Mountains through northern Europe into Greenland migrate south and then molt at staging sites and on the wintering grounds; these differences have been related to the distribution of food resources.[156] The extent of first-cycle molts in Sanderlings varies with wintering latitude: birds that winter in the temperate Northern Hemisphere replace no primaries in the first cycle, whereas first-cycle birds that winter in the Southern Hemisphere can replace all of their primaries.[157] The differences in the extent of first-cycle molts presumably reflect the need to counter increased feather bleaching and wear caused by higher levels of ultraviolet radiation in tropical and southern latitudes.

As with most rules, there are exceptions. In western North America, numerous hawks and falcons migrate while undergoing wing molt.[158] Conversely, in eastern North America few molting hawks are seen during migration, and the reasons for this east–west difference are not known. Might it relate to broad-scale patterns of food abundance, as with songbirds?

Some adult Herring Gulls along the Pacific Coast undergo wing molt while migrating. These birds probably can find food at almost any point along their migration route, and thus they undergo a gradual molt of their outer primaries while drifting southward. More sedentary populations of Herring Gulls, such as those in New England, often exhibit more intensive molt, with two or three outer primaries growing simultaneously. Their relatively impaired flying capability at such times could handicap sustained migratory flight, but presumably it doesn't adversely affect local foraging. Some Black Terns also migrate while undergoing a slow rate of wing molt and can feed as they migrate.[159] Another exception is found with swallows. These are un-

Figure 62. These Elegant Terns may undergo their prebasic wing molt in three stages as they track food resources. The inner primaries are molted on the breeding grounds in Mexico, then the birds migrate north to California and continue their molt. Often they don't complete primary molt in California, so they suspend it until reaching the nonbreeding grounds in Peru and Chile. The last stage of prebasic primary molt in mid- to late winter may overlap in timing with the prealternate molt of inner primaries. *Marin County, CA, 16 Oct. 2007. Steve N. G. Howell.*

usual among songbirds in being daytime migrants, but some species undergo a slow, protracted molt of their flight feathers as they migrate, and, like terns, they can also feed as they migrate. Among songbirds that migrate at night, an overlap of prebasic wing molt with active migration has been found in the Rose-breasted Grosbeak and Red-eyed Vireo, although not in other species.[160]

PATTERNS OF MOLTING

Once you start to think about why, when, and where birds molt, some interesting patterns emerge. We can always find exceptions, but identifying the broad-scale patterns can be informative in understanding the evolution of molt strategies in relation to life-history traits, both past and present. Some of these patterns are noted here, and the family accounts discuss numerous others.

PREALTERNATE MOLT AND MIGRATION. Something that soon becomes apparent is that prealternate molts are more common among migrants than residents. Why would this be? It may have started off that migrants exposed their feathers more to the elements during migration, and some also moved to sunnier areas where their feathers were more likely to fade and wear. These migrants also tend to be on the breeding grounds for relatively short periods and need to find mates quickly. Prealternate molts that originally developed to replace worn feathers may have incidentally permitted striking changes in appearance, especially for males, perhaps by happening at the same time that breeding hormones developed. This in turn may have helped attract mates and may have reinforced the need for prealternate molts. Conversely, many resident bird species form longer-lasting pair bonds and are not constrained for time—and the sexes often look alike or are relatively dull-plumaged. Think about a human couple that's been together for years and how they dress relative to folks hanging out at a singles bar; it's not that different.

PREALTERNATE MOLT AND HABITAT. Now let's add habitat to migration, or the lack of it. Another pattern is that resident forest species, whether tropical or temperate, typically lack prealternate molts; examples include such diverse groups as woodpeckers, trogons, antbirds, manakins, kinglets, thrushes, and chickadees. Birds living in forests tend to be protected from persistent exposure to sunlight, so their feathers remain relatively protected, and extra molts aren't needed. Conversely, many species that inhabit open or scrubby habitats, where they are exposed to the elements and harsh vegetation, often have prealternate molts, such as sparrows and tyrant-flycatchers, even if these molts bring about no marked change in appearance.

So, given that most tropical songbirds are nonmigratory and many live in forested habitats, we would expect prealternate molts to be rare in these birds—and they are. The only well-documented examples of Neotropical songbirds with prealternate molts are nomadic or partly migratory species of open and semiopen habitats, including some seedeaters and grassquits, and the Red-legged Honeycreeper (*Cyanerpes cynaeus*).

Figure 63. Conventional wisdom states that birds do not molt and migrate at the same time. However, many raptors in western North America, such as this adult Cooper's Hawk, undergo active molt of their flight feathers while migrating. *Salt Lake County, UT, 21 Sept. 2008. Jerry Liguori.*

MOLT AND TIME. Another pattern seen among diverse families of birds is that the extent of molt varies with environmental conditions and life-history traits. Not surprisingly, prealternate molts tend to be less extensive (or even absent) when less time is available for molt, as occurs with residents or short-distance migrants, which usually start breeding earlier than do long-distance migrants. This can even be seen within a single species. Thus, northern-breeding populations of Chipping Sparrows (which migrate long distances) have more-extensive prealternate molts than do southern-breeding populations (which start breeding earlier).[161] Something similar may even occur between age classes of a species. Thus, first-year male Baltimore Orioles, which often

Figure 64. The Red-legged Honeycreeper is one of a handful of Neotropical songbirds that undergo seasonal changes in their appearance. In fall this alternate-plumaged male will replace his intense blue plumage for a drab green, much as a male Scarlet Tanager does. *Suchitepéquez, Guatemala, 5 Mar. 2008. Steve N. G. Howell.*

start breeding later than adult males, have a more extensive prealternate molt than adult males,[162] perhaps because of more time available for molt.

Preformative molts vary greatly in extent, even within a species or between related species. This variation may reflect the time available for a molt, with later-hatched young tending to have less extensive preformative molts than earlier-hatched young.

Migrants can effectively "create time" for molt by moving south to warmer climates in winter. Thus, first-cycle Common Terns, which breed in the northern summer, are long-distance migrants that have protracted and complete preformative molts over the course of a Southern Hemisphere summer. Forster's Terns, however, are short-distance migrants that have a partial preformative molt before the onset of the temperate northern winter they experience, and they may even complete their molt before Common Terns start theirs (see the family account for terns). Another way of looking at this is that, all else being equal, molt may expand to fill the time available for it.

WING MOLT AND GEOGRAPHY. The example of the two terns leads us to the patterns that link migration distance, wintering latitude, and the location of wing molts (this subject is expanded upon in the family account for sandpipers). Among birds breeding in North America, the prebasic molts of residents and short-distance migrants (those wintering mainly north of the tropics) occur in late summer and fall, between the end of the breeding season and the onset of winter. In most songbirds these molts occur on or near the breeding grounds, but some species migrate in fall to molt in food-rich areas of the Southwest and northern Mexico.[163]

Among long-distance migrants (those wintering mainly in South America), some species molt before migrating whereas many others migrate to South America before molting. What this means is that any North American breeding species that does not molt its wings in North America is likely to be a long-distance migrant, be it a Buff-breasted Sandpiper, a Sabine's Gull, a Common Nighthawk, a Black Swift (see page 55, Life-history Predictions), an Eastern Kingbird, or a Cliff Swallow. This trend can even be seen in species that winter across wide spans of latitude.[164] For example, adult Greater Yellowlegs, Whimbrels, and Surfbirds that winter in California arrive on their wintering grounds and quickly undergo their prebasic wing molt in late summer and autumn, before the onset of winter. Individuals of these same species that winter in Chile wait until their arrival there to molt, and their prebasic wing molts may continue well into the "winter" (that is, the southern summer) because they do not have to worry about cold northern weather. Are such differences genetically programmed? Or do they simply reflect an individual bird's responses to the environment in which it winters?

COMPLETE PREFORMATIVE MOLTS. In most species the preformative molt is partial, but in some species it is complete. In other species, such as the Northern Mockingbird, Phainopepla, Northern Cardinal, and House Finch, the extent of this molt varies among individuals and populations, presumably reflecting factors such as the time available for molt and how worn the plumage may become. But what characteristics unite the species in which all individuals typically have complete preformative molts?

In some species, complete molts appear to be an ancestral trait that sometimes seems to have no obvious benefits today. Such species are mainly residents and short-distance migrants, such as gamebirds (grouse, quail, and turkeys), Horned Lark, Bushtit, Wrentit, European Starling, some blackbirds, and House Sparrow. The great increase in size that occurs in growing young gamebirds explains why a complete molt might occur (as discussed in the family account for turkeys), but in the other species there seem to be no compelling reasons why they should have complete preformative molts and other species sharing their habitats should not. Interestingly, most of these families have their origins in the Old World, suggesting that conditions there may have favored the evolution of complete preformative molts.

Another group of species with complete preformative molts comprises birds that rely on flight to feed: the Wilson's Storm-Petrel, jaegers, some terns, Black Skimmer, hummingbirds, tyrant-flycatchers, and swallows. For these species, a gradual molt of the primaries is necessary so they can maintain their powers of flight to forage; a hummingbird or swallow that couldn't fly wouldn't live very long. If these species followed the usual pattern of waiting until they were 1 year old before starting their first wing molt (that is, as part of the complete second prebasic molt), the necessarily gradual nature of this molt in aerial feeders would mean that their outermost primaries might not get replaced until about 18 months of age, by which time they could be too worn to function effectively.

A third group of birds with complete preformative molts consists of long-distance migrants that return north in their first summer, such as the American Golden-Plover and Baird's Sandpiper. For these species, having their wings in good con-

Figure 65. Whimbrels that migrate as far south as Chile undergo their prebasic wing molt over the "winter," which is summer in Chile. This individual is starting its primary molt on a date when Whimbrels wintering in California have usually completed their primary molt. *Region V, Chile, 7 Nov. 2007. Steve N. G. Howell.*

Figure 66. The poorly regarded House Sparrow is one of several species of resident songbirds that have a complete preformative molt by which first-cycle birds attain the adult plumage aspect. In many cases this complete molt may be an ancestral trait that conveys no obvious advantage to birds living today. Growing secondaries can be seen on this young male, as can a gap among the primaries indicative of a feather recently shed. *Marin County, CA, 31 Aug. 2008. Steve N. G. Howell.*

dition for long flights and exposure to year-round sunny conditions is important. Because prebasic wing molts occur on the nonbreeding grounds, the juvenile primaries would have to support three long-distance migrations before being replaced: north–south, then south–north, and then north–south again.

SIMPLE VERSUS COMPLEX STRATEGIES. When we view molt strategies as either simple or complex (see Four Fundamental Molt Strategies, pages 25–32), some patterns emerge. For example, all songbirds have complex strategies, whereas simple strategies occur mainly in larger birds that do not usually breed in their first year of life. The only North American bird families that have only the simple basic strategy are albatrosses and barn owls (see Table 1). Other families in which some to most species have this strategy include petrels, storm-petrels, and swifts. Most of these birds have relatively long nestling periods during which a strong juvenile plumage can develop. They also have few or no nest predators and nest either on remote islands or in places inaccessible to native predators.

The simple alternate strategy occurs almost exclusively among larger waterbirds, such as loons, pelicans, cormorants, ibises, spoonbills, large gulls, alcids, and possibly some waterfowl, rails, and large shorebirds (see Table 1). Is there something about frequent exposure to water, especially salt water, that predisposes birds to develop alternate plumages? Or are these birds just too big to manage two extensive molts of body feathers in their first cycle?

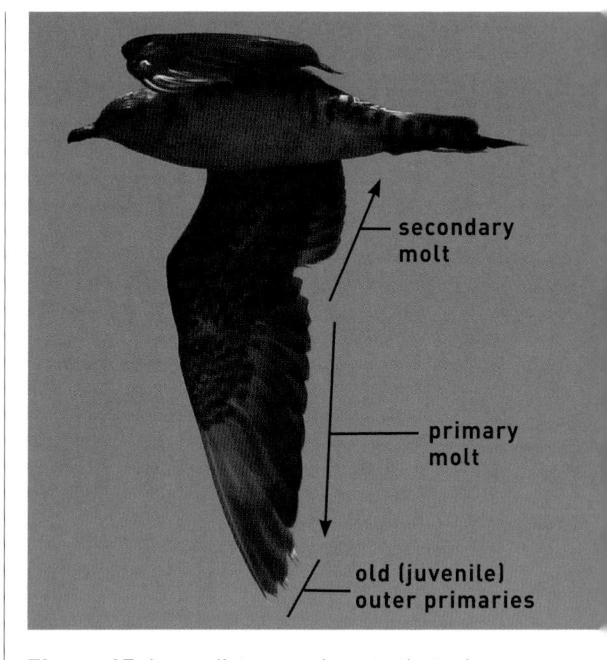

Figure 67. Long-distance migrants that rely on flight to forage, such as this Long-tailed Jaeger, undergo complete preformative molts on the nonbreeding grounds. Primary molt on this individual has reached p7 growing; p8–p10 are worn juvenile feathers that will be replaced by the summer, before the complete second prebasic molt starts in late fall. *At sea north of New Zealand (30°S, 169°E), 5 Apr. 2008. Steve N. G. Howell.*

LIFE-HISTORY PREDICTIONS

Recognizing patterns of molting, and understanding how they relate to other life-history traits such as migration distances and breeding seasons, allows us to make predictions about some of the many unknowns that remain in the world of birds (see page 51, Patterns of Molting).

For example, where do Black Swifts winter? The American Ornithologists' Union[165] in 1983 noted that this species "winters in Mexico (presumably), [south] through the breeding range . . . to Costa Rica." Yet nobody has ever seen a Black Swift in wing molt in North America, or in Central America to the best of my knowledge. This fact points to the species being a long-distance migrant that likely winters somewhere in South America, a prediction that was subsequently supported by specimens collected migrating south through Colombia in fall.[166,167]

How about Cave Swallows? As recently as 1998, the nonbreeding range for northern-breeding populations of Cave Swallows was considered "unknown."[168] However, unlike Black Swifts and Cliff Swallows (the latter another long-distance migrant that winters in South America), Cave Swallows are often seen in wing molt in the U.S. and Mexico. This suggests they are not long-distance migrants, so we could predict that they winter somewhere in Mexico or Central America. Thus, it came as no surprise when thousands of Cave Swallows of the subspecies that breeds in the U.S. were found wintering in El Salvador.[169]

It's not just the nonbreeding grounds but sometimes even the breeding grounds of a species that remain unknown. For example, nobody knows exactly where Hornby's Storm-Petrel (*Oceanodroma hornbyi*) nests, although it is presumed to be somewhere in the vast barren deserts of northern Chile. But when would you search for it breeding? Well, other storm-petrels in the genus *Oceanodroma* exhibit little or no overlap of wing molt and breeding, and first-year birds molt earlier than do adults. Thus, if Hornby's Storm-Petrels are seen year-round at sea but birds in wing molt are seen only during certain periods (say, April to November), it would make sense to look for active colonies during periods when birds are not molting or when wing molt is in its early stages (say, December to June). Not enough observations of molting Hornby's Storm-Petrels have been accumulated, or at least published, to predict when it breeds, but an appreciation of molt timing could be helpful in such a case.

An example nearer to home involves the Black-capped Petrel, which actually may comprise two or more species.[170] One of the differences between the types of Black-capped Petrel is that white-faced birds molt a month or so earlier than black-faced

Figures 68–69. The Black-capped Petrel may actually comprise two distinct species that have not been formally recognized. Besides differences in head pattern, the two types differ in their wing-molt timing, suggesting they have different breeding schedules. Thus, molt timing lends support to the idea of two species. The "white-faced" types are typically well into their primary molt in late May, at which point most "black-faced" types are in worn plumage but have not yet started primary molt. *Dare County, NC, 21 May 2008 (black-faced) and 28 May 2008 (white-faced). Steve N. G. Howell.*

birds, suggesting that the breeding season of white-faced birds also might be a month or two earlier. This information should be useful when searches are made for the as-yet-unknown breeding grounds of the white-faced Black-capped Petrels.

SEEING MOLT IN THE FIELD

How can you tell if a bird you are looking at is molting? Sometimes it's easy, but at other times you simply can't tell. Bigger birds have bigger feathers, so it's usually easier to see molt on, say, a pelican than on a wren. And remember, most birds are molting only at certain times of year and usually not when they're migrating. People often mistakenly refer to mottled first-year male Summer Tanagers or Indigo Buntings as "molting" when in fact these birds have stopped molting and simply exhibit a patchy appearance.

There are two fairly easy ways to see molt, and evidence of molt. The first is by noting gaps caused by missing or growing feathers, such as in the wings. The second is by observing the contrast in wear and color between new and old feathers. Because molt is defined as feather growth, technically it can really only be an active phenomenon. However, there is some value to distinguishing between "active molt" (which is ongoing) and "interrupted molt" (which may simply be suspended, as for migration, or may be arrested, not to be continued); these two terms are often used by birders and field ornithologists.

ACTIVE MOLT. The easiest place to observe active molt is in the wings of large birds in flight. If you watch Turkey Vultures through the summer in North America, it is often difficult to find one that isn't molting its flight feathers, and when the first juveniles appear, usually in July or August, they stand out by having uniformly fresh wings with no molt. After a while, you may also start to notice evidence of wing molt on perched birds, where gaps in the primaries can be quite easy to detect once you know what to look for (compare Figures 76 and 77).

Because molting birds are often vulnerable when they have an incomplete coat of feathers, they can be somewhat retiring during this period. For example, how often do you see a Red-tailed Hawk molting from its barred juvenile tail feathers to the red adult tail feathers? Young Red-tails seem to lie low during this time and probably hunt mainly from perches in the woods, rather than getting up and flying around.

Figure 70. Active wing molt is evident on this Anna's Hummingbird, which is growing its inner primaries (visible as blacker feathers projecting past the tips of the tertials), has shed its middle primaries (hence the big gap in the middle of the primaries, exposing much of the base of p7), and still has its four juvenile outer primaries, which will molt slowly once the rest of the wing has filled in (see Figures 12 and 59). The fine whitish tips to the tertials, and the relatively fresh condition of the outer primaries, point to this being a first-cycle bird and thus undergoing its complete preformative molt. *Marin County, CA, 24 Aug. 2008. Steve N. G. Howell.*

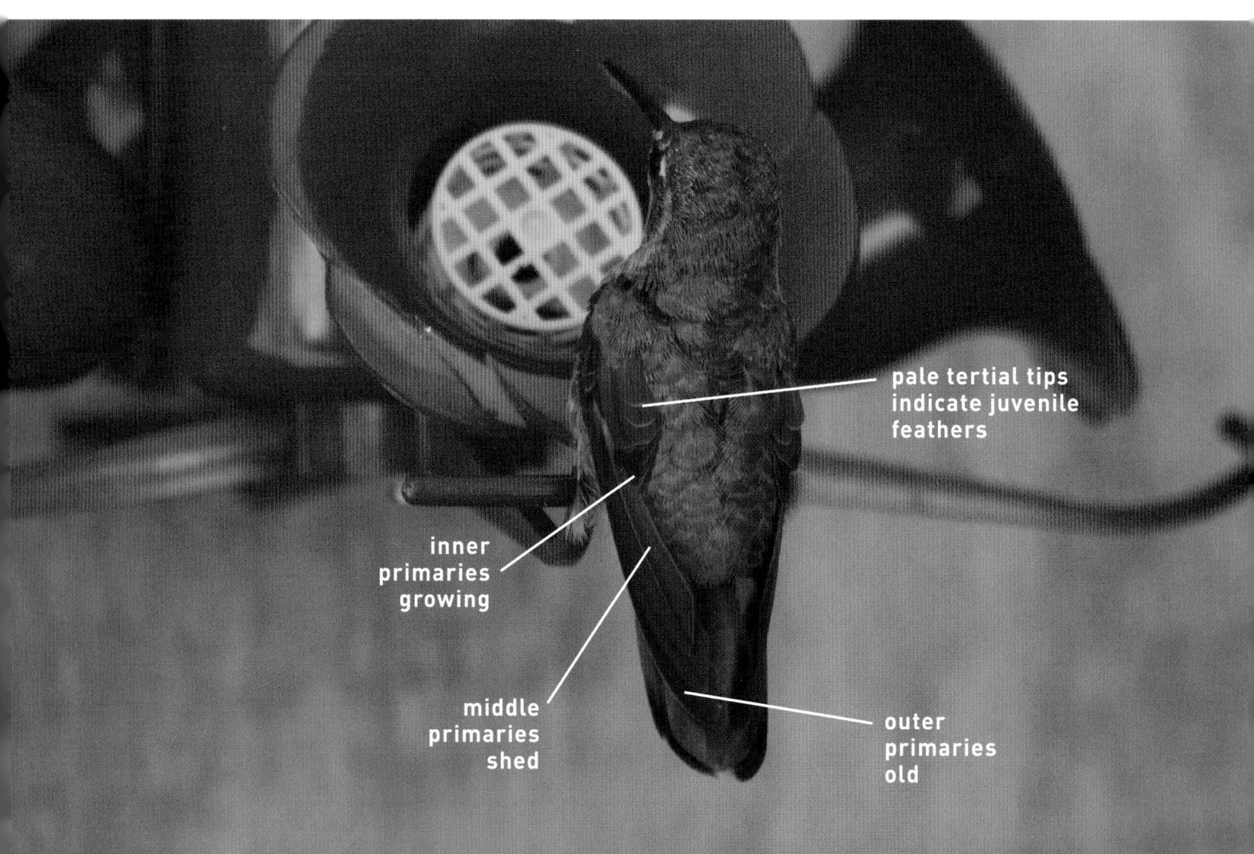

LEFT: **Figure 71.** On this Forster's Tern, the molt contrast between the fresher silvery inner primaries and the older darker outer primaries is obvious, and it indicates that the prealternate molt included six inner primaries. Adult Forster's typically don't complete their primary molt until October–November, when there is little or no time for a prealternate wing molt. Immatures, however, complete their second prebasic molt in August–September and thus have time for an extensive prealternate wing molt in September–October. This molt can create a contrast in the wing, suggesting the pattern of a Common Tern. The age of this bird can thus be inferred as a second cycle, based on the extent of its prealternate molt. *Jalisco, Mexico, 24 Jan. 2007. Steve N. G. Howell.*

Figure 72. Turkey Vultures have extensive overlap between molt and breeding, so almost any individual you look at in North America in summer is likely to show obvious wing molt. On this adult, p8 is growing on both wings, and p9–p10 are old; p10 is unusually worn, which may indicate it was not molted in the last cycle. On the top wing, p1 has been shed at the start of the next wave; on the bottom wing, p1 is growing and p2 has been shed. There is also some molt going on among the secondaries, and the outer tail feathers are growing in. *Marin County, CA, 5 Aug. 2008. Steve N. G. Howell.*

Figure 73. When juvenile Turkey Vultures, such as this bird, first appear in the skies after fledging in July–August, they stand out from older age classes by having uniform-generation wings with no evidence of molt. Compare with Figure 72. *Marin County, CA, 1 Sept. 2008. Steve N. G. Howell.*

Figure 74. This fresh-plumaged first-cycle House Finch (the sex of this individual, which is mostly or entirely in its juvenile plumage, can't be determined) provides a useful baseline from which to follow molt in subsequent ages (see Figures 75–77). Note the uniform-generation upperwing coverts, with two evenly edged pale wingbars and narrow pale edgings to the tertials. *Marin County, CA, 4 Aug. 2008. Steve N. G. Howell.*

Figure 75. Here's a male House Finch in worn plumage in late summer. All of the wing feathers appear to be similarly worn and thus are presumably of a similar age. There is no step in the greater coverts, and the most exposed inner coverts are the most worn, as we would expect. This is why these feathers are replaced in the preformative molt rather than the less exposed outer coverts (see Figure 76). The age of this bird is uncertain. It could be an adult, or it could be a first-cycle bird that had a complete preformative molt. *Marin County, CA, 8 June 2008. Steve N. G. Howell.*

Sparrows, wrens, and other small songbirds can also "go to ground" during their molt. In some habitats they seem to disappear for a while, only to reappear a few weeks later in fresh plumage, when they often announce that they're back by a burst of fall singing.

If birds are habituated to people and have a steady supply of food, they can be much easier to observe. Thus, bird feeders and city parks are great for observing molt. For example, Canada Geese and Mallards undergo their synchronous wing molt at many city parks, and they are flightless in full view (see Figure 43). Yet geese and ducks "in the wild" find remote areas or seek dense cover for this molt, which is when they are most vulnerable to predators. Close-range birds at feeders can often be studied easily, and you may see more subtle evidence of molt.

Molt is most easily seen on birds that change their appearance when they change their feathers. Thus, detecting the molt from juvenile to formative plumage can be easy with species in which these two plumages look quite different, but is difficult when juveniles and adults are similar in appearance. Given good views, though, you can start to appreciate how the soft, lax look of juvenile plumage differs from the stronger formative or adult plumage.

Following the progress of wing molt on hummingbirds is fairly easy if individuals return regularly to your feeder, and keeping track of this can even provide new and interesting information[171] (compare Figures 12, 59, and 70). Gulls are often easy to observe; just take a little bread and you can follow their molts and changes of appearance fairly easily. In fact, it was a study of molt in the Western Gull that precipitated a revision in 2000 of the H-P system.[172,173] So don't ignore those common birds—there's still much to be learned.

Detecting molt can be difficult if not impossible, however, on many birds in the field, especially if they are simply undergoing a low-level renewal of body feathers where the incoming plumage looks just like the plumage being replaced. Conversely, birds undergoing an intense molt in a short period are usually obvious (see Figure 11). Once you open your eyes to molt, it can often be surprisingly easy to see, and it adds another dimension to your everyday birding. When do the House Sparrows at my feeder usually start molting? How many of these adult Greater Yellowlegs are in wing molt, and does this reflect how many will winter here? Are the goldfinches or House Finches molting earlier than usual this year? And so on.

Paying attention to molt can also help you get more out of your birding and add value to your field notes. For example, how many Turkey Vultures counted at your local hawk-watch are molting their primaries? Are these more likely to be local

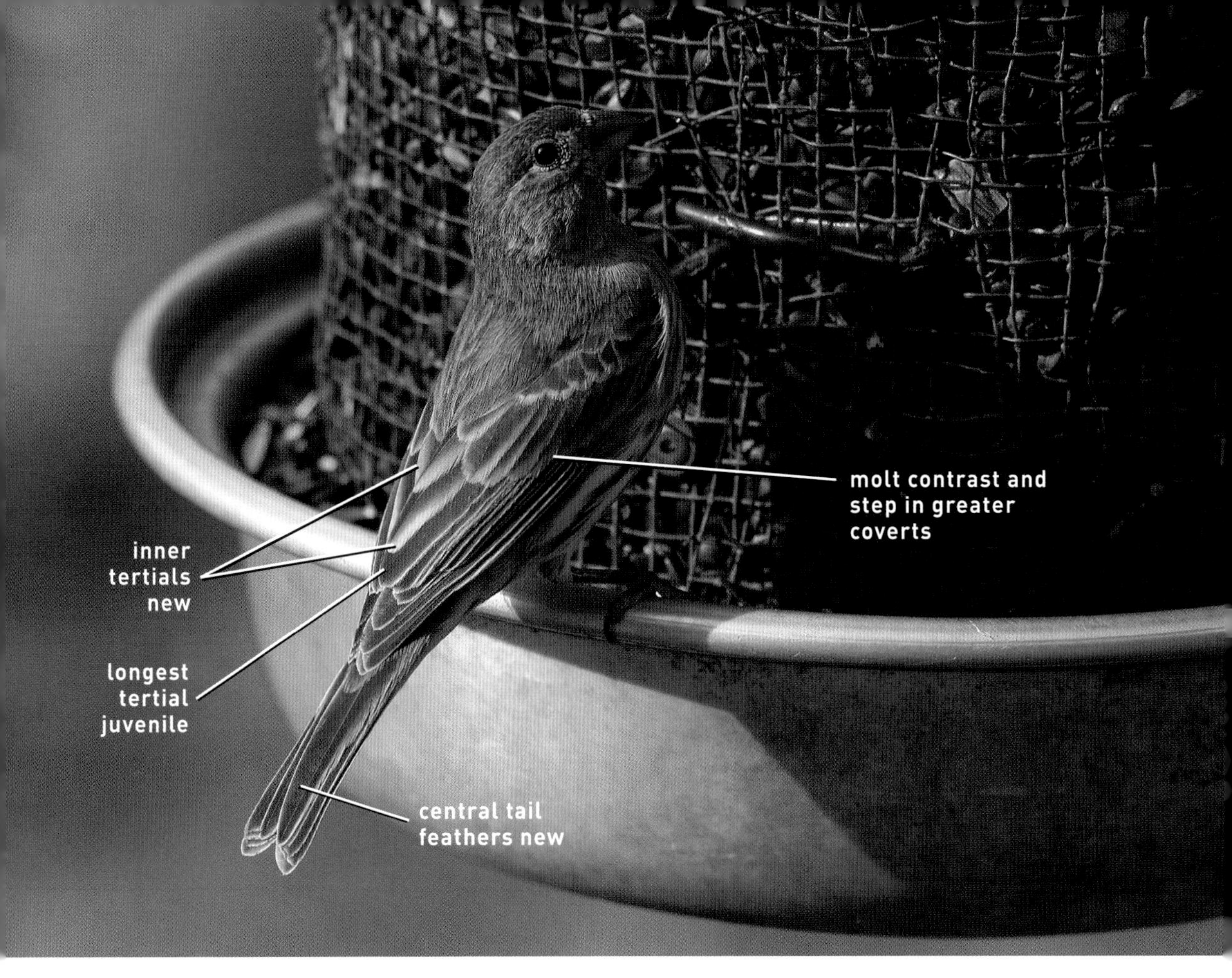

Figure 76. There's a male House Finch at your feeder, so what? Well, let's take a closer look. This bird provides a good example of how molt contrasts can be subtle. The inner two tertials are fresher and have broader edgings than the longest tertial, which is a retained juvenile feather, like the secondaries and primaries. Looking carefully, we can also see that the inner five greater coverts have broad fresh tips much like the inner two tertials; the outer five greater coverts are a little hidden in shadow, but they are obviously shorter than the inners. Even though by this time of year there is no appreciable difference in wear among the greater coverts, a step such as this is a clue that two generations of coverts may be present. For whatever reason, formative greater coverts are often slightly longer than juvenile greater coverts. Compare these greater coverts with those of a fresh juvenile (Figure 74) or uniformly worn adult (Figure 75), and note how these other ages don't show this step. Let's look again. The central tail feathers look a little darker and more neatly edged than the other tail feathers, yet the central feathers should experience the most fading and wear, so it looks like they've also been replaced. So, this male House Finch shows molt contrasts in the tertials, greater coverts, and tail feathers, indicating that it is a first-cycle bird. *Frederick County, VA, 26 Nov. 2006. Steve N. G. Howell.*

birds rather than the true long-distance migrants, which may suspend molt for migration? If you see 150 first-cycle migrant Western Sandpipers at your marsh one day in late August, and then 2 days later you see 150, are they the same birds? If you noticed that 50 percent of them had new gray back feathers (that is, formative plumage) on your first visit, but only 10 percent of them had such feathers on your second visit, then you could say confidently that there had been some turnover. Noting the wing molt patterns of individual birds can also help you keep track of different birds, such as on a pelagic trip when albatrosses may follow the boat for much of the day, or when South Polar Skuas appear at different times—hey, is this the same bird? Chances are, the wing molt might be different enough to answer this question.

INTERRUPTED MOLT. In birds with molts that are suspended or simply incomplete (such as with many preformative molts), it is sometimes possible to see a point of contrast between newer, recently molted feathers and the older feathers that were not

Figure 77. Here's a male House Finch undergoing a complete molt; it may even be the same individual as in Figure 75. Compare the feather tracts of this individual with birds not in molt (Figures 74 and 76). What do you see? The inner two tertials and the middle and outer greater coverts have been shed or are starting to grow in, and the inner greater coverts have mostly grown in. Songbirds often molt the greater coverts early on in the sequence to provide protection for the growing secondaries. Thus, in this instance, the incoming tertials will be protected. Molt has not started among the other secondaries, which show up as a stack of frayed whitish edgings. In the primaries, molt has reached p6 being shed: you can count inward from the frayed whitish edges of p9, p8, and p7, and also note the dark area on the newly exposed portion of p7, which used to be covered by p6 (compare this molting wing with that of the bird in Figure 76, where p6 is clearly visible). *Marin County, CA, 3 Aug. 2008. Steve N. G. Howell.*

Figure 78. Active molt can be obvious on species such as the American Goldfinch, which changes its colors seasonally. In this instance, the incoming yellow alternate feathers contrast with the old pale gray basic feathers. But if this female molted simply from pale gray to pale gray, we would have a hard time knowing whether or not it was molting. *Marin County, CA, 18 Mar. 2008. Steve N. G. Howell.*

Figure 79. Molt contrasts are pretty obvious in the wings of this Great Blue Heron, or are they? Looking again, we can see that p6 is growing, and so is s1, the innermost secondary. When looking for molt limits, it is important to determine whether or not a bird is actively molting, since any bird in molt—like this one—is likely to show a contrast between new and old feathers. *Marin County, CA, 29 Aug. 2008. Steve N. G. Howell.*

molted. This point of contrast is known as a molt contrast, or a molt limit. In some species, molt contrasts can be obvious, such as between the alternate and basic primaries on terns (see Figure 71). But they can also be subtle, although with close-range views, and especially with sharp photos, molt contrasts can be appreciated on many species (see Figure 75). The first thing to do when looking for molt contrasts is to make sure that the bird isn't actively molting, since any bird in molt may have a contrast between old feathers and incoming new feathers.

Why do molt contrasts occur, and where do we look for them? Molt contrasts occur when molts are incomplete, such as the preformative molts of many songbirds and shorebirds. In these species, the adult prebasic molts are complete, so the presence of a molt contrast on a bird in winter identifies it as being in its first plumage cycle. The areas where preformative molt tends to vary most in its extent are among the upperwing coverts and sometimes the tertials, less often in the flight feathers (see Figure 41). On a closed wing, the inner greater coverts tend to be more exposed than the outers, and so it is the inner coverts that tend to be replaced. Hence, any molt contrast in the greater coverts tends to be between fresh inner feathers and worn outer feathers (see Figure 76).

Many larger birds such as hawks, owls, and petrels can have incomplete prebasic molts of the flight feathers, whereas juveniles have uniform-generation flight feathers for their first cycle. Thus, the presence of a molt contrast in the flight feathers of a large bird indicates it is in at least its second plumage cycle.

In spring, molt contrasts can also indicate a first-cycle bird, but be aware that adults of some species (notably sandpipers and some wood-warblers) replace wing coverts and tertials in their prealternate molt. Thus, adults of these species can also show molt contrasts in spring between worn basic and fresh alternate feathers. However, if you can distinguish three generations of feathers (juvenile, formative, and alternate), then you can still say a bird is in its first cycle.

What if we can't see any molt contrasts? While the presence of molt contrasts on a North American songbird in fall and winter indicates it is a first-cycle bird, the lack of contrast does not necessarily

Figure 80. In general, on smaller species such as songbirds and sandpipers, molt contrasts occur most often among the body feathers or wing coverts (also see Figure 77). This first-cycle Sanderling shows a contrast between grayish formative scapulars (which are already quite worn) and bleached whitish juvenile upperwing coverts. Perhaps easier to appreciate is the contrast between two fresher and darker formative tertials and a browner, more faded juvenile tertial. Compare these contrasts with the uniformly fresh plumage of the adult in Figure 81. *Monterey County, CA, 28 Dec. 2008. Steve N. G. Howell.*

Figure 81. Compare the overall uniformly patterned upperparts of this adult Sanderling completing its prebasic molt (the outer primaries are not fully grown) with the first-cycle bird in Figure 80. *Marin County, CA, 21 Oct. 2008. Steve N. G. Howell.*

Figure 82. On large species such as hawks, owls, and petrels, molt contrasts are often most conspicuous among the flight feathers and indicate that a bird is more than one cycle old; first-cycle birds have uniform-generation flight feathers. This Great Gray Owl shows obvious contrasts in its primaries and secondaries between older, browner, and more faded flight feathers and fresher, grayer, and darker feathers. Counting in from the outermost primary (p10) at the leading edge of the wing, we can see that p7 is contrastingly fresh. This replacement pattern accords with the second prebasic molt (see Figure 58), indicating that this is a bird in its second winter. Further support for this age designation is how brown and faded the older feathers are; worn adult feathers tend to be grayer overall, and the contrast between two generations of adult primaries is less distinct. *Lake County, MN, 24 Jan. 2005. Christopher L. Wood.*

ABOVE: **Figure 83.** This singing Wilson's Snipe shows a molt contrast on the upperwing coverts between the faded, fairly plain-looking old feathers and some new and brightly patterned inner greater coverts and median coverts. Both adult and first-cycle birds can replace coverts in the prealternate molt, however, so at this season a molt contrast alone is unhelpful for determining age. *Plumas County, CA, 10 June 2007. Lyann A. Comrack.*

RIGHT: **Figure 84.** The male Yellow-rumped Warbler is a handsome bird that many take for granted. But let's take a closer look at this individual (of the northern and eastern population known as Myrtle Warbler). As well as the head and body feathers, the median coverts and four inner greater coverts have been renewed in the prealternate molt: note the molt contrast between the outer and inner greater coverts. As in snipe (Figure 83), this molt pattern can occur in both adult and first-cycle Myrtle Warblers, so at this season a molt contrast alone is unhelpful for determining age. *Tompkins County, NY, 29 Apr. 2007. Christopher L. Wood.*

Figure 85. An awareness of molt can help in everyday birding with the common species around your home. For example, a raven molting its tail in fall can be missing or growing the longest feathers and thus show a squared-off tail shape that might suggest a crow. *Marin County, CA, 31 July 2008. Steve N. G. Howell.*

mean it is an adult. It could be, but it also could be a first-cycle bird that has replaced all of the feathers where molt contrasts might be visible. For those interested in pursuing this aspect of identification, a good source on the extent of molts, and on where molt contrasts can occur, is Peter Pyle's two-volume *Identification Guide to North American Birds.*[174,175] Always be aware, though, that preformative molts (called first prebasic molts in part 1 of Pyle) are inherently variable in extent, even within a species.

USING MOLT FOR IDENTIFICATION

Having some knowledge of molt can help you with field identification and improve your birding skills. Not too many years ago, the field separation of adult Common and Arctic terns in summer was considered a challenging identification problem. Today, many birders routinely distinguish these two species by the color of the outer primaries, perhaps unaware that in Common Terns what they are seeing is a molt contrast.

At one level, an awareness of molt may simply mean appreciating how it can change the appearance of a common species, such as a raven molting its tail, which can show a tail shape that might suggest a crow. At another level, it might offer insight into identifications that are usually considered intractable, such as the separation at sea of Fea's and Zino's petrels, which breed at different seasons—and thus presumably molt at different seasons.[176]

The most obvious way of using molt in identification is simply by noting whether or not a bird is undergoing wing molt. This is because the timing and location of wing molt can reflect factors such as breeding seasons, migration distances, and the age of the bird. The presence or absence of wing molt can often be seen when other more conventional field marks are not discernable, and it might be your first clue in an identification. It is important to realize, though, that while the *presence* of wing molt can be a useful field mark, the *absence* of wing molt does not necessarily help. This is because first-cycle birds are often in fresh plumage and not molting when older age classes are molting.

For example, Leach's Storm-Petrels breed in summer and molt their wings mainly from fall into winter. Conversely, most of the Band-rumped Storm-Petrels that visit waters off the eastern U.S. breed in the northern winter and are in wing molt during the summer. Thus, if you see a Leach's/Band-rumped storm-petrel in spring off North Carolina and it's obviously in wing molt, then it's almost certainly a Band-rumped even if it's too distant for you to see any of the traditional field marks. But remember, a nonmolting bird is not necessarily a Leach's, because first-cycle Band-rumpeds won't be in wing molt. Taking this one step further, as many as four species of "Band-rumped" storm-petrels breed on different islands in the eastern Atlantic.[177] Because some of these species breed at somewhat different seasons, the timing of their wing molts may be one clue to help distinguish them at sea.

Another example would be if you thought you saw a Cliff Swallow in wing molt in North America in the fall. Cliff Swallows are long-distance migrants that winter and molt in South America, but Cave Swallows molt in North America—so you'd be more likely to be seeing a Cave Swallow, even if you were in Ontario or Maine! The same thing applies with Common Nighthawk (a long-distance migrant that molts in South America) and Lesser Nighthawk (a short-distance migrant that molts in North America).

If you see a golden-plover in California that is in wing molt, you can be pretty sure it's a Pacific Golden-Plover, some individuals of which winter along the West Coast, as opposed to an American Golden-Plover, which winters (and molts its wings) in South America. Jaegers mostly undergo their wing molt on the nonbreeding grounds, so if you see a jaeger in obvious wing molt off northern California in August or September, it's probably a Pomarine Jaeger, since Parasitic Jaegers winter mainly to the south and don't start wing molt until late September or later.

Both adult and first-cycle Hammond's Flycatchers molt in late summer and fall, before migration, so all Hammond's seen on fall migration will be in bright fresh plumage. Conversely, all other western flycatchers in the genus *Empidonax* molt mostly on their nonbreeding grounds. This means that if you find a ratty-plumaged *Empidonax* flycatcher on fall migration, it isn't a Hammond's. But again, remember, while adults of the other *Empidonax* flycatchers will be in worn plumage during fall migration, the first-cycle birds of these same species will be in relatively fresh plumage.

How do you identify a silent Fish Crow? Well, most people don't, but it turns out that wing-molt timing can be helpful in separating Fish Crows from American Crows, at least in some places. For example, in Cape May, New Jersey, American Crows usually complete their wing molt by mid-August, whereas Fish Crows are still undergoing wing molt

Figure 86. Learning to see wing molt can be helpful for many aspects of birding, and never more so than when sea-birding, or pelagic birding. You're on a Memorial Day pelagic trip off North Carolina and a Leach's or a Band-rumped storm-petrel flies quickly by the boat—you snap a few pictures and the bird has gone. Your photos aren't the greatest, and you can't be certain about the tail shape or how much white is on the uppertail coverts. But look at that wing molt: the inner primaries are new and black, there's a gap at p6 on both wings, and the outer primaries are paler and older than the new inners. Most of the Band-rumped Storm-Petrels visiting North American waters are from winter-breeding populations in the eastern Atlantic, whereas Leach's Storm-Petrels breed in the summer. These opposite breeding seasons are reflected in opposite wing-molt timings, with Band-rumped molting in summer and Leach's in winter—so your bird is almost certainly a Band-rumped Storm-Petrel. *Dare County, NC, 1 June 2008. Steve N. G. Howell.*

Figure 87. Pelagic trips off California in fall often produce lots of jaegers, which can be some of the most challenging birds to identify at sea. A jaeger flies by the boat on a fall pelagic trip off California and you get a reasonable photo, but it's molting the long central tail feathers. It's also in heavy wing molt, with p4 growing and p5 shed. Hmm, not so easy to identify. However, because jaegers undergo their wing molt on the non-breeding grounds, and because Pomarine Jaeger is the only species that winters commonly off central California, there's a good chance that any jaeger with obvious wing molt in fall is going to be a "Pom," like this individual. *Monterey County, CA, 26 Sept. 2008. Steve N. G. Howell.*

in September and October.[178] Thus, any crow in the Cape May area showing obvious wing molt in September or later, which is much easier to see than conventional field marks, should be a Fish Crow. Careful observations are needed to see if these timing differences are useful in other areas where both species occur—and this is something almost any birder could do.

You see a dull-plumaged female or immature hummingbird in late summer in the West. Is it a Black-chinned or an Anna's? You check the shape of the inner primaries, which the field guides say will help you separate these two species, but the bird is in heavy wing molt and these feathers are mostly missing (see Figure 70). Well, you've probably just answered your question! Anna's Hummingbird is a short-distance migrant that molts its wings in summer and fall, whereas Black-chinned is a relatively long-distance migrant that mostly molts in fall and winter on the nonbreeding grounds in Mexico. Any presumptive Black-chinned or Anna's in heavy wing molt in late summer in the West is almost certainly an Anna's.

All of these examples are fine, but where can you find out when and where different species molt? A handicap to pursuing this approach to birding has long been that information on molt is scattered widely in the literature, or not even published. Fortunately, Pyle's *Identification Guide to North American Birds* contains information on the timing, location, and extent of molts for all species occurring regularly in North America. These two volumes are invaluable references for birders as well as biologists, but note that they should be taken as a starting point rather than as definitive, even for common species you'd think might be well known. For example, American Crows in both California and New Jersey start their wing molt in May (at least in California these are mainly first-year birds) and breeding adults are molting in June, yet the Pyle guide indicates that crows molt from July through September.

We've now addressed several introductory questions about molt, covered some background theory, and discussed some practical applications for your growing knowledge of molt. The following family accounts review the molting strategies of North American bird families and relate them to different aspects of their life history and ancestry, showing how molt offers new insights into the fascinating world of birds.

REFERENCES

1. Pyle 1997b; **2.** Pyle 2008; **3.** Poole and Gill 1992–2002; **4.** Cramp 1985; **5.** Cramp 1988; **6.** Cramp 1992; **7.** Cramp and Simmons 1977; **8.** Cramp and Simmons 1980; **9.** Cramp and Simmons 1983; **10.** Cramp and Perrins 1993; **11.** Cramp and Perrins 1994; **12.** Marchant and Higgins 1990; **13.** Marchant and Higgins 1993; **14.** Higgins and Davies 1996; **15.** del Hoyo et al., 1992–2007; **16.** Pyle 1997b; **17.** Pyle 2008; **18.** Sibley 2000; **19.** Voitkevich 1966; **20.** Höst 1942; **21.** Nolan et al. 1992; **22.** Howell 2001a; **23.** Battley et al. 2006; **24.** Wetmore 1936; **25.** Middleton 1986; **26.** Murphy and King 1986; **27.** Voitkevich 1966; **28.** Lucas and Stettenheim 1972; **29.** Nice 1962; **30–31.** Marchant and Higgins 1990; **32.** Chapman 1905; **33.** Wetmore 1936; **34.** Knappen 1932; **35.** Ammann 1937; **36.** Stresemann 1963; **37.** Meijer 1991; **38.** Richardson and Kaminski 1992; **39.** Murphy et al. 1988; **40.** O'Briain et al. 1998; **41.** Ainley and Boekelheide 1990; **42.** Tickell 2000; **43.** Snow and Snow 1967; **44.** Ashmole 1963; **45.** Ginn and

Melville 1983; **46.** Jenni and Winkler 1994; **47**. Stresemann and Stresemann 1966; **48.** Jenni and Winkler 1994; **49.** Prevost 1983; **50.** Wood 1950; **51.** Grubb 2006; **52.** Langston and Rohwer 1996; **53.** Edwards and Rohwer 2005; **54.** Dawson 2003; **55.** Lenton 1984; **56.** Veit and Jones 2004; **57.** Wood 1950; **58.** Murphy et al. 1988; **59.** Burtt 1979; **60.** Howell 2001b; **61–62.** Murphy 1999; **63.** Ankney 1979; **64.** Howell 2001c; **65.** Snyder et al. 1987; **66.** Stiles and Wolf 1974; **67.** Prys-Jones 1982; **68.** Stutchbury and Morton 2001; **69.** Willoughby 1991; **70.** Miller 1933; **71.** Michener and Michener 1940; **72.** Dorsch 1993; **73.** Jongsomjit et al. 2007; **74.** Earnst 1992; **75.** Lande 1980; **76–77.** Voitkevich 1966; **78.** Beebe 1914; **79.** Dwight 1900a; **80.** Chandler and Marchant 2001; **81.** Pyle 2008; **82.** Miller 1933; **83.** Johns 1964; **84.** Boss 1943; **85.** Howell 2007; **86.** Pyle and Howell 2004; **87.** Burtt 1986; **88.** Hill and McGraw 2006; **89.** McGraw and Hill 2000; **90.** McGraw et al. 2002; **91.** Tickell 2003; **92.** Humphrey and Parkes 1963a; **93.** Hardy 2003; **94.** Hudon and Brush 1990; **95.** Humphrey and Parkes 1959; **96.** Humphrey and Parkes 1963b; **97–98.** Howell et al. 2003; **99.** Dorsch 1993; **100.** Jenni and Winkler 1994 and references therein; **101.** Sutton 1935; **102.** Rohwer 1986; **103.** Willoughby 1986; **104.** Thompson and Leu 1994; **105.** Howell et al. 2004; **106.** Pyle 2008; **107.** Howell 1999; **108.** Willoughby 1991; **109.** Howell and Corben 2000b; **110.** Howell et al. 2003; **111.** Stresemann and Stresemann 1966; **112.** Ginn and Melville 1983; **113.** Hedenström and Shunada 1998; **114.** George 1973; **115.** Brown and Saunders 1998; **116.** Sullivan 1965; **117.** Marks 1993; **118.** Dorward 1962; **119.** Stresemann and Stresemann 1966; **120.** Filardi and Rohwer 2001; **121.** Zann 1985; **122.** Rogers et al. 2005; **123.** Wyndham 1981; **124.** Potts 1971; **125.** Ginn and Melville 1983; **126.** Prevost 1983; **127.** Shugart and Rohwer 1996; **128.** Filardi and Rohwer 2001; **129.** Pyle 2005c; **130.** Pyle 2006; **131.** Stresemann and Stresemann 1966; **132.** Schreiber et al. 1989; **133–134.** Prevost 1983; **135.** Dorward 1962; **136.** Stresemann and Stresemann 1966; **137.** Snyder et al. 1987; **138.** Pyle 2008; **139.** Stresemann and Stresemann 1966; **140.** Rohwer and Johnson 1992; **141.** Howell and Corben 2000c; **142.** Howell et al. 1999; **143.** Ainley et al. 1976; **144.** Pitelka 1958; **145.** Ainley et al. 1976; **146.** Marchant and Higgins 1990; **147.** Baltosser 1995; **148.** Pyle et al. 1997; **149.** Dittmann and Demcheck 2006; **150.** Stresemann and Stresemann 1966; **151.** Pyle 1997b; **152.** Murphy and King 1991; **153.** Dietz et al. 1992; **154.** Boyd and Jehl 1998; **155.** Young 1991; **156.** Greenwood 1983; **157.** Myers et al. 1995; **158.** Sullivan and Liguori 2009; **159.** Zenatello et al. 2002; **160.** Cannell et al. 1983; **161.** Willoughby 1991; **162.** Rohwer and Manning 1990; **163.** Rohwer et al. 2005; **164.** Pyle 2008; **165.** AOU 1983; **166.** Howell and Webb 1995; **167.** Stiles and Negret 1994; **168.** AOU 1998; **169.** Komar 1997; **170.** Howell and Patteson 2008; **171.** Dittmann and Demcheck 2006; **172.** Howell and Corben 2000a; **173.** Howell and Corben 2000b; **174.** Pyle 1997b; **175.** Pyle 2008; **176.** Howell and Patteson 2007; **177.** Robb et al. 2008; **178.** Michael O'Brien, pers. comm.

FAMILY ACCOUNTS

ANATIDAE
Swans, Geese, and Ducks
(CBS, CAS, SAS?; 49 species)

Few birds in North America are as handsome as a male Mallard in full finery, yet how many of us take the time to look at this common species? And how many of us could guess whether the male's bright "breeding" plumage is alternate or basic? Looking at duck molts opens up a Pandora's box of fascinating but often unanswered questions about how molts in waterfowl have evolved. This cosmopolitan family of aquatic birds is familiar throughout the world as a traditional food source and as a common feature of ornamental bird collections in many city parks. Waterfowl have also spawned a fantastic conservation legacy—for the simple reason that you can't kill them if you don't have any to kill. (If only the Ivory-billed Woodpecker had been a viable game species, it might still be with us!)

Recent studies have supported the view proposed by the pioneering genetic work of Charles Sibley[1] that waterfowl (order Anseriformes) and upland gamebirds (Galliformes) form an ancient lineage, the Galloanseres, which is why these groups have been moved to the front of many recent field guides. Within waterfowl, various subdivisions have been recognized, and North American species include three subfamilies: the more primitive whistling-ducks (Dendrocygninae) and geese and swans (Anserinae), and the more advanced typical ducks (Anatinae). Almost all North American waterfowl are migratory to some degree, although most favor temperate habitats and only a handful of species migrate in any numbers south of central Mexico.

The sexes of whistling-ducks, geese, and swans look alike and have no seasonal changes in appearance, whereas the sexes of most typical ducks in North America look different, and many exhibit seasonal variation in their appearance. All waterfowl young are precocial and downy, able to swim and walk shortly after hatching, and they mostly feed themselves. Other features are discussed in the separate accounts that follow for geese and swans and for typical ducks.

One well-known aspect of waterfowl molt is that birds become flightless for a few weeks when the primaries and secondaries are shed at the same time. This strategy has been attributed to having large, fat bodies set against relatively small wings, so that a short period of no flight is better than a prolonged period of labored or impaired flight (see page 36). It is also an efficient way to complete wing molt in a short time. These synchronous wing molts generally occur in late summer and fall, when food resources are plentiful, and individuals can even gain weight while molting.[2] Because birds are vulnerable when flightless, these wing molts typically occur in areas with ample cover (as in most dabbling ducks) or in remote areas or large open water bodies (as in most diving ducks) relatively free from predators. Waterfowl may undergo fairly long-distance molt migrations to find safety at this time in their life cycle, and some species stage in large numbers at traditional molting grounds.[3]

The molts of ducks and geese have long attracted attention. Indeed, the puzzles they present were instrumental in the development of the original Humphrey-Parkes system. Not until recently, however, were the prebasic molts of North American

The colorful male Mallard in basic (or formative) plumage is a stunning sight. Ducks provide a good example of how plumage coloration can confound our understanding of molt sequences. *Frederick County, VA, 24 Nov. 2006. Steve N. G. Howell.*

Whether the cryptic eclipse plumage of ducks, such as shown by this male Mallard, is an alternate or a supplemental plumage remains to be elucidated. Following growth of its flight feathers, this bird will molt its head and body plumage, along with its tail, and thus acquire its colorful basic plumage by fall. *Sonoma County, CA, 14 July 2008. Steve N. G. Howell.*

waterfowl clarified, in a paper[4] that reinforced the idea that the processes controlling plumage coloration are independent from those controlling molts (see pages 17–18). By comparing the simpler molts of geese with the more complex molts of ducks, it follows that the bright and colorful plumages worn by many male ducks (such as Mallard, Harlequin Duck, and Hooded Merganser) are actually their basic plumages. The molts of geese and ducks all occur at about the same time and are comparable, but in ducks the hormones controlling color simply have been activated early because of unique life-history traits (see below, Why Are Ducks Different?). Ironically, Humphrey and Parkes[5] considered these bright plumages as alternate, presumably through association (bright = breeding = alternate), even though their system was proposed to free the mind from such preconceptions. This means that much literature over the past 50 years has considered the bright plumages of ducks to be alternate plumages, but from an evolutionary standpoint they are in fact basic plumages. If you're used to the old system, then simply switch basic and alternate; if you didn't know the old system, don't worry—it's easier to start afresh without preconceptions.

The partitioning of molting periods in waterfowl, as in all birds, is a convenience that allows patterns to be discerned; the reality may be far more complicated. Thus, individual Canada Geese may be molting some head and body feathers almost year-round, although their main molting period is during July–November.[6] Does this simply represent a protracted prebasic molt, or might some geese have a prealternate molt? Canvasbacks also molt head and body feathers for much of the year, and it has been suggested that the energetic costs of this molting are not sufficient to interfere with migration, courtship, or even the early stages of egg formation.[7]

Let's review the molt strategies of the two main subfamilies of waterfowl represented in North America: geese and swans versus typical ducks. The whistling-ducks (Dendrocygninae) exhibit the complex basic molt strategy, and their molts are broadly similar to those of geese.

ANSERINAE
Geese and Swans
(CBS; SAS?)

Geese and swans apparently exhibit the complex basic molt strategy typical of many birds, although some geese may have the simple alternate strategy. They form long-lasting pair bonds sometime in their second or third year of life, usually breed for the first time at 2 to 3 years of age, and often stay together in family groups through the winter. The young are protected by their parents, grow a relatively durable juvenile plumage, and fledge in 1.5 to 3 months. Adult plumage aspect is typically attained within a year, either with the formative plumage or with the second basic plumage.

PREBASIC MOLTS. Prebasic molts occur mainly in late summer and fall on the breeding grounds or at staging grounds and may continue at a low level through the winter. As in other waterfowl, the primaries and secondaries of geese and swans are shed synchronously, mainly during mid- to late summer (often while tending young), followed by molt of the head and body feathers, which occurs mainly from late summer to fall.[8] Among breeding pairs of swans there can be an offset in the timing of wing molt so that one member of the pair is always fully winged; this has been attributed to the need for territory defense and protection of the young, rather than to insufficient food not allowing both parents to molt together.[9]

PREFORMATIVE MOLTS. Preformative molts in geese and swans are variably protracted and mainly involve head and body feathers (sometimes also tertials and tail feathers). These molts start in fall, often on or near the breeding grounds, and may continue through winter into early spring. The molts are generally more rapid in northerly-wintering species and smaller species, such as Emperor Goose and Ross's Goose, and more protracted in more southerly-wintering species and larger species, such as Brant and Snow Goose. Geese and swans undergo their second prebasic molt in late summer and fall (often slightly earlier than the prebasic molt of breeding adults), and their molt schedules then become aligned with those of adults. An interesting sideline to the plumage aspect of first-cycle Mute Swans is discussed in the family account for wood-warblers, under Delayed Plumage Maturation (page 223).

PREALTERNATE MOLTS? Although the conventional wisdom is that geese don't have prealternate molts, this matter has never been satisfactorily resolved, and several species do molt head and neck feathers in the late winter and spring.[10] Because geese have lots of feathers, though, might it simply be that their prebasic molt continues for much of the year, perhaps with a suspension in the coldest midwinter periods? In an experiment on Canada Geese, feathers clipped in September from the head and neck were renewed in January, whereas those clipped on the breast were not renewed until the next complete molt, in late summer.[11] This strongly suggests that a second molt occurs, although it brings about no change in appearance.

Interestingly, this late-winter "prealternate" molt of head and neck feathers has been reported only in geese with bold head and neck patterns (the White-fronted Goose and the genus *Branta*, which includes Canada Goose and Brant[12]), which suggests it may serve to freshen up their appearance. If this molt does occur, might it be more extensive in second-cycle immatures that have yet to form pair bonds, and less extensive in adults that have paired for life and have less need to impress one another? There appears to be only one molt in the first cycle, a variable and protracted molt of head and body feathers. Thus, if a prealternate molt does occur in some geese, then their molt strategy would be simple alternate.

The White-fronted Goose undergoes its protracted preformative molt through the winter, by which it gradually acquires adultlike features such as the eponymous white front and black mottling on the belly. While this White-fronted Goose can be aged easily as a first-cycle bird, the age of the Canada Goose behind it is more difficult to ascertain; given what can be seen here, it could be either an adult or first-cycle bird. *Marin County, CA, 2 Jan. 2008. Steve N. G. Howell.*

ANATINAE
Typical Ducks
(CAS; SAS?)

The four groups of typical ducks in North America are dabbling ducks (the genus *Anas*, including Mallard, teals, wigeons, etc.), scaups and allies (the genus *Aythya*), sea ducks (including eiders, scoters, goldeneyes, and mergansers), and stifftails (including the Ruddy Duck). Despite our familiarity with ducks, and a relative wealth of literature about them, their plumage sequences are still not well understood. Most species in all four groups have the complex alternate molt strategy, but some species may have the simple alternate strategy.

Unlike geese and swans, typical ducks have distinct alternate plumages, and the sexes usually look quite different. Linked to this, the breeding strategy of migratory Northern Hemisphere ducks differs from that of geese and swans in several ways: ducks form new pair bonds each year, the males tend to abandon females when the young have hatched, and families separate before winter. Adult plumage aspect is attained between a few months of age (as with the formative plumage of most dabbling ducks) and about 2 years of age (as in some eiders). Dabbling ducks, scaups, and stifftails first breed at 1 to 2 years of age, sea ducks at 2 to 3 years of age. Young fledge at 1 to 2 months of age.

The prebasic wing molt of ducks occurs on the breeding grounds or at staging grounds in late summer and fall, followed by the head and body molt, which often completes on the wintering grounds. In earlier and longer-distance migrants (such as Northern Pintail and Blue-winged Teal), the prebasic head and body molt occurs mainly on the nonbreeding grounds. Preformative molts occur from fall through winter into spring, with the timing varying somewhat among species. These molts are usually partial (but sometimes include tail feathers) and thus resemble the preformative molts of geese in both extent and timing. The timing of the partial prealternate and possible presupplemental molts in ducks depends on how these molts are defined (see below).

PREBASIC MOLTS. When we ignore the colors and complex plumage patterns of ducks and simply look at the sequence and extent of their molts, the complete molt of head and body feathers in fall follows the synchronous wing molt and appears directly comparable to, or homologous with, the prebasic molt in geese and swans.[13] These complete molts are thus best viewed as prebasic molts. A hindrance in recognizing this for many people has been that these molts often produce bright and colorful plumages that have been equated with alternate plumages—despite the intent of the H-P system to remove such associations.

Duck molts also illustrate well why terms such as "post-breeding molt" and "breeding plumage" should be avoided when trying to compare molts of different species. The use of these terms can even produce incongruities, such as stating that the White-winged Scoter undergoes its "post-breeding" molt in April–May,[14] which is immediately *before* the breeding season!

In Ruddy Ducks, it has been suggested that birds molt their flight feathers twice a year, although how frequent this is among wild birds (versus captives) and whether it occurs in all age classes and both

Unlike most North American ducks, the adult male Ruddy Duck molts into a bright alternate plumage in late winter and then back into a dull basic plumage in fall, after breeding. *Sonoma County, CA, 14 Dec. 2006. Steve N. G. Howell.*

Whistling-ducks, such as these Black-bellied Whistling-Ducks, have a complex basic molt strategy, as do geese and swans. Within a few months, the dusky-billed first-cycle bird at the front of the group will look much like an adult. *Nayarit, Mexico, 15 Jan. 2007. Steve N. G. Howell.*

Like many northern-breeding geese and gulls, the Surf Scoter migrates south in a fairly strong juvenile plumage and then has a protracted molt over the winter, by which time males acquire adultlike black feathering. But whether one or two molts occur over a scoter's first winter and spring remains unresolved. If only a single molt occurs before the complete second prebasic molt in late summer, then the molt strategy of scoters would be simple alternate. But if a second molt occurs in spring, they would have the complex alternate molt strategy. *Sonoma County, CA, 26 Feb. 2008. Steve N. G. Howell.*

sexes remains unclear.[15] This putative double wing molt recalls a similar wing-molt pattern in some grebes, which is discussed in the family account for grebes.

PREFORMATIVE MOLTS. Preformative molts in dabbling ducks generally occur in fall and early winter and produce a plumage aspect much like that of a basic adult. The same is true of scaups, although their preformative molt averages later and may not complete until late winter or spring, when in females it may overlap with the prealternate molt. Sea ducks (and apparently also Ruddy Ducks[16]) tend to retain much of their juvenile plumage longer and undergo a protracted preformative molt that occurs throughout the winter and which, in males, produces a plumage aspect variably intermediate between the basic adult male and female.

In Ruddy Ducks, the preformative molt may be complete in some individuals, but why might this be? Ruddy Ducks lay large eggs, and the juvenile wings are smaller than adult wings, with poorer-quality primaries. Thus, a female Ruddy Duck producing eggs but having juvenile wings could have difficulty flying, and it has been suggested that preformative wing molts enable first-year birds to attempt breeding.[17]

In scoters there is little evidence for another first-cycle molt before the second prebasic molt in late summer into adultlike plumage, in which case the molt strategy of scoters would be defined as simple alternate. However, a limited additional molt in spring would be difficult to distinguish from the protracted overwinter molt, and the jury is still out on how many molts occur in scoters during their first cycle.

A combination of earlier breeding and more exposure to stronger sunlight at lower latitudes may contribute to greater wear of the juvenile feathers of dabbling ducks versus those of sea ducks, which could then be reflected in the earlier molt timings of the former. Dabbling ducks and scaups mostly breed earlier in the year and winter at mid- to low latitudes. Sea ducks mostly breed later and winter at higher latitudes. And because sea ducks do not breed in their first summer, there is no pressure to quickly acquire a bright plumage for display—or perhaps they don't breed in their first year *because* they cannot physiologically molt quickly and produce the bright plumages required for winter courtship. It also may be that juvenile sea ducks simply grow stronger feathers than do dabbling ducks, something that could be tested fairly easily. Might it be that sea ducks eat better-quality food? Or because they tend to breed in higher latitudes, perhaps they have longer periods of daylight in which to forage, which can result in stronger feathers being grown?

PREALTERNATE MOLTS. Prealternate molts in most species of typical ducks that breed in the Northern Hemisphere have been considered unusual in that the timing apparently differs between the sexes. For example, in dabbling ducks and scaups the partial prealternate molt of females occurs during late winter and spring[18] and tends to produce a browner, more cryptically patterned aspect that provides better camouflage for nesting. This molt averages more extensive in paired birds[19, 20] and involves mainly feathers of the head, back, chest, and sides (and sometimes the tertials and tail feathers). In males, however, it has been argued that the molt in late summer, after breeding, should be considered the prealternate molt.[21] This partial molt in males produces a duller (often femalelike) aspect, often termed the "eclipse" plumage, which makes a bird less conspicuous during the wing molt that follows.

Most sea ducks apparently follow a similar pattern, but in both sexes of Long-tailed Duck the prealternate molt may start early in spring.[22,23] And in scoters, Jonathan Dwight[24] reported that all birds he examined in spring were molting, although the molt caused no outward change in appearance, and he found no evidence of an eclipse plumage; thus it may be that both sexes of scoters have their prealternate molt in spring, at the same time.

Curiously, the pattern of a complete fall molt and partial spring molt is shown by both males and females of several Southern Hemisphere ducks, both in species where the sexes look alike and in species where they look different.[25] This raises some interesting questions. For example, given that waterfowl, including typical ducks, probably originated in the Southern Hemisphere,[26] a partial spring molt may be the ancestral condition for male ducks. Might this molt still occur to varying degrees in Northern Hemisphere ducks?

Well, yes, it might. Several critical studies have found that males of a number of northern ducks do have a limited spring molt of head and body feathers, although it produces no appreciable change in appearance. These molts occur mainly in February–March, when females are undergoing their prealternate molt, and they have been found in the Mallard,[27] Gadwall,[28] American Wigeon,[29] and Greater Scaup,[30] but not in the Northern Pintail.[31] The spring molt in the Greater Scaup has been considered simply a continuation of the interrupted complete molt started in fall, which might be the case. But could this interpretation also reflect putting the molt into the box where it "should" fit in order to conform to the system? Whatever the answer, these largely overlooked molts seem remarkably similar to those of males in Southern Hemisphere ducks, and critical study is still needed to figure out what is going on.

PRESUPPLEMENTAL MOLTS? If both males and females of northern ducks do have a spring prealternate molt, then the summer molt of males into their eclipse plumage would represent a novel, presupplemental molt. This scenario suggests a remarkable parallel to the situation found in ptarmigans (see the family account for grouse and pheasants). And if a presupplemental molt exists in male ducks, does a comparable but overlooked late-summer molt occur in females? A limited or vestigial molt of cryptic brown feathers in females would be difficult to detect. Careful molt studies mapped onto a robust family tree could help address these questions.

Supplemental plumages have been reported for the Long-tailed Duck, which has been described as having up to four "plumages,"[32,33] although no feather follicle may produce more than three generations of feathers.[34] These descriptions, however, have been based on the patterns and shapes of feathers (which can vary seasonally within a single molt, particularly if it is protracted or interrupted) and not necessarily on the number of molts involved. It has also been argued that Long-tailed Ducks undergo only a protracted partial (prealternate) molt from spring through summer, during which time the appearance of incoming feathers can change, and a complete molt in fall (into basic plumage).[35] Some authors have argued that there are three molts per cycle, but that the only feathers molted three times are the longest scapulars.[36] Further study is required to establish how many molts Long-tailed Ducks, and indeed all ducks, undergo in their annual cycle, and how extensive these molts might be.

The Problem with First-cycle Molts

Given that the molts of adult ducks are not well understood, you might not be surprised to learn that first-cycle molts also may not fit neatly into the boxes that have been suggested for them. For example, many authors have reported an additional molt, immediately following juvenile plumage and preceding the preformative molt. Indeed, for Humphrey and Parkes to term the bright plumage an alternate plumage, their system required the existence of a preceding "first prebasic" molt. However, the few studies that have looked critically at post-juvenile molts in ducks have found this molt to be lacking (as in scoters[37]), or at best vestigial and ephemeral (as in the Gadwall[38] and Redhead[39]). Notwithstanding these careful studies, some authors have invoked, for example, a "first prebasic" molt of all head and body feathers for scoters during fall[40] —yet no such molt occurs! Based on extensive examination of museum specimens, it has now been concluded that this molt "does not appear to exist in most genera of [ducks]."[41] So what is going on?

At least four possible explanations exist, which are not mutually exclusive. First, some individuals, or some species, do indeed have an additional pre-

Putting names to the molts of Northern Shoveler remains something of a puzzle. If first-cycle birds, like this male, are to look adultlike by spring, then surely they have two molts following their juvenile plumage: a fall molt into this distinctive plumage, which resembles that of an eclipse male, and then a late-winter molt into a bright plumage similar to that of a breeding adult. The logical interpretation of this pattern would be of a preformative molt in fall and a prealternate molt in late winter. Could it be this simple? See the text for discussion of this conundrum. *Colima, Mexico, 14 Feb. 2008. Steve N. G. Howell.*

formative molt, but its frequently ephemeral nature makes it difficult to detect. Second, because ducks grow considerably in size during their first month or two of life, it may be that the delayed activation of juvenile feathers (as the skin's surface area expands and needs to be covered) has been mistaken for an extra molt. A parallel is found in some passerines, in which there can be two or even three phases of juvenile feather growth (see page 21). Third, because the processes controlling feather pigmentation and molt are independent, the earliest-molted formative feathers may come in with juvenile-like patterns but later-molted feathers may have adult-like patterns (as has been found in the Blue-winged Teal[42]). This process could be misinterpreted as two different molts.

The fourth possibility is more intriguing. At least in some longer-distance migrants such as the Blue-winged Teal and Northern Shoveler, it does appear that two molts have been added into the first cycle. Given how the molt strategies of ducks have been misunderstood and misrepresented, however, a critical reexamination of first-cycle molts in these species is desirable. For now, though, it is thought that the Blue-winged Teal and Northern Shoveler have a partial molt that replaces some of their juvenile plumage in fall, prior to migration; the plumage produced by this molt in males looks variably intermediate between the adult male and female basic plumages (and much like the male eclipse plumage). Then, on the nonbreeding grounds in late winter and spring, birds undergo a second molt into a plumage that, in males, resembles the adult male's basic plumage.

Now, if the ancestral pattern of duck molt involved a partial prealternate molt by both sexes in late winter, as hypothesized above, then might the second molt by the Blue-winged Teal and Northern Shoveler actually be a first prealternate molt and not an additional preformative molt? Could this situation parallel the molts of some migrant songbirds (such as the Baltimore Oriole and Indigo Bunting), in which first-cycle males have more-extensive and later prealternate molts than the adults?

Why Are Ducks Different?

So how do the molts and plumages of ducks relate to their life history? For example, why do most North American ducks acquire bright plumages in fall and not wait until spring? Presumably it is because they court and form pairs during the winter and spring, a strategy that is unusual among migratory birds. But why would they do that? Why not wait until they arrive on the breeding grounds? Surely their strategy adds a degree of difficulty in the need to remain together during migration, and also adds extra time during which a mate might die. Other species, such as grebes and shorebirds, typically wait until they arrive on the breeding grounds to start singing and displaying.

Interestingly, the Ruddy and Masked ducks are exceptions to the "typical" duck pattern: they molt into bright alternate plumage, display, and form pairs on the nesting grounds. This may relate to their being part of the stifftail group, which likely originated in low- to mid-latitudes of the Southern Hemisphere. In such areas, breeding might occur at any time of year, whenever favorable conditions exist. Thus, these species may be programmed to respond at short notice to suitable water levels.

For "typical" Northern Hemisphere ducks, various theories have been proposed to explain early pairing.[43] There are appreciably more males than females in duck populations, which has been attributed to greater female mortality during parental care. So could the pressure on many males to secure mates from a pool of few females have pulled the courting period forward in time? In other words, males can't risk waiting until they are on the nesting grounds, when all the females might be taken. However, given that the high mortality of females can be linked to their being single parents, might the sex ratio be more balanced if males stayed with the females?

An unbalanced sex ratio may be part of the answer, but a more important factor simply may be the nature of a duck's habitat—water. Ducks display mostly while swimming. Those species that court in winter and in early spring breed mostly in higher northern latitudes, where water can freeze. Furthermore, in early spring the northern nesting grounds may offer relatively little in the way of food—not enough to feed the female and allow her to gather the resources needed to produce eggs. Consequently, females gather nutrients in late winter and spring and store them for egg production until they reach the nesting grounds. In winter and spring, paired females benefit from protection by their mates while foraging, and within flocks pairs may have higher status than single birds; paired females thus have access to better feeding conditions.[44] Pairing on the nonbreeding grounds, where large flocks form, also offers a large choice of potential mates.

This may explain winter pairing in ducks, but does it explain the bright plumage colors? Not really. In South America there are species of teal, pintails, and wigeons (both resident and migratory) in which the sexes look similar or even identical. Both sexes are femalelike in appearance, and both may stay with the young.[45,46] And even in North America there are Mottled Ducks, Mexican Ducks, and Black Ducks, in which both sexes are dull-plumaged. As a rule among birds, bright male plumages occur mainly in migratory species breeding at higher latitudes in the Northern Hemisphere, such as some wood-warblers and orioles, and this is true also in ducks. In the songbirds, the time constraints for mate choice in a short but seasonally predictable summer breeding season may help explain the existence of bright plumages. This may also be the case with Northern Hemisphere ducks, except that their life-history characteristics require that pairs be formed in winter and spring.

Another possibility that has been suggested is that striking sexual differences in ducks may be promoted where numerous species occur together and court in the same region. Thus, from the point of species-isolating mechanisms, it may be that the differences between males of species occurring together are more significant than the differences between the sexes.[47]

Whereas adult male Northern Shovelers and Blue-winged Teal wintering in California typically acquire a fully colorful basic plumage in fall, birds migrating longer distances may acquire a duller basic plumage with varying degrees of an immature-like aspect. For these longer-distance migrants to attain a breeding aspect by spring, they will have to undergo an additional molt. Whether more-northerly wintering birds have a comparable molt (which brings about no appreciable change in their appearance) remains unknown. Answering such questions is made challenging by a lack of clear information on how to determine the age of ducks in the field. For example, is this Northern Shoveler an adult male, or an immature male with an advanced plumage aspect? *Colima, Mexico, 14 Feb. 2008. Steve N. G. Howell.*

More Questions Than Answers

Clearly, much remains to be learned about a group as "well studied" and familiar as ducks. For example, how have molts in northern-breeding ducks evolved to be different from those of the presumed ancestral pattern still shown by species in South America? Do male northern ducks have a prealternate molt in spring, homologous to the female's spring prealternate molt? Could this molt in males be more extensive in first-cycle birds, as with the Blue-winged Teal? Has it been completely lost in some species, at least in adults? Do females have a supplemental plumage in late summer, which could be homologous to the male's eclipse plumage?

Do all species of ducks pair anew in winter? Or do the "female-plumaged" species such as the Mottled Duck form longer-lasting pair bonds? At least in the Southern Hemisphere, the species in which sexes look alike may maintain year-round pair bonds,[48] and it has been argued that this is the ancestral condition in ducks.[49] How do the two strategies compare? For example, are male Northern Pintails that desert the young more successful in breeding than male Yellow-billed Pintails (*Anas georgica*) in South America that stay with the young?

Did the southern ducks originate from colorful northern ancestors and some males subsequently lose their bright colors? Or did the bright northern

The sexes look similar in numerous species of ducks that breed in the Southern Hemisphere, which is where waterfowl are believed to have arisen. Both sexes may be bright (like this pair of Southern Wigeon [*Anas sibiliatrix*]) or dull and femalelike in appearance. And in these southern ducks both sexes have a complete prebasic molt in fall and a partial prealternate molt in spring; no evidence for a summer molt into eclipse plumage has been found. *Magallanes, Chile, 23 Oct. 2006. Steve N. G. Howell.*

species originate from southern ancestors? That most groups of waterfowl probably originated in the Southern Hemisphere may argue in favor of the latter. But secondary, southward invasions by northern species could have resulted in the loss of sexual dimorphism, as has been suggested for presumed derivatives of the Mallard and Northern Pintail in Hawaii and on subantarctic islands of the Indian Ocean.[50] To throw another spoke into the wheel of thought, it has been found that the bright "male" plumage of ducks such as the Mallard is actually the "default" option, and that females, in the absence of estrogen, will acquire a bright "malelike" plumage.[51] It has even been suggested that the ancestral state in waterfowl may have been for both sexes to be brightly colored and that subsequent selection occurred for cryptic coloration.[52]

Why do males of Northern Hemisphere ducks desert the females after the eggs have hatched? Is it simply because the bright male plumages might attract predators? If male desertion causes higher female mortality, does this increase pressure on males to form pairs earlier in the season? Did bright male plumages or desertion of young come first? Is this another chicken-and-egg (well, duck-and-egg) conundrum?

The next time you see a Mallard, take a second look at this iconic duck. And perhaps ponder what might be going on with its molts and with the molts of other waterfowl, and how this may relate to their life-history strategies and perhaps to their southern ancestry.

REFERENCES

1. Sibley and Alquist 1990; **2.** Ankney 1979; **3.** Salomonsen 1968; **4.** Pyle 2005a; **5.** Humphrey and Parkes 1959; **6.** Gates et al. 1993; **7.** Lovvorn and Barzen 1988; **8.** Pyle 2005a; **9.** Earnst 1992; **10.** Cramp and Simmons 1977; **11.** Delacour 1954:156; **12.** Cramp and Simmons 1977; **13.** Pyle 2005a; **14.** Cramp and Simmons 1977:650; **15.** Hohman 1996; **16.** Pyle 2005a; **17.** Hohman 1996; **18.** Pyle 2005a; **19.** Lovvorn and Barzen 1988; **20.** Rohwer and Anderson 1988; **21.** Pyle 2005a; **22.** Sutton 1932; **23.** Stresemann 1948; **24.** Dwight 1914; **25.** Weller 1968; **26.** Livezey 1991; **27.** Combs and Fredrickson 1995; **28.** Paulus 1984; **29.** Wishart 1985; **30.** Billard and Humphrey 1972; **31.** Miller 1986; **32.** Salomonsen 1941; **33.** Palmer 1976; **34.** Salomonsen 1949; **35.** Sutton 1932; **36.** Stresemann 1948; **37.** Dwight 1914; **38.** Oring 1968; **39.** Weller 1970; **40.** Palmer 1976; **41.** Pyle 2005a:216; **42.** Greij 1973; **43–44.** Rohwer and Anderson 1988; **45.** Weller 1968; **46.** S. N. G. Howell, pers. obs.; **47.** Sibley 1957; **48.** Weller 1968; **49.** Livezey 1991; **50.** Sibley 1957; **51.** Voitkevich 1966:217–218; **52.** Kimball and Ligon 1999.

CRACIDAE
Plain Chachalaca
(CBS; 1 species)

"Chachalaca" is an onomatopoeic rendition of these birds' rollicking, ear-splitting choruses, which can pass through the woods from group to group with a domino-like effect that drowns out any other sounds. In south Texas brushlands, the Plain Chachalaca is the northernmost member of an exclusively New World family of primitive, mostly arboreal gamebirds known as cracids, which is an abbreviation of their family name. Cracids occur throughout the Neotropics south to Argentina, in habitats ranging from lowland rain forests to highland cloud forests. All species are nonmigratory, and the sexes look alike (as in chachalacas) or strikingly different, with males averaging larger. Adult plumage aspect is usually attained with the formative plumage. Young cracids are precocial and downy, able to run and even fly within a few days of hatching, but they are fed by their parents. Chachalacas first breed at 1 to 2 years of age.

As far as is known, the molting of all cracids follows the complex basic strategy. Perhaps because chachalacas inhabit tropical lowlands, where there is no seasonal pressure to molt before winter sets in, their molts tend to be protracted, and there may even be some overlap between molting and breeding. The preformative molt is often complete, and sometimes the second prebasic wing molt can start before the preformative wing molt has completed.[1] This can set up opportunistic patterns of stepwise primary molt (discussed more fully in the family account for hawks), and some birds may require two cycles to replace all of their primaries.[2]

As in other gamebirds, the juvenile flight feathers of chachalacas enable very young birds to fly and escape danger, but such feathers would not support a full-grown bird. Thus, the complete preformative molts allow these juvenile flight feathers to be replaced with feathers better suited to support the weight of an adult bird.

REFERENCES

1. Stresemann and Stresemann 1966; **2.** Pyle 2008.

The Plain Chachalaca differs from most other native North American gamebirds in having a complete preformative molt. *Hidalgo County, TX, 25 Apr. 2004. Bill Schmoker.*

TETRAONIDAE AND PHASIANIDAE
Grouse and Pheasants
(CBS, CAS; 16 species)

Although often lumped by birders under the mundane moniker of "chickens," gamebirds include a fantastic diversity, from a gobbling group of male turkeys and the white-on-white of a winter ptarmigan in the snow, to the spectacular dawn dances of prairie-chickens and the handsome patterns of the elusive Mountain Quail. But how closely are all of these groups related? The taxonomy of gamebirds has had a checkered history, with from one to five families being recognized by different authors. The New World quail are now usually treated as a distinct family,[1] and it seems reasonable to view the other four groups also as distinct families.[2,3] The five main groups of "chickens" are the grouse of northern Eurasia and North America (family Tetraonidae); the New World quail, found from southern Canada to northern Argentina (Odontophoridae); the pheasants and partridges of the Old World (Phasianidae); the guineafowl of Africa (Numididae); and the turkeys of North America (Meleagrididae). Given similarities in their molting strategies, the nonnative pheasants and partridges found in North America are here grouped into a single account with the native grouse.

North American "chickens" occur mainly in temperate habitats ranging from forested to open country and are largely terrestrial in their lifestyles. All New World species are ostensibly nonmigratory, although some species undergo short-distance (sometimes elevational) movements in winter. In some species, the sexes look alike, and in others they are strikingly different. Males average larger than females, particularly in species that engage in communal displays. Except for ptarmigans, no species show appreciable seasonal changes in their appearance. Adult plumage aspect is usually attained with the formative plumage. The young are precocial and downy, able to run, feed, and even fly within days of hatching. The age of first breeding is 1 to 2 years of age.

Despite the great external diversity within the grouse and pheasants, the molting strategy of all species is complex basic or complex alternate, the difference between these two strategies in most species being a limited prealternate molt of some head and neck feathers. Ptarmigans are a notable exception, discussed below. Prebasic and preformative molts of all North American species occur from summer into fall, before the onset of winter. In the migratory Common Quail (*Coturnix coturnix*) of the Old World, the preformative and prebasic wing molts can be suspended for migration, to be continued on the nonbreeding grounds.[4]

PREFORMATIVE MOLTS. A distinctive feature of the preformative molt of gamebirds is that it is complete except for the outer two juvenile primaries and primary coverts, which typically are held through the first year.[5] An exception occurs in pheasants, which usually have a complete preformative molt.[6] Complete preformative molts are more typical of tropical gamebirds (such as chachalacas and guineafowl) than of temperate gamebirds,[7] so perhaps this trait in pheasants points to a tropical ancestor.

As in other gamebirds, the juvenile flight feathers of grouse and pheasants enable very young birds to fly and escape danger, but such feathers would not support a full-grown bird. Thus, the juvenile flight feathers are replaced in the preformative molt with feathers that can support the weight of an adult bird. But why would the outer two primaries be retained? That this also occurs in turkeys and New World quail suggests it may be an ancestral trait in gamebirds. Grouse and quail typically have rounded wings in which the long middle primaries provide more lift than the relatively short outer primaries. In order for juveniles to fly and escape predators as soon as possible, it appears sufficient, and presumably more cost-efficient, to invest initial energy in only the inner 8 primaries rather than in all 10; the outer 2 primaries can grow in later, at which time they are close to the adult size and don't need to be replaced. Also, since complete preformative molts are more typical of tropical gamebirds,[8] perhaps the incomplete molts developed in an ancestral species of temperate habitats, where less time was available for molt before the onset of winter.

PREALTERNATE MOLTS. An interesting obstacle to the study of prealternate molts in gamebirds is that shooting them is not generally permitted in spring. Thus, few specimens exist in museum collections from periods when a prealternate molt might occur. Studies to date suggest that most species lack such molts,[9,10] but a limited prealternate molt (in late April–June) has been detected in prairie-chickens,[11] Greater Sage-Grouse,[12] and a few Old World species of grouse.[13]

These prealternate molts are limited mainly to head and neck feathers and often serve to make males more cryptic by breaking up their bright throat patterns. These plumages generally follow, rather than precede, the courtship and display periods and thus recall the eclipse plumages of some male ducks, which serve to make birds less conspicuous during wing molt. Unlike ducks, however, gamebirds have sequential wing molt and do not become flightless. Prealternate molts occur in both sexes of Old World grouse but have been reported only for male sage-grouse. This may reflect the lack of specimen material, combined with a less noticeable change of appearance in females. In the Ruffed

Grouse, some body feathers may be shed in spring but not replaced until fall, which may relate to temperature regulation or may represent early onset of the prebasic molt.[14] Clearly, study is needed on the extent, timing, and even the existence of prealternate molts in most North American gamebirds.

Molts in Ptarmigans

The molts of ptarmigans have long attracted attention and engendered confusion, although they were outlined clearly more than 100 years ago.[15] Ptarmigans are unusual among birds in having three molts a year (prebasic, prealternate, and presupplemental), during which their appearance can go through an almost bewildering series of changes, as detailed by Finn Salomonsen.[16] It may be that high predation pressure (such as from foxes and raptors) has driven the continually changing summer appearance of ptarmigans and that they can fuel these molts because their vegetable food is abundant in the Arctic summer. But what caused them to develop three molts in the first place? The resident British population of Willow Ptarmigan (known as the Red Grouse) does not have a white plumage phase and reportedly has only two molts a year;[17] a partial molt in females occurs during March–May, and in males during April–June, when it runs into the complete prebasic molt. But how carefully have the molts of that population been studied? Could it have three molts?

FIRST-CYCLE MOLTS. There appear to be two preformative molts in ptarmigans. The first is a partial molt producing patterned brown feathering in late summer, the second a near-complete molt (apparently homologous to the preformative molt in other gamebirds) producing white feathering in fall. As in some ducks, determining the extent (or even confirming the existence) of the first preformative molt is made difficult by the possibility that some or all of it may represent delayed activation of juvenile feather follicles as the young bird increases in size. However, careful studies of first-cycle molts in the Wild Turkey[18] and Common Quail[19] have found

Greater Sage-Grouse are unusual among North American grouse in that males have a limited prealternate molt of head and neck feathers in summer, perhaps to render them less conspicuous when molting. In display like this, they are anything but inconspicuous. *Jackson County, CO, 26 Apr. 2008. Bill Schmoker.*

Ptarmigans, like this pair of White-tailed Ptarmigans (the male at left) are among the few species of birds with well-documented supplemental plumages. In late fall, birds molt into a white or mostly white basic plumage that serves them well through the snowy winter. In late winter and spring they have a partial molt and in summer another partial molt, both of which serve to upgrade their plumage and provide changing colors and patterns that provide fine camouflage amid the changing vegetation of the tundra. *Larimer County, CO, 3 July 2006. Steve N. G. Howell.*

that two preformative molts occur in these species, so two molts likely do occur in ptarmigans.

Which plumage is basic? In adult ptarmigans, difficulties in naming the molts lie in identifying molt homologies between ptarmigans and other grouse and in determining the homologies of molts within ptarmigans—which molt is prebasic, which is prealternate, and which is presupplemental? The timing of the molts and the pigmentation of the feathers reflect an incompletely understood interplay of day length and various hormones.[20,21,22] Because molts can overlap in timing, and because the appearance of feathers can seemingly change continuously over several months of molting, it is difficult to say where one molt stops and the next one starts. Thus it is not possible to certainly assign every feather to one of the three molts. Some authors have considered the white nonbreeding plumage as basic, others as supplemental.

When we ignore the colors and patterns of plumages and look simply at the sequence and timing of molts in relation to other grouse, in particular how the wing molt relates to the body molts, it follows that the white plumage is best considered a basic plumage.[23] Among more southerly populations of ptarmigans that do not live in areas with solid snow cover, the basic plumage can be mottled with dark, and the amount of dark feathering may vary from year to year depending on temperature.[24,25] The basic plumage may even lack white completely in populations living in areas not prone to persistent snow cover, as with the appropriately named Red Grouse. An interesting feature of the white basic feathers of ptarmigans is that the barbules have numerous air-filled cavities, which may help with insulation or may make the plumage look more like snow.[26,27]

ALTERNATE AND SUPPLEMENTAL PLUMAGES. The two extra molts of male ptarmigans involve a partial molt, mainly of head and neck feathers in March–May, and a more extensive molt of head and body feathers in July–September. Females have a more extensive molt of head and body feathers in April–June, and a more limited molt in July–September. It has been proposed recently that the molts in ptarmigans have evolved separately in each sex, such that the spring molt of males is presupplemental and the spring molt of females is prealternate.[28] This novel interpretation stems from comparisons between the present-day molts of ptarmigans in terms of their extent, and between molts in ptarmigans and sage-grouse. Given that homologous characters in animals have high genetic correlations between the sexes,[29] it is more parsimonious to consider that the ptarmigan ancestor simply developed two homologous molts rather than both sexes evolving two molts that are not homologous.

Although the extent of these two molts in present-day species sometimes varies between the sexes, in some populations of Rock Ptarmigan both the timing and extent of these two molts are similar in males and females.[30] And if the Red Grouse in Britain really does have only two molts per cycle, this simpler situation suggests that the spring molt in both sexes of ptarmigans would be prealternate, like that found in British populations. The summer molt would thus be the novel, presupplemental molt. The prealternate molt of sage-grouse may just as easily have evolved independently from those of ptarmigans.

A similar situation may occur in Northern Hemisphere ducks (see the family account for waterfowl), in which both sexes may have a prealternate molt in spring (more extensive in females) and a presupplemental molt in summer (perhaps reduced to the point of being absent in some females). Thus, despite the relative wealth of literature on waterfowl and gamebirds, their molts remain to be satisfactorily elucidated. The above discussion highlights the difficulty (impossibility?) of attempting to trace the evolutionary history of molts. Do the seeming parallels between the molts of ptarmigans and ducks indicate some common ground, or are these just the logical developments that molts follow given certain life-history traits?

REFERENCES

1. AOU 1998; **2.** Cramp and Simmons 1980; **3.** del Hoyo et al. 1994, **4.** Cramp and Simmons 1980; **5.** Dwight 1900a; **6.** Cramp and Simmons 1980; **7–8.** Stresemann and Stresemann 1966; **9.** Dwight 1900a; **10.** Pyle 2008; **11.** Dwight 1900a; **12.** Pyle 2007; **13.** Cramp and Simmons 1980; **14.** Rusch et al. 2000; **15.** Dwight 1900a; **16.** Salomonsen 1939; **17.** Cramp and Simmons 1980; **18.** Leopold 1943; **19.** Lyon 1962; **20.** Höhn and Braun 1980; **21.** Höst 1942; **22.** Stokkan 1979; **23.** Pyle 2007; **24.** Jacobsen et al. 1983; **25.** Watson 1973; **26.** Dyck 1979; **27.** Holder and Montgomerie 1993; **28.** Pyle 2007; **29.** Lande 1980; **30.** Jacobsen et al. 1983.

As well as being simply white, the body feathers of a basic-plumaged ptarmigan contain numerous air-filled cavities. These structures may increase the insulation value of the plumage or may help it to appear more like snow—certainly these White-tailed Ptarmigan blend beautifully with their winter surroundings. *Clear Creek County, CO, 25 Nov. 2008. Bill Schmoker.*

MELEAGRIDIDAE
Wild Turkey
(CBS; 1 species)

Large and leggy, the Wild Turkey is best known in its domesticated form as the traditional food of Thanksgiving feasts. Turkeys compose a small North American family that is sometimes lumped into an expanded family that includes other gamebirds (as discussed in the account for the grouse and pheasant families). Turkeys live mainly in open wooded habitats and are nonmigratory. The sexes look similar, but males are appreciably larger and somewhat more brightly colored. Adult plumage aspect is attained in 1 to 2 years. Like other gamebirds, young turkeys are precocial and downy, able to run and feed within a day or two of hatching, and they can make short flights at 1 to 2 weeks of age. The age of first breeding is 1 to 2 years.

In their molting, turkeys apparently follow the complex basic molting strategy. Prebasic and preformative molts occur from summer into fall, before the onset of winter. A prealternate molt of body plumage has been reported in captive birds[1] but has not been found in wild birds.[2]

PREFORMATIVE MOLTS. A study by Aldo Starker Leopold[3] found that Wild Turkeys have two preformative molts. The main preformative molt follows on quickly from, and at times overlaps with, the acquisition of juvenile plumage. In most populations this molt is almost complete except that the outer two juvenile primaries are retained, as in many other gamebirds. In turkeys of the southeastern U.S., however, it is usually only the outermost primary that is retained, and up to 5 percent of birds in that region replace all of their juvenile primaries.[4]

This first preformative molt produces a fairly dull plumage in which the sexes appear similar, although later-molted feathers tend to be brighter, and it is completed when birds reach about 14 to 15 weeks of age. The extent of this molt, which includes the primaries, suggests it is homologous with the complete or near-complete preformative molts of other gamebirds. A second preformative molt, which runs on directly from the first, starts with the central tail feathers at about 14 weeks of age. This second molt does not include primaries but appears to include body feathers, with the males acquiring squared, black-tipped feathers and the females acquiring rounded, buff-tipped feathers.

Wild Turkeys undergo a great increase in size in their first few months of life, which may be why they have developed two preformative molts. *Sonoma County, CA, 5 Dec. 2008. Steve N. G. Howell.*

Young Wild Turkeys, like most young gamebirds, can fly at an early age to escape danger, here with the adult female running rather than flying. *McHenry County, ND, 16 June 2008. Steve N. G. Howell.*

Turkeys undergo a great increase in body size during their first few months of life, and the second preformative molt serves to replace feathers (such as the central tail feathers) that the bird outgrows as it gains in mass. As with the first-cycle molts of ducks, ptarmigans, and other birds, however, it is difficult to separate overlapping body molts and determine the extent of each. Might some of the "second" preformative body molt in turkeys simply be a continuation of the first molt as birds increase in size and new follicles develop?

An interesting sidelight to Leopold's study was that domestic turkeys have more extensive first-cycle molts than do wild turkeys: after the first preformative molt, domestic birds retain only the outermost juvenile primary, and in the second preformative molt they often replace more tail feathers than do wild birds. Variations in the extent of preformative molts, between wild and domestic birds and between most wild populations and those in the Southeast, have not been explained. Might they reflect environmental conditions (past or present), time of breeding, or . . . ?

REFERENCES

1. Leopold 1943; **2.** Williams and McGuire 1971; **3.** Leopold 1943; **4.** Williams 1961.

ODONTOPHORIDAE
New World Quail
(CAS, CBS?; 6 species)

Members of this family are familiar over much of the U.S., from the loud whistles of bobwhites in the Southeast to the striking plumage patterns in the West of California's state bird, the California Quail. Although New World quail have traditionally been lumped with grouse and pheasants into the family Phasianidae, they apparently are not closely related to other gamebirds despite many similarities (see the introduction to the account for the grouse and pheasant families). Most tropical quail are forest-based, but the North American species inhabit mainly open woodland and scrubby areas. Quail are nonmigratory and undergo no seasonal changes in appearance. The sexes usually look obviously different, and adult plumage aspect is attained with the formative plumage. The young are precocial and downy and able to run, feed, and fly within days of hatching. New World quail first breed at 1 year of age.

The molting strategy of North American quail appears to be complex alternate, although prealternate molts are not well known, in part because hunting quail in spring is not permitted. Preforma-

Did this male Northern Bobwhite undergo a prealternate molt in spring? Prealternate molts of gamebirds in general are poorly known, and among bobwhites they may be more extensive in southern and tropical populations than in northern populations. *Yuma County, CO, 27 July 2005. Bill Schmoker.*

New World quail, such as these California Quail, differ from other gamebirds in retaining all of their primary coverts in the preformative molt. *Marin County, CA, 19 Mar. 2008. Steve N. G. Howell.*

tive and prebasic molts occur in summer and fall, before the onset of winter, and prealternate molts occur in spring.

PREFORMATIVE MOLTS. Preformative molts in quail are much like those in other gamebirds, which typically replace all feathers except for the outer two juvenile primaries and their coverts. Quail, however, retain all of the juvenile primary coverts, not just the outer two.[1,2] Replacing most of the juvenile primaries makes sense; the feathers that support a chick in flight would not support a full-bodied bird (as discussed in the account for grouse and pheasants). But why quail don't also replace the accompanying juvenile primary coverts is a mystery. Presumably these feathers are adequate to protect the new primaries and there is no need to replace them. At least two tropical-forest quail, the Singing Quail (*Dactylortyx thoracicus*) and the Marbled Wood-Quail (*Odontophorus gujanensis*), also retain their juvenile primary coverts through the first year,[3,4] suggesting that this trait occurs throughout the family. So why are primary coverts replaced in grouse and pheasants? Was the ancestor of New World quail a small bird that could get by for a year with juvenile primary coverts? And were the ancestors of grouse and pheasants much larger birds whose juvenile primary coverts would not have been sufficient to protect the large new formative primaries?

PREALTERNATE MOLTS. Prealternate molts were documented by Jonathan Dwight[5] in all North American species except Mountain Quail, for which he had limited material; however, the existence of a prealternate molt in California Quail has been questioned.[6] Prealternate molts in quail occur during March–May and are usually very limited in extent, mainly involving feathers on the head and throat. Unlike the prealternate molts of several grouse species, those of quail occur before, or at the start of, the breeding season and bring about no appreciable change in appearance; they simply may be facultative molts of feathers that become abraded by scrubby vegetation, as occur in many songbirds such as wrens and sparrows.

Among Northern Bobwhites in Cuba, the prealternate molt is more extensive than in populations of temperate northern latitudes and includes feathers on the head, neck, chest, upper back, and underparts.[7] And in the related Spot-bellied Bobwhite (*Colinus leucopogon*) of Central America, the prealternate molt includes several tail feathers as well as head and body feathers.[8] These findings mirror a pattern found commonly among various groups of birds, whereby the preformative molt is often more extensive in southerly populations, which are exposed to stronger sunlight. An obvious, but unanswered, question that follows is whether tropical quails of shady forest habitats lack prealternate molts.

REFERENCES

1. Van Rossem 1925; **2.** Leopold 1939; **3.** Dickey and Van Rossem 1938; **4.** Petrides 1945; **5.** Dwight 1900a; **6.** Raitt 1961; **7.** Watson 1962; **8.** Dickey and Van Rossem 1938.

GAVIIDAE
Loons
(SAS; 5 species)

The melancholy wailing tremolos of loons on warm summer evenings are a sound of the North, and in most of the Lower 48 states loons are associated with winter. Known as "divers" in the Old World, loons are a small family of large, heavy-bodied aquatic birds that live in temperate northern latitudes. They are of uncertain taxonomic affinities and may be part of an assemblage that includes penguins, frigatebirds, and petrels.[1] All species are short- to medium-distance migrants that withdraw south to winter on open water. The sexes look alike, but there are strong seasonal and age-related differences in appearance. The precocial downy young are able to swim shortly after hatching but are fed and guarded by their parents, and young loons fledge in 2 to 3 months. Adult plumage aspect is attained in about 2 years, and loons rarely breed before 3 years of age.

The molting strategies of loons have long been debated[2,3] and have posed challenges to the Humphrey-Parkes system for those who focus on the molts of adults. By carefully tracing molts starting with the juvenile plumage, however, it can be seen that loons have the simple alternate molt strategy.[4] Most molting occurs on the wintering grounds, and wing molts are synchronous (see page 36). The differences in wing-molt timing among species show interesting parallels with the molts of puffins (discussed below).

PREBASIC MOLTS. There are only a few times in their annual cycle when loons can molt their wings, and these time constraints are overcome by having a synchronous wing molt. By shedding their primaries and secondaries at the same time, birds become flightless for a few weeks while the new wing feathers grow. Another advantage of this strategy for

In their second summer, loons such as this Yellow-billed Loon typically acquire a plumage that is duller and less fully patterned than that of a breeding adult. Although it may be tempting to say the whitish feathers on the head and neck of this individual are retained basic feathers, they are in fact likely to be alternate feathers that did not attain breeding aspect—as is true of all the other head and neck feathers, which are either sooty brown or have only a dull oily black gloss (unlike the deep oily green gloss of an adult). Similarly, all of the back feathers appear to be of one generation and thus are alternate: these range from all-black to having pale brownish triangles to having bold white rectangles, presumably reflecting changes in hormones during the course of the molt. *Sonoma County, CA, 11 July 2009. Steve N. G. Howell.*

heavy-bodied birds with small wings is that a short period of no flight may be better than a prolonged period of labored flight.

There is, however, a striking difference among adults of different species in the timing of this synchronous molt: Red-throated Loons undergo wing molt in late fall (mainly September–November), whereas the four other species, which are all larger, molt their wings in early spring (mainly February–April). Unlike geese and swans, which undergo synchronous wing molt while tending young, both members of a breeding pair of loons need to remain fully winged for territory and brood defense, as well as for commuting to feeding areas. The traditional view has been that the larger loons don't have time for wing molt before northern waters freeze up, so they postpone it over the winter until favorable conditions return, in early spring.[5] This sounds reasonable enough, but if we trace wing molts starting with the first cycle (see below), it appears that, in comparison with the Red-throated Loon (which has a "normal" molt strategy), the adult wing molts of the larger loons are actually being advanced in time rather then being delayed.[6]

For high-latitude breeding species that do not migrate far south in winter, the balance between breeding and molt can be a tight one. In some cases it may be that more resources for molt are available in the longer spring window before breeding (as also occurs with Horned and Atlantic puffins and the Ivory Gull) than in the short window after breeding and before winter sets in. Synchronous wing molts have the advantage of compressing molt into a short period, and perhaps they are also more easily shifted in timing than are typical, protracted molts.

It is unclear why nonbreeders or failed breeders of the larger loons don't undergo their wing molt in fall, as they don't appear to have time constraints. However, the late-winter wing molt coincides with hormone levels that produce bold black-and-white spotting on the upperwing coverts. A wing molt in fall may not produce these patterns, which might be enough to select against an occasional fall wing molt in adults.

FIRST-CYCLE MOLTS. First-cycle molts in all loons are broadly similar. Young fledge in a strong juvenile plumage, which can be retained well into or even through the winter. A variable and protracted molt of head and back feathers starts in mid- to late winter and continues through spring and perhaps into summer. Most feathers produced by this molt are dull-patterned, like adult basic plumage, but sometimes a few white-spotted back feathers are produced. In mid- to late summer, all species undergo a complete second prebasic molt, including a synchronous wing molt.

After this, the molts of the Red-throated Loon are "normal," with partial prealternate molts in spring and complete prebasic molts in late fall. The larger

Common Loons wear their juvenile plumage into the winter and have a variably extensive molt of head and body feathers in late winter and spring, before a complete second prebasic molt in summer. This first-cycle individual has acquired some new, mostly gray scapulars that contrast with the worn, whitish-fringed, and mostly blackish brown juvenile feathers. Also note the projecting primaries, which many adults at this season have shed. *Sonoma County, CA, 20 Feb. 2007. Steve N. G. Howell.*

There are relatively few times in the annual cycle of a loon when it can molt its flight feathers. Adult Common Loons, like this individual, renew their flight feathers in late winter, prior to northward migration. This represents the first stage of their interrupted prebasic molt, which is completed in fall. Late winter and spring are also when they undergo their prealternate molt of head and body feathers. Thus, the dark flecks on this bird's chin are alternate feathers, whereas the white-spotted upperwing coverts are basic feathers! Note that the primaries have been shed—there is no wing projection above the thighs, as would be shown by a full-winged bird. *Sonoma County, CA, 26 Feb. 2008. Steve N. G. Howell.*

loons, however, molt their wings again in spring, which appears to be the first stage of their seasonally split prebasic molt (see above). Given that only a single first-winter molt appears to occur in all loons (much like the single first winter molt of some ibises and large gulls), their molt strategy is considered simple alternate.

PREALTERNATE MOLTS. Prealternate molts in adult loons involve head and body feathers plus the tail, and in all species they occur mainly in February–April. In the larger loons, these molts overlap in timing with the first stages of the prebasic molt (of upperwing coverts and flight feathers), which effectively means the birds may have a complete "prebreeding" molt in spring, although the homology of the wing molt lies with the prebasic molt.

REFERENCES

1. Sibley and Alquist 1990; **2.** Sutton 1943; **3.** Woolfenden 1967; **4.** Howell and Pyle 2005; **5.** Woolfenden 1967; **6.** Howell and Pyle 2005.

PODICIPEDIDAE
Grebes
(CAS, SAS?; 7 species)

Grebes are a cosmopolitan family of small to medium-large aquatic birds of mainly temperate habitats. They are of uncertain taxonomic affinities, and their frequent placement in field guides next to loons reflects little more than association through similar life-history adaptations. Some North American grebes are residents, whereas others are medium-distance migrants. The sexes look alike, but most species show distinct seasonal differences. Adult plumage aspect is attained within 1 year, and grebes first breed at 1 to 2 years of age. The precocial downy young are able to swim shortly after hatching but are fed and guarded by their parents. They fledge at 6 to 11 weeks of age.

Grebes have the complex alternate and perhaps also the simple alternate molt strategies. Prebasic and preformative molts in migratory species may start on the breeding grounds but occur mainly on the wintering grounds, or at staging sites such as Mono Lake, California, where well over a million Eared Grebes stage in fall,[1] or on Boundary Bay, British Columbia, where thousands of Western and Red-necked grebes gather.[2] Prealternate molts occur mostly or entirely on the nonbreeding grounds.

One aspect of grebe feather replacement that falls outside the box of molt terminologies is caused by the continual plucking and ingesting of feathers, mainly from the flanks and thighs. This practice presumably helps protect the stomach lining from damage by sharp fish bones, and renewal of feathers in these tracts is thus an ongoing, year-round process.

PREBASIC MOLTS. Like loons and waterfowl, grebes shed their primaries and secondaries at the same time and thus are flightless for a few weeks while the new wing feathers grow in (see page 36). Wing molt overlaps with, or slightly precedes, most of the head and body molt. Unlike loons, grebes do not use their wings for brood and territory defense, and rarely or not at all for commuting to feed. So why don't grebes molt their wings while tending their young, as do geese and swans? The answer may be that having full wings is important for protecting the young, which often ride, and are effectively brooded, on their parents' backs and under their wings. It also may be that limited food resources, at least on smaller lakes and ponds, are insufficient to fuel both the growing young and an adult wing molt.[3]

The main wing-molt period for most grebes is in fall (July–October), but some Eared, Western, and Clark's grebes have been found molting their wings in late winter (January–March[4,5,6]). Because grebes use their wings relatively little, it seems unlikely that any individual would regularly undergo two wing molts per year, and why birds molt at different seasons has yet to be resolved. The age of birds undergoing this winter wing molt is not known, although it has been suggested for Western and Clark's grebes that this molt occurs only in first-cycle birds.[7] Thus, if the second prebasic wing molt follows in fall, the late-winter molt would be part of the preformative molt.

If different wing-molt timings occur among adults, might this simply reflect different nesting times? Whereas early nesters have time to molt in fall, after breeding, late breeders might not have time for wing molt before the onset of winter, so perhaps they postpone it until the next period when it is energetically feasible. Or perhaps the wing molt in some birds (or populations?) is actually occurring before the breeding season, as in larger loons. To establish this would require following wild individuals through their first year or two of life, which is problematic at best.

The late-winter wing molts in North American grebes occur only in colonial or semi-colonial species that breed mainly in mid-latitudes, where water levels at breeding lakes are prone to strong fluctuations. Thus, might synchronous wing molt be inherently flexible in its timing from year to year? This could allow breeding birds to take advantage of suitable water levels when they occur, and potentially postpone wing molt when water levels allow late-season nesting. Interestingly, a late-winter wing molt also occurs among some Ruddy Ducks, a species that shares several life-history traits with grebes.

FIRST-CYCLE MOLTS. The smaller species of grebes apparently show the "textbook" complex alternate pattern, with a preformative molt of head and body feathers in fall, after which birds look much like adults, and then a partial prealternate molt in spring. After this comes the complete prebasic molt and entry into the definitive molt cycle. The two larger species, Western and Clark's grebes, may also follow this strategy, but it is unclear whether one or two molts occur over their first winter. If only one molt occurs, as in loons, then the molt strategy of these large grebes would be simple alternate.

PREALTERNATE MOLTS. Prealternate molts in adult grebes involve head, neck, and body feathers and are perhaps most extensive (or at least bring about the most striking changes in appearance) in more-northerly breeding species. This may be related, as in so many groups of birds, to shorter periods available for pair formation and breeding at higher latitudes. It might also be that the less-striking changes apparent in lower-latitude species involve less extensive molts, which may in turn enable them to breed more quickly on the ephemeral wetlands often found in lower latitudes.

REFERENCES

1. Boyd and Jehl 1998; **2–3.** Stout and Cooke 2003; **4.** Sibley 1970; **5.** Storer and Neuchterlein 1985; **6.** Storer and Jehl 1985; **7.** Pyle 2008.

The inconspicuous Least Grebe has the complex alternate molt strategy typical of most, if not all, grebes in the region. This bird is in presumed basic plumage and has a nonbreeding aspect, with a whitish throat (which becomes black in breeding aspect). *Nayarit, Mexico, 16 Jan. 2007. Steve N. G. Howell.*

DIOMEDEIDAE
Albatrosses
(SBS; 3 species)

Synonymous with the windy southern oceans, albatrosses are large, long-winged seabirds that can fly almost effortlessly and commute thousands of miles on foraging trips. All albatrosses are migratory or nomadic, and the three North Pacific breeding species range regularly into North American waters to feed—and molt. Along with the petrels and storm-petrels, albatrosses form the group of seabirds known as tubenoses, so named for their tubelike nostrils. The sexes look similar, but males of some larger species attain their adult plumage aspect several years earlier than do females; there are no seasonal changes in appearance. Ages differ to varying degrees, and it can take anywhere from 1 to 20 years to attain adult plumage aspect, depending on species and sex. Young albatrosses are semi-altricial and downy and fledge in 5 to 10 months. The age of first breeding is usually 6 to 12 years.

Albatrosses appear to have the simple basic molt strategy. It is difficult, though, to establish whether the head and body molt that starts within 6 months or so of fledging represents the start of a protracted second prebasic molt, or perhaps may be a limited preformative molt. The novel patterns of wing molt in albatrosses are not well understood but appear to be an adaptation to differential rates of wear in the primaries (see below). Breeding adults molt almost entirely outside the breeding season, although limited head and body molt may start during the chick-feeding period.

The Unique Wing Molts of Albatrosses

Albatrosses rely on their remarkable powers of flight to survive in a challenging marine environment, and presumably for this reason they have developed a unique strategy of wing molt.[1,2,3,4] Basically, the primary molt is split into two series: p8–p10 molt outward as a series (annually in Black-footed and Laysan albatrosses but every other year in Southern Hemisphere species), whereas the middle primaries appear to molt inward from p7, perhaps in stepwise waves, such that p7 (the most exposed of the inner series) is replaced more frequently than the relatively protected inner primaries.

The second prebasic molt of Black-footed and Laysan albatrosses involves only two to four outer primaries (usually p8–p10), but subsequent molts include a variable number of the middle and inner primaries, plus the outer primaries. The reason the first molt involves few primaries probably relates to the time needed for young birds to learn to live in their environment. Another species in this genus, the Galapagos Albatross (*Phoebastria irrorata*), also replaces its outer primaries each year.[5]

Why would these northern-breeding species, but not the southern ones, replace their outer primaries every year? It seems likely that the frequency of outer primary molt in albatrosses reflects environmental conditions.[6] Thus, in Black-footed and Galapagos albatrosses, the effects of strong tropical sunlight, and perhaps abrasion on the nesting islands, may necessitate annual replacement of the outer three primaries (whose tips are exposed on the closed wing). The annual molt in Black-footed and Galapagos albatrosses is facilitated by occurring in the productive waters of the California and Humboldt currents, respectively. By contrast, the high-latitude lifestyles of southern albatrosses do not bring them into contact with intense sunlight, and their primaries molt during midwinter, a potentially less favorable time;[7] hence, they have a 2-year replacement schedule for their outer primaries.

Time Constraints and Molt-Breeding Tradeoffs

Like many large seabirds, albatrosses have prolonged breeding seasons. Thus, adult Black-footed and Laysan albatrosses arrive at the colonies in October–November, and their young don't fledge until June–July. This means breeding birds have only 3 to 4 months available to them for molting. In contrast, some nonbreeding birds can replace all of their primaries in one molt season, which for them spans April to October. Given their prolonged period of immaturity, why haven't albatrosses developed the standard stepwise patterns of wing molt like other large seabirds such as boobies, frigatebirds, and cormorants? It may simply reflect ancestral traits in the development of wing-molt patterns, and it "just didn't happen" in albatrosses. Also, in stepwise molt the outer primaries are not replaced until two or three molt waves have developed, which might take too long for an albatross that needs its outer primaries to be in good condition for long-range flying in strong winds.

The short periods available for wing molt mean that sooner or later an albatross may be unable to replace enough primaries between breeding seasons to allow it to fly with maximal efficiency. In such cases it would have to miss out on a breeding season in order to catch up with its molt. This tradeoff between molt and breeding may explain why up to 25 percent of "breeding adult" Laysan Albatrosses do not breed each year; instead, they skip a year to "recharge" their flight feathers.[8] In this regard, it has been found that Laysans with higher parasite loads replace fewer primaries and begin molt later than

The age at which Steller's (or Short-tailed) Albatross attains this handsome "adult" plumage aspect is unknown and may differ between the sexes, with females taking several more years than males. *Off Torishima Island, Japan, 1 May 2008. Steve N. G. Howell.*

do healthier birds and are thus more likely to skip a breeding season.[9]

Arguments for molt-breeding tradeoff have also been made for a few other species of large birds.[10] For example, populations of Pelagic Cormorants breeding in eastern Asia experience colder sea temperatures than do populations breeding in western North America, so the former need to migrate farther and have less time available for molt.[11] Most North American birds apparently have complete primary molts each year, but Asian birds have incomplete primary molts. Will Asian Pelagic Cormorants eventually skip a year of breeding to catch up with their molt, as appears to happen with albatrosses?

REFERENCES

1. Edwards and Rohwer 2005; **2.** Howell 2006; **3**. Langston and Rohwer 1995; **4.** Rohwer and Edwards 2006; **5.** Harris 1973; **6.** Howell 2006; **7.** Prince et al. 1993; **8.** Langston and Rohwer 1996; **9.** Langston and Hilgarth 1995; **10.** Rohwer 1999; **11.** Filardi and Rohwer 2001.

PROCELLARIIDAE
Petrels
(SBS, CBS; 20 species)

Petrels (which include shearwaters and fulmars) occur throughout the world's oceans, from the Arctic to the Antarctic. Like albatrosses and storm-petrels, they are a family of "tube-nosed" seabirds; that is, species that have their nostrils encased in tubes on the bill. All petrels are migratory or at least nomadic, ranging widely over the ocean in search of food. They come to land only to breed. Of the 20 species occurring regularly in North American waters, only 2 (the Northern Fulmar and Manx Shearwater) breed in North America north of Mexico. The remainder are migrants from as far afield as New Zealand, Chile, and the Mediterranean, and many of them visit North America to molt in food-rich waters such as the Bering Sea, California Current, and Labrador Current. The sexes and ages of petrels look similar in all species that occur in North America, and there are no seasonal changes in appearance. Young petrels are semi-altricial and downy, and they fledge in 1.5 to 5 months. The age of first breeding is usually 5 to 6 years.

Most petrels have the simple basic molt strategy, but a few species apparently have the complex basic strategy, with a preformative molt limited to head and body feathers. Molting generally occurs at sea, and most species do not overlap wing molt and breeding, although the molt of head and body feathers often starts when birds are still nesting. Preformative molts are not well known in petrels but occur in some species that breed and molt in subtropical latitudes, such as the Buller's Shearwater.

Petrels of tropical latitudes tend to have lower wing-loading and range widely over warmer, food-poor environments. Conversely, those of temperate habitats tend to have higher wing-loading and inhabit windier, food-rich environments.[1] Reflecting patterns of food abundance, tropical tubenoses tend to have more-protracted wing molts during which they maintain good flight capabilities, whereas temperate tubenoses often have relatively rapid wing molts during which they may be temporarily almost flightless but can dive for concentrated prey that fuels a quick molt.

PREBASIC MOLTS. Like many species, from crows and hummingbirds to gulls, petrels often drop their inner few primaries almost simultaneously, so that there is an obvious gap in the wing. The middle and outer primaries then tend to be molted gradually, one or two at a time, except for species that have time constraints on their molt—and have the concentrated food resources needed to fuel a rapid molt. Thus, long-distance migrants such as the Sooty Shearwater are often seen growing their outer two or three primaries at the same time. Species in food-poor environments, such as the Black-capped Petrel, tend to molt their primaries gradually, as do some Sooty Shearwaters. The latter are presumably nonbreeding individuals not rushed for time.

In some species, such as the Black-capped Petrel, the "sequential" wing molt can start with p2

This Great Shearwater, over unusually mirror-calm seas, is nearing the end of its second prebasic molt, which was likely started in the Southern Hemisphere and then suspended to finish in the Northern Hemisphere; note the pointed juvenile outer primary, soon to be molted. *Dare County, NC, 26 May 2008. Steve N. G. Howell.*

Adult Sooty Shearwaters, like this bird, are typically well into their wing molt by May. Note how the greater coverts have been shed synchronously, revealing a white stripe across the exposed bases of the secondaries. *Monterey County, CA, 11 May 2008. Steve N. G. Howell.*

Sooty Shearwaters breed in the Southern Hemisphere but migrate to the Northern Hemisphere to molt, where concentrations of birds in food-rich waters can number in the hundreds of thousands. These birds formed part of a mass of shearwaters swarming over bait fish right off the beach, and at this season most adults would be finishing their prebasic wing molt. *Monterey County, CA, 24 July 2008. Steve N. G. Howell.*

and move outward;[2] p1 is usually replaced before the molt has reached the middle primaries. It is not known how frequent this pattern is, whether it relates only to certain ages, and whether individuals can start with p1 in some years and with p2 in other years.

In general, wing molt in petrels does not overlap with breeding, but it does in two shearwaters, Cory's and Black-vented, both of which breed in mid-latitudes.[3,4] Cory's Shearwater is the largest shearwater and undertakes long-distance migrations, which could limit the time available for its wing molt. By starting wing molt while still feeding young, it can fit a complete molt into the annual cycle. Black-vented is a small shearwater that "should" have time to molt in its nonbreeding period, but it starts wing molt in June–July, during the chick-rearing period, and completes it before the onset of winter. Why this might be is unclear. Perhaps it's a safeguard strategy against El Niño events that periodically affect the waters it inhabits and, through winter storms and food shortages, might compromise a complete molt in some years. It has also been suggested that because shearwaters flap their wings underwater while pursuing prey, a reduced wing-surface area might facilitate deeper diving, which may be advantageous during the chick-rearing period.[5]

Molt and Migration

In petrel species that molt regularly in North American waters, four groups can be identified in terms of their breeding latitudes and migration distances.[6]

NORTHERN RESIDENTS. The one species—Northern Fulmar—that remains year-round in temperate and subtropical northern oceans has a protracted wing molt. Its molt starts in late summer and continues gradually over the winter, or it may suspend in midwinter before completing in spring. The second prebasic molt occurs earlier, during the northern summer when food is plentiful, and is usually completed before winter.

NORTHERN MIGRANTS. These species breed in temperate or subtropical northern latitudes and migrate south to tropical or southern latitudes in the nonbreeding season. Most molting occurs in the winter, but wing molt can start when birds are nesting and can be suspended over migration to complete in winter, as occurs with Cory's Shearwater.[7] The second prebasic molt averages earlier in timing and starts during the northern summer food peak, but it may be suspended and complete on the nonbreeding grounds.

SOUTHERN MIGRANTS. These are species that migrate from southern temperate or subtropical regions (where they breed) to northern temperate or subtropical waters to molt in the food-rich northern summer. They generally have a relatively rapid wing molt, as in Sooty, Short-tailed, and Great shearwaters. The second prebasic molt strategy is variable: some birds molt at about 1 year of age in the Southern Hemisphere summer, before migrating back north; some may spend their first year in the Northern Hemisphere and molt there slightly earlier than the adults; and some start molt in the Southern Hemisphere and suspend it to complete in the Northern Hemisphere summer.

TROPICAL MIGRANTS. These species breed in tropical or subtropical latitudes and often range to food-rich areas in the tropics or subtropics to molt. They usually molt in the "summer season" of either hemisphere, such as Black-capped and Cook's petrels molting from March to September in the Northern Hemisphere. An exception is Murphy's Petrel, which appears to molt in the subtropical North Pacific during midwinter. Why, nobody knows. The second prebasic molt in these species generally occurs a month or two ahead of the adult prebasic molt.

REFERENCES

1. Spear and Ainley 1998; **2.** S. N. G. Howell, unpub. data; **3.** Monteiro and Furness 1996; **4–5.** Keitt et al. 2000; **6.** S. N. G. Howell, unpub. data; **7.** Monteiro and Furness 1996.

For whatever reason, Sooty Shearwaters don't seem to molt in the North Atlantic. The Sooties that arrive off the East Coast in May and June are juveniles in fresh plumage, which are not molting their wings. Thus, an all-dark petrel in wing molt should draw attention. This is a Trinidade Petrel, individuals of which are often in wing molt during May and June off the East Coast. *Dare County, NC, 23 May 2007. Steve N. G. Howell.*

Petrels offer many examples of birds breeding at opposite times of year having opposite molt schedules. Note how the summer-breeding Cory's Shearwater (left) is just starting primary molt, with p1–p3 shed, while the winter-breeding Black-capped Petrel (right) is finishing primary molt, with p10 growing. *Dare County, NC, 29 July 2007. Steve N. G. Howell.*

OCEANITIDAE
Storm-Petrels
(SBS, CBS; 9 species)

These tiny seabirds are found worldwide, from temperate waters of the northern oceans to tropical islands and the Antarctic. They come ashore only to breed. An increasing body of evidence points to there being two families of storm-petrels, the Southern Hemisphere Oceanitidae and the Northern Hemisphere Hydrobatidae, which traditionally have been lumped as "storm-petrels" based simply on their small size.[1,2] The two families differ in several structural and behavioral ways, but their molt strategies are similar. Storm-petrels range from long-distance, transequatorial migrants to largely sedentary species of mid-latitudes. The ages and sexes look alike, and there are no seasonal changes in appearance. Young are semi-altricial and downy and usually fledge at 2 to 4 months of age wearing a strong juvenile plumage. The age of first breeding is usually 4 to 5 years, reflecting how long it takes to learn to live in the tough marine environment.

Most storm-petrels have the simple basic molt strategy, but at least one species has the complex basic strategy. Prebasic molts are protracted and usually complete, but some Fork-tailed Storm-Petrels retain a few secondaries after molting,[3] presumably a reflection of the constraints placed on them by residing in the stormy North Pacific. The molt of body feathers, tail feathers, and sometimes even inner primaries can start during the later stages of breeding, and some species may suspend wing molt for migration. The second prebasic molt usually occurs a few months earlier than the prebasic molt of breeding adults, and molt timings then synchronize with the adult cycle.

Perhaps because of its relatively weak juvenile plumage, in combination with its long migrations, Wilson's Storm-Petrel has a complete preformative molt. This preformative molt spans July to December or later, whereas prebasic molt spans April to September.[4] The strategy of Wilson's Storm-Petrel and the relative timing of its preformative and prebasic molts parallel those of other transequatorial migrants that rely on flight to forage, such as the South Polar Skua, jaegers, and swallows, showing how shared life-history features can shape patterns of molting.

Molt and Migration

The patterns of molt shown by storm-petrels occurring in North American waters mirror those of petrels (see details in the family account for petrels).[5] Northern residents include the Fork-tailed and Ashy storm-petrels; northern migrants include Leach's, Black, and Least storm-petrels; southern migrants include Wilson's Storm-Petrel; and tropical migrants include the White-faced Storm-Petrel.

Studies at the Farallon Islands off central California contrast the molts of the two storm-petrel species breeding there in relation to their life histories: the Ashy Storm-Petrel, a short-ranging sedentary species of the California Current, and Leach's Storm-Petrel, a long-ranging migrant that winters in the equatorial Pacific.[6,7]

The Ashy Storm-Petrel has a protracted breeding season, with eggs laid from late April to mid-July and young fledging from late August to November.

The famous storm-petrel flocks found in Monterey Bay in fall usually contain thousands of birds, many of which are molting. Whereas a nonmolting Ashy Storm-Petrel has a long and deeply forked tail like the bird at left, molting birds like the individual at right can look "sawn-off" at the rear and are often mistaken for Least Storm-Petrel. Besides being paler and grayer overall than a Least, note this bird's relatively broad wings with obvious molt (most Leasts that reach central California are not molting) and its large blocky head with a relatively small slender bill (Least has a smaller, rounded head and a thicker bill). *Santa Cruz County, CA, 1 Oct. 2006. Steve N. G. Howell.*

Wilson's Storm-Petrel is a transequatorial migrant that has a relatively rapid molt in the Northern Hemisphere before returning south to its breeding grounds around Antarctica. This bird is in the early stages of its prebasic molt, with the greater coverts and the inner primaries growing. *Dare County, NC, 28 May 2007. Steve N. G. Howell.*

Leach's has a more synchronized season, with eggs laid from early May to mid-June and young fledging from late August to September. In both species, some body molt begins around the time when the first young hatch in mid-June; a few weeks later tail molt starts, and a few weeks after that the wing molt starts. While the timings of these events in both species are similar, their timings in relation to breeding events are rather different. Leach's complete tail molt by the time their young fledge, and most do not start wing molt before migrating south. In contrast, Ashy Storm-Petrels continue to molt their tail and wing feathers while feeding young through the late fall. Leach's probably suspend wing molt for migration, and Ashy likely suspend wing molt in the northern winter. When birds return to the islands in March, some adults of both species are still completing their wing molt.

Thus, the overall period required for a complete molt by both species is similar, despite their very different life-history strategies, and reflects how molt may expand to fill the time available for it.

REFERENCES

1. Nunn and Stanley 1998; **2.** Hackett et al. 2008; **3.** Pyle 2008; **4–5.** S. N. G. Howell, unpub. data; **6.** Ainley et al. 1974; **7.** Ainley et al. 1976.

PHAETHONTIDAE
Tropicbirds
(CBS; 3 species)

Much sought-after in North American waters, tropicbirds are spectacular streamer-tailed inhabitants of tropical oceans around the world. Traditionally they have been one of six families united in the order of birds known as Pelecaniformes (or totipalmates), the other five being boobies, pelicans, cormorants, anhingas, and frigatebirds. These waterbird families share several features, most conspicuously their totally palmated (that is, fully webbed) feet with all four toes connected by webbing—hence, the common name of totipalmates for this group. Tropicbirds, however, are rather distinct from the other totipalmates. Indeed, a 2008 genetic study suggests that tropicbirds are only distantly related to other totipalmates and instead share an ancestor with flamingos and pigeons![1] All tropicbirds spend most of their lives at sea, where they range widely, and come to land only to nest. The sexes look alike and have no seasonal changes in appearance. Ages differ, and the adult plumage aspect is attained in about three plumage cycles. The semi-altricial young hatch with a downy covering and

This first-cycle Red-billed Tropicbird has shed its innermost primary in the start of its first wave of primary molt. Yellow-billed birds such as this can be mistaken for the more lightly built White-tailed Tropicbird (might molt timing differ between the two species?), but note the black outer primary coverts, diagnostic of Red-billed in all plumages (these feathers are white overall on White-tailed). *Dare County, NC, 22 May 2007. Steve N. G. Howell.*

Tropicbirds, such as this adult White-tailed Tropicbird, often seem to materialize overhead—only to vanish as quickly as they appeared. The molts of tropicbirds remain poorly known because immature birds spend their first year or two of life at sea. The extensively white tips to the outer primaries distinguish this bird from a Red-billed Tropicbird. *Dare County, NC, 26 May 2007. Steve N. G. Howell.*

fledge in 3 to 4 months. The age of first breeding is 3 to 4 years.

Because tropicbirds spend their first year or two of life at sea, little is known of their molts during this period, although they appear to exhibit the complex basic molt strategy,[2,3] with accelerated patterns of stepwise wing molt. Identifying plumage cycles is also complicated by the fact that some populations breed year-round and that cycles may be less than annual. For example, on Ascension Island in the tropical Atlantic, successfully breeding White-tailed Tropicbirds have a 9- to 10-month cycle and Red-billed Tropicbirds an 11- to 12-month cycle.[4] Preformative molts have been assumed to involve only head and body feathers,[5] but the first wave of wing molt also appears to be part of the preformative molt (see pages 38–45 and the family account for boobies and gannets).

Prebasic molts in tropicbirds apparently fill the period birds are at sea between breeding, but they sometimes impinge at either end of the breeding season. Thus, some birds at the start of the breeding season may still be molting body feathers, and occasionally primaries, and birds feeding large chicks late in the season can begin molting their primaries.[6,7,8] In White-tailed Tropicbirds the prebasic molt requires 4 to 5 months, and in the larger Red-billed Tropicbird it requires 5 to 7 months.[9]

One exception to the separation of molt and breeding occurs with the central tail streamers, which can project more than half a meter beyond the rest of the tail. These specialized feathers appear to be molted alternately in adults of all species, such that one feather is almost always growing.[10,11] Why might this be?

A study of the timing of tail-streamer growth in the Red-tailed Tropicbird[12] found that the wirelike streamers of this species grow about 2 millimeters per day, thus requiring almost 6 months to become full-length. Individual birds have only two full-length streamers for about a month before one is shed, which coincides with the prebreeding courtship period. Red-billed and White-tailed tropicbirds also have two full-grown streamers at the start of the season but rarely thereafter.[13] The possession of two full streamers thus appears to serve as an important cue in coordinating the pairing of birds ready to breed, something that may be important in species that can initiate breeding in any month of the year within a single colony. Among the many questions that remain unanswered: Do birds with unequal-length streamers pair to breed? And are such pairings, if they occur, less successful than those of pairs in which both courting birds have full-length streamers?

REFERENCES

1. Hackett et al. 2008; **2.** LeValley and Pyle 2007; **3.** Pyle 2008; **4.** Stonehouse 1962; **5.** Pyle 2008; **6.** Diamond 1975a; **7.** Schreiber and Ashmole 1970; **8–9.** Stonehouse 1962; **10.** Kinsky and Yaldwyn 1981; **11–12.** Veit and Jones 2004; **13.** Stonehouse 1962.

SULIDAE
Boobies and Gannets
(CBS; 5 species)

From squadrons of Northern Gannets plunge-diving into icy waters of the Labrador Current to lightweight Red-footed Boobies snatching flying fish over tropical blue waters, these distinctive seabirds are collectively known as sulids, an abbreviation of their family name. Sulids are one of six families traditionally united in the order of birds known as totipalmates (see the introduction to the family account for tropicbirds). The molting strategies of totipalmates are not especially well known, which may seem surprising given that many are large, conspicuous, and often common birds. Most ornithologists looking at molt prefer to study songbirds, which are quick and easy to prepare into study specimens and don't take up much room in museum cabinets. But wouldn't you rather have 2 boobies than 100 cardinals?

Sulids are large, powerfully built seabirds that occur almost worldwide, with boobies living in tropical oceans and gannets in temperate oceans. They are migratory or nomadic and come to land to nest and roost. The sexes look similar in most cases and have no seasonal changes in appearance. Ages look different, and the adult plumage aspect is attained in 3 or 4 cycles. Young sulids are altricial and hatch virtually naked, but they quickly grow a downy coat and fledge in 3 to 4 months. The age of first breeding is usually 4 to 6 years.

The molting strategy of sulids is complex basic, which reflects how their stepwise wing molts are interpreted (see below, Which Path Did Stepwise Molt Take?). Molts are generally protracted, and wing molts occur year-round or almost year-round, particularly in nonbreeding immatures of tropical species. In breeding adults, wing molt is usually interrupted for a few months during breeding, but it can also overlap completely with breeding.[1] This variation may reflect energy constraints and interannual variation in food resources. Wing molts proceed in accelerated stepwise patterns, with the first wave of wing molt starting when a bird is 6 to 8 months of age (see pages 38–43).

On Ascension Island in the tropical Atlantic, both the Masked and Brown boobies require about 6 months between laying their eggs and fledging their young. Yet the Masked Booby breeds on a 12-month cycle, whereas the Brown Booby has a cycle of only 8 to 10 months.[2] Could this difference

The gradual stepwise wing molt of Red-footed Boobies, like this white-morph adult, allows them to maintain flying proficiency, which is important when chasing flying fish. *Western Pacific (18°N, 147°E), 27 Apr. 2008. Steve N. G. Howell.*

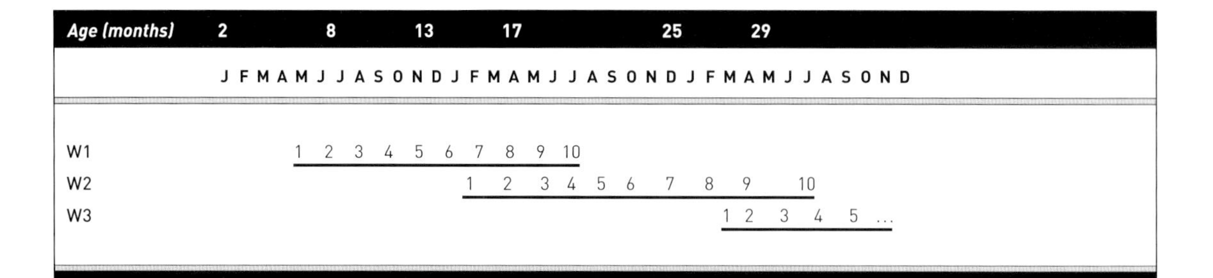

Figure 88. Diagrammatic representation of the first three successive waves (W1, W2, W3) of stepwise primary molt in a Masked Booby hatched in November on Ascension Island (the numbers 1–10 represent approximate timing for molt of individual primaries 1-10). Ages represent months that correspond to those reported by Dorward (1962). Note how each new wave starts before the preceding wave has ended and that the timing of waves becomes slower after the first wave has completed.

reflect the smaller size of the Brown Booby, which may need less time to complete its wing molt?

Stepwise Wing Molts in Tropical Seabirds

Stepwise patterns of wing molt are one way that large, long-winged birds can renew all of their flight feathers once per cycle (see pages 36–45). This approach to overcoming time constraints was first recognized, and described carefully, in a study of Masked Boobies, in which wing molt could be traced in birds of different ages.[3] In nonbreeding immatures, successive cycles, or waves, of primary molt can start before the preceding waves have completed (Figure 88). Thus, the first wave of primary molt starts with p1 at about 6 to 7 months of age and moves outward sequentially through the primaries, with a replacement rate of about one primary per month. The second wave starts again with p1 at about 14 to 15 months of age, when the first wave has reached the shedding of p7–p8. After this point the feathers grow more slowly, at a rate of about three-quarters of a primary per month, perhaps because two or more primaries are growing at the same time. The first wave completes at about 18 months of age, and a third wave starts anew with p1 at about 27 months of age when the second wave has reached about p7 being shed.[4]

By the time a bird has reached its breeding age, at about 4 to 6 years, all of the primaries can be replaced in one episode of molting courtesy of two or three waves of molt. These waves can involve, for example, p1–p6 and p7–p10, or p1–p3, p4–p8, and p9–p10. In breeding adult Masked Boobies, the primaries grow at an average rate of about half a feather per month. Thus, in a breeding adult booby all of the primaries are effectively of a similar age, even though they pertain to two or three different cycles. The next molting period restarts where the previous one left off. In the examples given above, this would mean two waves restarting at p7 and again at p1, or three waves restarting at p4, p9, and again at p1. In this way, it can appear that the primaries are being molted simply in two or three sets; inner, middle, and outer. This is effectively what is happening in an adult booby, but the process is established by the development of accelerated wing-molt cycles in immatures.

In an aseasonal environment such as the tropical ocean, life-history theory predicts that breeding and molt should alternate as quickly as possible so that a bird can produce as many young as possible dur-

Despite being large and conspicuous, the Northern Gannet remains poorly known in terms of the timing of wing molt in immatures. Primary molt typically starts in late winter or early spring. This first-cycle bird had shed p4 in its first wave of primary molt. *Dare County, NC, 29 May 2008. Steve N. G. Howell.*

ing its lifetime. But the tropical ocean is generally a food-poor environment in which seabirds need unimpaired flight to forage successfully. A standard sequential wing molt in such conditions, with only one primary growing at a time, could require a year or more to complete, such that breeding might have to occur at intervals of 18 months or even longer.

Courtesy of stepwise patterns of wing molt, however, tropical seabirds such as the White-tailed Tropicbird, Brown Booby, and White Tern (*Gygis alba*) can all have cycles of less than 1 year,[5,6,7] and larger species such as the Masked Booby can have an annual cycle. Because these species all require a few years to reach sexual maturity, the young can spend their time setting up accelerated waves of wing molt that will allow them as adults to renew all of their primaries in one molting period.

Which Path Did Stepwise Molt Take?

When stepwise patterns in other bird species are examined, the first wave of wing molt may be considered either as the second prebasic molt or as a preformative molt that kick-starts the process (this phenomenon is discussed more fully in the family account for hawks). In temperate-region species with annual breeding regimes, it is easy to distinguish these two pathways to the development of stepwise molt. In tropical seabirds, however, it is more difficult to establish whether the first wave of wing molt represents an early start to the second prebasic molt, or whether it is a novel molt inserted to get the stepwise patterns started.

One approach to solving this riddle may be to relate the timing of early molt waves to the adult cycle. Thus, if Masked Boobies breed and molt on a 12-month cycle, the frequency of their prebasic molts would be expected to average 12 months. By the time they are about 2 years old, however, the Masked Booby, as well as the Osprey[8] and European Shag (*Phalacrocorax aristotelis*),[9] are starting their third post-juvenile primary molt. Conversely, in similar-sized species with temperate annual cycles, such as the Red-shouldered Hawk, Golden Eagle, and Western Gull, birds at this age are starting their second post-juvenile primary molt (see Figure 89 in the family account for hawks). The discrepancy between these two groups of birds suggests that the accelerated stepwise waves set up by boobies and other totipalmates involve a preformative wing molt, which means their molt strategy is considered to be complex basic.

REFERENCES

1–4. Dorward 1962; **5.** Ashmole 1968; **6.** Dorward 1962; **7.** Stonehouse 1962; **8.** Prevost 1983; **9.** Potts 1971.

PELECANIDAE
Pelicans
(SAS; 2 species)

Unmistakable and huge, pelicans are a small family of waterbirds found worldwide in warm climates. They include two marine species (brown pelicans, found only in the New World) and six freshwater species (white pelicans, found on all continents except Antarctica). Traditional classification has placed pelicans with other totipalmates (see the family account for tropicbirds), but the closest relatives of pelicans remain unclear. Most species of pelicans are migratory, and their long, broad wings are well suited for soaring and gliding flight. The sexes look similar, and adults have seasonal changes in appearance. Ages look different, and adult plumage aspect is attained in 3 or 4 cycles. Young pelicans are altricial and hatch virtually naked, quickly growing a downy coat and fledging in 3 months. The age of first breeding is usually 3 to 5 years.

Pelicans have the simple alternate molt strategy, but the homologies of their molts are unresolved. Prebasic molts are protracted, and in adults they mostly follow breeding. At least in Brown Pelicans, the wing molts appear to develop in accelerated stepwise patterns, with the first wave of wing molt usually starting when a bird is about 8 months old (see pages 38–45). Nonbreeding immatures usually replace all of their primaries in one episode of molting courtesy of two molt waves, but adults can develop three or more waves in the primaries.

An interesting aspect of the plumage of Brown Pelicans is how many tail feathers they have. In most bird species the number of tail feathers does not vary, but in Brown Pelicans it varies from 19 to 24, with the greater number of feathers almost always being on the right side. Plunge-diving pelicans twist to the left as they hit the water, and it has been suggested that an asymmetrical tail may facilitate this rotation,[1] presumably by providing more lift on the right side.

How Many Molts—and Why?

Both the Brown Pelican and the American White Pelican appear to have homologous molts and seasonal plumage changes, but how many molts they undergo in the adult cycle is unclear. The usual view has been to recognize three molts,[2,3] but it may be that only two occur. Because American White Pelicans mostly molt from white to white, it is difficult to see how many molts they have. Sadly, a detailed descriptive study of molts and plumages in the Brown Pelican[4] failed to precisely describe actual molts as opposed to changes in appearance, and the authors concluded that they didn't know if pelicans had two or three molts per cycle.

After breeding, pelicans have a protracted prebasic molt that continues into winter; by late winter both North American species have acquired yellowish crown or nape feathers that intensify in color as the spring proceeds, apparently through pigment applied from the uropygial gland.[5] It is unclear whether the yellowish crown of the Brown Pelican and the elongated yellow nape and chest plumes of the American White Pelican represent the last stages of a protracted prebasic molt or are produced by a prealternate molt that occurs in late winter. Brown Pelicans, at least, molt their neck feathers in late winter as part of what is presumably a prealternate molt. At about the time the young hatch, both species molt their crown feathers and develop blackish freckling on a white crown, which is also when American White Pelicans shed the fibrous "horn" on their bill. Why would birds have such a molt at this point in their annual cycle?

American White Pelicans nest on low-lying islands, and the young gather into sometimes large crèches at about 3 weeks of age. It has been suggested that the adults' newly acquired dark head freckling, which is highly variable in pattern, may help the young to recognize their parents.[6] This explanation does not work, however, for the tree- and cliff-nesting Brown Pelicans, in which, by necessity, the young remain in or near the nests. But perhaps the molt is a residual trait carried over from a common ancestor with the white pelicans. The summer head molt of breeding pelicans shows an interesting parallel to the notably early phase of prebasic head molt that occurs in Royal, Elegant, and Sandwich terns, which all develop white forehead spotting at about the time their young hatch. Interestingly, the young of these terns also form crèches that can number hundreds of densely packed birds, suggesting that variable adult head patterns might aid the young in recognizing their parents.

The question still remains, though: do pelicans have a complete prebasic molt in late summer to winter, a prealternate molt in late winter, and a presupplemental molt of head feathers in summer? Or might the summer head molt simply represent the first stages of the prebasic molt? Another possibility is that the molt to dark-freckled head plumage is a delayed phase of the prealternate molt![7] One approach to solving this problem might be by comparison with other species of pelicans, yet the molts of pelicans in general appear to be poorly known, so an answer will have to await critical study, preferably by following known-age birds over time.

REFERENCES

1. Schreiber et al. 1989; **2.** Palmer 1962;
3–5. Schreiber et al. 1989; **6.** Knopf 1975; **7.** Pyle 2008.

The molts of the American White Pelican remain something of a puzzle, as is so often the case with many large, conspicuous birds. Do pelicans have two or three molts per cycle? More work is needed to address this question. *Yucatán, Mexico, 28 Nov. 2007. Steve N. G. Howell.*

PHALACROCORACIDAE
Cormorants
(SAS; 6 species)

These large, black, long-necked waterbirds occur worldwide in coastal marine habitats and also locally inland at lakes and along rivers. Numerous species of cormorants in other parts of the world are boldly patterned in black and white, with brightly colored bills and feet, but in North America we have six species of mostly black, "standard-issue" models. Some species are short-distance migrants that withdraw south or away from inland areas in winter, while others are hardy residents of temperate marine habitats. The sexes look similar, and adults display slight to moderate seasonal changes in appearance. Ages look different, and adult plumage aspect is attained in 2 to 3 years. Young cormorants are altricial and hatch virtually naked, but they quickly grow a downy coat and fledge within 2 months. The age of first breeding is usually 3 to 5 years.

The molting strategy of cormorants appears to be simple alternate, but published accounts of molts in this family are often vague. Prebasic molts are protracted and occur mainly after breeding, from late summer into winter. Wing molts proceed in accelerated stepwise patterns, often with three or more waves in breeding adults (see pages 38–45). The extent and timing of prealternate molts are not well known, which is compounded by the fact that they may overlap in timing with the protracted prebasic molts! Prealternate molts apparently involve some head and body feathers grown from winter into spring, although ornamental "breeding" plumes on the head and neck may simply be slow-growing basic feathers (see the family accounts for Anhinga and herons for more on this subject).

FIRST-CYCLE MOLTS. The first-cycle body molt in cormorants is variable in extent and perhaps even skipped in some individuals, perhaps those hatched latest. It occurs over the first winter, starting sometime between late fall and late winter. Whether this molt represents a preformative molt (which it may resemble in early-molting birds) or a prealternate molt (which it resembles in late-molting birds) cannot be resolved at present, and it may run straight into the complete second prebasic body molt, which starts in spring or summer. Claims that both a preformative molt and a first prealternate molt occur in the Neotropic Cormorant[1] require verification, and may have resulted from a failure to appreciate the variable timing of molt and the variable appearance of feathers acquired over the first winter.

For such large and conspicuous birds, cormorants have been rather neglected in terms of critical observations about their molting. That said, it was a study of a small cormorant, the European Shag (*Phalacrocorax aristotelis*), that helped elucidate the development of stepwise patterns of molt in the primaries.[2] Now, critical observations are needed to determine the extent and timing of first-cycle molts and of prealternate molts.

The outer two primaries of this Brandt's Cormorant are obviously worn and faded, as are some of the middle secondaries. Given how ragged these feathers appear, they are likely to be juvenile feathers that are a little over a year old (postjuvenile generations of feathers tend to be of better quality and not so prone to fading). On this individual, p7 is growing and p8 has been shed, presumably as part of the first wave of primary molt that began in the previous winter; the second wave of primary molt is likely to start soon, setting up stepwise patterns that allow adults to replace all of their primaries in one episode of molting. *Monterey County, CA, 26 Sept. 2008. Steve N. G. Howell.*

Winter Wing Molt?

Although wing molt in cormorants is most intensive from late summer into early winter, some species also appear to molt flight feathers slowly over the winter.[3] In other cormorants of temperate and mid-latitude regions, there is a midwinter molt suspension,[4] and wing molt also overlaps with the breeding season.[5,6,7] Thus, wing molt in most cormorants is essentially a gradual, year-round activity interrupted for periods of bad weather, food shortages, and peaks in breeding activity.

So why would cormorants molt throughout the winter, at a time when very few species in North America are molting? Is wing molt during the breeding season not favored because of constraints on food resources at that time? Or does most of the

The first-cycle plumage aspect of the Double-crested Cormorant is rather variable, as shown by these three birds, one of which had already shed its innermost primary in the start of its first wave of primary molt. Cormorants are unusual among North American birds in molting flight feathers through the winter. *Sonoma County, CA, 23 Dec. 2008. Steve N. G. Howell.*

wing molt actually occur from late summer to early winter and again in late winter and spring, and thus not really throughout the winter?

Interestingly, among tropical totipalmates with accelerated patterns of stepwise wing molt, such as tropicbirds, boobies, pelicans, and frigatebirds, wing molt also occurs throughout the nonbreeding season. This is easier to understand for birds living in a warm tropical environment than for cormorants living in cold temperate waters. Of North American breeding species having stepwise molt, most that reside in temperate regions, such as condors and hawks, have normal stepwise molts that overlap extensively with the breeding season and are interrupted in winter. With respect to winter molting in cormorants, it has been suggested that stepwise molts may have developed in an ancestor in a tropical environment and have been retained by all descendent species.[8] Thus, might the year-round or winter molting patterns of cormorants represent an ancestral trait?

REFERENCES

1. Telfair and Morrison 1995; **2.** Potts 1971; **3.** Filardi and Rohwer 2001; **4.** Potts 1971; **5.** Bernstein and Maxson 1981; **6.** Cooper 1985; **7–8.** Rasmussen 1988.

ANHINGIDAE
Anhinga
(CBS, SAS?; 1 species)

Sometimes known as snakebirds, anhingas often swim with only their long, curved, snake-like head and neck above the water. They are part of a small family of totipalmate diving birds found worldwide in warm tropical regions. Although anhingas superficially resemble cormorants, the two families differ in several ways, including how they molt their wings. Also unlike most cormorants, anhingas soar readily and migrate in wheeling flocks resembling kettles of hawks, with which they often mix during migration. The sexes and ages look different, and adult plumage aspect is attained in 2 to 3 years. Like cormorants, anhingas are altricial and hatch naked, but they quickly grow a downy coat and fledge in about 2 months. The age of first breeding is likely about 3 years, corresponding to attainment of the adult plumage aspect.

Anhingas are yet another family in the totipalmate assemblage whose molting strategy is not well known. There appears to be a preformative molt of head and body feathers but no prealternate molt in the first cycle.[1] Breeding adults have filamentous white plumes on the head and neck that may be alternate feathers or might be specialized basic feathers that grow slowly, like the plumes of herons or the white filoplumes found on breeding petrels.[2] Claims that the prealternate molt is complete "except for the wing quills"[3] are in error. Thus, the molting strategy of anhingas is complex basic or simple alternate. As is so often the case, further study is needed.

PREBASIC MOLTS. Prebasic molts follow the breeding season and involve a synchronous or near-synchronous loss of the primaries and secondaries, so that birds are flightless for a few weeks. During this period anhingas are understandably wary, and they slip stealthily into the water at any sign of danger. But why would anhingas have a synchronous wing molt and cormorants a stepwise wing molt? Anhingas are relatively lighter-bodied than cormorants, so the explanation that a short period of flightlessness for heavy-bodied birds is better than a long period of labored flight doesn't seem to apply.

Synchronous wing molt also has the straightforward advantage of "getting it done" in a short period, so perhaps time constraints are the drive. Anhingas often occur in seasonal wetlands and at small bodies of water, and they spend a lot of time soaring in search of feeding areas and while migrating, for which having a full wing area is important. In many waterbirds, breeding is timed so that the young fledge in times of plenty, which may also be a good time to complete a synchronous molt quickly.

Unlike cormorants, the Anhinga has a synchronous wing molt, during which birds are flightless but wary. When fully winged, however, anhingas tend to be conspicuous, like this adult female. Note the enigmatic corrugations on the central tail feathers. *Dade County, FL, 5 May 2007. Lyann A. Comrack.*

Thus, birds can be full-winged for the rest of the year, which allows for easier mobility.

Other nomadic and migratory waterbirds such as pelicans, herons, and ibises often share the same wetlands with anhingas, but none of these (except apparently the Least Bittern) undergoes a synchronous wing molt. However, these other species all tend to be large and conspicuous and might have a hard time avoiding predators if they were flightless. The Anhinga's diving ability gives it an edge over these other species, and synchronous wing molt is the strategy that works for its lifestyle. Our initial question then becomes turned around: why don't cormorants have synchronous wing molts?

The Riddle of the Corrugations

An intriguing feature of the Anhinga's plumage is that the outer webs of the longest scapulars and central tail feathers have transverse corrugations, which are absent or at best weakly defined on juveniles but strongly defined on adults, especially males. No convincing explanation has been advanced for these corrugations. It has been suggested that they might

The Anhinga is sometimes known as the "snakebird" for its habit of swimming with only its neck and head projecting above the water surface. After being submerged like this, the wings can be waterlogged, and birds may have difficultly taking off. *Nayarit, Mexico, 21 Jan. 2009. Steve N. G. Howell.*

reflect light, which might relate to display, or that those on the tail might produce a mechanical sound in a putative flight display,[4] but the latter suggestion does not account for the corrugated scapulars.

It is notable that these corrugated feathers align down the midline of the upperparts, which suggests they could relate to channeling the flow of water. Most diving birds that swim underwater in pursuit of fish, such as loons, grebes, and cormorants, have small tails that do not offer much resistance to the water, yet the Anhinga has a big tail that would appear to be unwieldy underwater. Could the corrugations help birds swim faster underwater by reducing turbulence? Some high-tech clothing designed for swimmers these days has small surface corrugations that reduce turbulence and improve speed through the water, so it would not be surprising if the Anhinga came up with this idea a long time ago.

The age- and sex-related development of these corrugations, which are unlike anything found in any other North American bird, suggests that they may relate to courtship or breeding status. Perhaps adult males with deeper corrugations are somehow showing that they can swim faster in pursuit of prey and thus might be better at provisioning food. For now, though, the riddle of the Anhinga's ridged feathers remains unsolved.

REFERENCES

1. Pyle 2008; **2.** Imber 1971; **3.** Palmer 1962:359; **4.** Owre 1967.

FREGATIDAE
Frigatebirds
(CBS, SBS?; 1 species)

Looking like they were created by Disney as birds of evil, the black angular shapes of frigatebirds (often known simply as frigates) hang in the wind over tropical beaches worldwide. They are another small distinctive family of totipalmate birds (see the introduction to the family account for tropicbirds) that lack obvious close relatives. Frigates are extremely light-bodied relative to their big wingspans, and they can soar easily even in calm conditions. The sexes and ages of frigatebirds differ in appearance, and females are appreciably larger than males. Adult plumage aspects may require 8 to 10 years or longer to develop, and there are no seasonal changes in appearance. The altricial young hatch naked but quickly grow a downy covering; they can fly in 4 to 5 months but benefit from about a year or more of further parental care, during which time they return to the nest to be fed.[1,2] The age of first breeding is not well known but appears to be around 10 years.[3]

Frigates spend most of their life in the air and much of their first few years at sea; hence, little is known of their molts. They probably exhibit the complex basic molt strategy, or perhaps the simple basic strategy. Interpreting plumage cycles in frigates is complicated by potentially year-round breeding and by cycles that may be up to 2 years long. Prebasic molts of adults likely fill the periods birds are at sea between nesting cycles. Flight feathers apparently are not molted during the breeding season,[4,5,6] and the wing molt proceeds in stepwise waves (see pages 36–45). Because the length of molt cycles is uncertain (see below), it is unknown whether stepwise waves develop on an accelerated schedule (as in other totipalmates) or on a normal schedule. The second wave of primary molt starts when the first wave has reached p8 or p9 being shed,[7] which is similar to the patterns found in boobies and tropicbirds. By the time frigatebirds reach breeding age, all of the primaries may be replaced in one cycle, courtesy of three waves.

A Unique Lifestyle— and a Unique Molting Strategy?

Frigatebirds have a unique lifestyle, and much about their molts remains to be learned. They have a remarkably long period of immaturity, which, among birds, is matched perhaps only by the largest albatrosses. And because young frigates receive up to a year or more of post-fledging care, adults that breed successfully cannot nest every year and probably do so every other year, as also happens in the big albatrosses. However, given that on average

Magnificent Frigatebirds, like this adult female (above), have a very gradual wing molt. This means their wings don't develop any appreciable gaps that might compromise their ability to outmaneuver Blue-footed Boobies and shake them up for food. *Nayarit, Mexico, 22 Jan. 2007. Steve N. G. Howell.*

only 20 percent of nesting attempts by frigates are successful, many adults may try to nest in successive seasons or perhaps move to other colonies that are on slightly different schedules.[8,9] How often they may keep trying to breed possibly reflects whether they need to molt. And deciphering how all of this might relate to their molt cycles is not an easy task.

The many years that frigates take to reach breeding age and attain their adult plumage aspect probably reflect their unique life history. The ocean is not a gentle or forgiving mother, and most seabirds require several years to learn how to feed themselves before they start breeding and have other mouths to feed. Frigates inhabit tropical oceans, which are generally food-poor regions where it is tough to make a living on the best of days, and the areas they inhabit are often prone to hurricanes, which can disrupt breeding attempts and life in general. As if this weren't enough, the plumage of frigates gets saturated easily, and their anatomy makes it difficult for them to take off from the sea.

To compensate for not being able to settle on the sea, frigates are agile and accomplished flying machines, and they feed by plucking fish from near the sea surface, by snatching up nestling seabirds, and by pirating food from boobies and other birds. These highly specialized feeding techniques take time to perfect, which helps explain the extremely prolonged post-fledging period during which young birds presumably learn how to start living in their environment. Such light-bodied birds that rely on such precision in their flight cannot afford large gaps in the wings. This, in combination with the challenges of finding food to fuel molt in a tropical marine environment, means that wing molt in frigatebirds is an extremely slow process, with a given primary not usually shed until the adjacent feather is almost fully grown.

So, how long does a frigatebird take to attain its adult plumage aspect? And how long are its molt cycles? Although eggs may be found in any month of the year, most egg-laying at a given colony usually occurs over a 3- to 6-month period, suggesting a seasonal, and often annual, cycle to breeding.[10,11,12] Given that five analogous plumage stages have been identified in four of the world's five frigatebird species,[13,14] and that the adult plumage aspect is apparently attained at about 10 years of age,[15] it is tempting to suggest that a frigatebird plumage cycle may be about 2 years. That juvenile head and body plumage is worn for over a year, perhaps even 2 years, before molt is apparent also suggests that

Frigatebirds pass through several immature plumage stages before attaining the adult plumage aspect at about 10 years of age. A knowledge of wing molt cycles can help unravel the ages and plumage sequences. This Magnificent Frigatebird has reached p8 growing in its first wave of wing molt, with the outer two primaries (p9–p10) being worn juvenile feathers. The inner primaries are quite worn relative to the new middle primaries, suggesting that the wing molt has been protracted and that a second wave of primary molt may start soon. The plumage pattern on the body resembles that of a juvenile, but some new white and black feathers are appearing. *Oaxaca, Mexico, 15 Dec. 2008. Steve N. G. Howell.*

This immature Magnificent Frigatebird shows an obvious contrast between the older outer four primaries and the fresher and blacker inner five, with p6 shed or just starting to grow in. The outer primaries appear relatively fresh and postjuvenile. In combination with an extensively white body, this indicates that the wave of primary molt that has reached p6 being shed is the second wave. *Oaxaca, Mexico, 15 Dec. 2008. Steve N. G. Howell.*

plumage cycles might be longer than annual.[16,17,18] In Great Frigatebirds (*Fregata minor*), it has been reported that some birds retain individual primaries for up to 4 years,[19] which might correspond to two molt cycles and, in turn, offer more support for the idea of 2-year molt cycles.

Might some birds molt on annual cycles and others on 2-year cycles? In the frigatebird species of open-ocean environments, both parents continue to feed the juvenile for several months after it can fly. But in the coastal-living Magnificent Frigatebird, the species we have in North America, males usually desert their youngster when it is 3 to 6 months old, while females stay with it for up to another year.[20,21,22] This has been attributed to relatively rich nearshore feeding grounds, which allow the female alone to provision the youngster. Thus, males can return to breed each year, but successfully breeding females can breed only every 2 years. It is usually assumed that male frigates leave to molt after deserting, whereas females do not molt while still feeding young.[23,24] Might it be that male frigatebirds have an annual molt cycle and females a biennial one, something that has not been found in any bird species? And how would this relate to the development of immature plumages? Although different molt cycles for different sexes is an intriguing possibility, it is more likely that both sexes of frigatebirds have either an annual or a biennial molting cycle. This is because homologous characters, such as molting strategies, have high genetic correlations between males and females and do not diverge quickly.[25] But then again, might frigatebirds be the exception to the rule?

Long-term studies involving hundreds of marked birds are required to answer the questions of molt cycles in frigatebirds. Sadly, one paper that could have helped address the puzzle of frigatebird molts and plumage sequences failed to describe wing-molt timings, did not identify plumage cycles, and was based on only 12 known-age birds (9 of them less than 6 years old). The authors came to the astonishing conclusion that the adult plumage aspect of Great Frigatebirds is attained via only a single "post-juvenile" head and body molt that takes place over 8 to 10 years, and they implied that the male's lanceolate back feathers require 3 to 4 years to grow.[26] Admittedly, frigatebirds are unusual in taking many years to acquire adult plumage, but there is no precedent for any bird taking that many years to complete a single molt, particularly a species living in a tropical, saltwater environment. The authors also overlooked two studies that carefully identified plumage stages in frigatebirds and showed that the chest sides change from black in juveniles to white in the next plumage stage to black again in adult males,[27,28] which could not be achieved by only one molt.

Why So Much Variation in Appearance?

Frigatebirds have complex sequences of age-related plumage patterns and are unusual among seabirds in that the sexes look quite different, a phenomenon known as sexual dimorphism, or sexual dichromatism. But why do they have so many immature plumages, and why are they strongly sexually dimorphic?

AGE VARIATION. In having so many immature plumages, frigates provide a singular example of a phenomenon known as delayed plumage maturation (a topic discussed more fully in the family account for wood-warblers). At least five hypotheses have been advanced to explain this phenomenon,[29] which has been explored mainly among songbirds. However, each hypothesis seems to overlook, or at best sidestep, the most obvious explanation, which might be termed the puberty hypothesis: plumage aspect simply signals physiological immaturity. Other attributes of immature plumages, such as being cryptic or signaling social status, may just be incidental attributes.

SEX VARIATION. Striking sexual dimorphism in appearance is found among diverse species of birds, although among seabirds it occurs consistently only in frigatebirds. Among birds in general it occurs most frequently in species that experience time constraints in their breeding cycle, such as northern-breeding ducks and warblers. It is also frequent among species in which males play little or no part in nesting and often form leks to attract females, such as grouse, hummingbirds, and the tropical cotingas, manakins, and birds-of-paradise. Why it should have developed to such an extent in frigatebirds is unclear, but aspects of both time constraints and leks can be identified in their breeding strategies.

The well-lit underwings of this adult male Magnificent Frigatebird allow us to distinguish where the preceding waves of primary molt ended. On the top wing there appear to have been three waves, ending with p4, p8, and p10; note how p4 and p8 look fresher and darker than the adjacent outer primaries. On the bottom wing the waves are slightly different and end with p2, p4, and perhaps p10 (it is difficult to evaluate the relative wear among the outer three primaries, all of which appear quite fresh). *Oaxaca, Mexico, 14 Dec. 2008. Steve N. G. Howell.*

Male frigates may be under pressure to find a mate quickly and start breeding, so they can desert their family and try to breed in the next season. One study[30] found that early-breeding males could desert earlier in the season and have 84 percent chick survivorship, and that 75 percent of these males returned to attempt breeding the next year. In contrast, among males that paired early or late but deserted late, chick survivorship was 64 percent or lower, and only 50 percent of these males returned in the next season. Thus, early-breeding males potentially produce more offspring in their lifetime, which suggests that sexual dimorphism may be driven partly by the time constraints of needing to pair quickly and breed.

The displaying system of frigatebirds, in which groups of males gather and inflate their big red throat pouches, resembles that of species that form leks, such as many grouse and cotingas. However, male frigates are then tied to the female that selects them and go on to participate in nest building, incubation, and feeding of the young—all things that "real lek males" don't do. Lek systems typically develop in areas with plentiful food, such that females can raise young as a single parent. But given the constraints of finding food in a tropical marine environment, a conventional lek simply wouldn't work. Still, male Magnificent Frigatebirds do desert their mate and young as soon as they can, to try and get back for the next breeding season, so they are a little like lek-wannabes.

TOTAL VARIATION. Another group of birds among which plumage dimorphism occurs frequently is pirates and hunters, such as jaegers and hawks. In these cases, however, the dimorphism is not related to sex or age. Still, might there be some advantage for hunters to have a varied appearance? Could it improve their chances of hunting if the victim has to process more than one search image?[31] Beware of all-black birds; no, beware of black birds with white heads; no, beware of black birds with white chests; and so on.

REFERENCES

1. Diamond 1975b; **2.** Nelson 1975; **3.** Valle et al. 2006; **4.** Diamond 1975b; **5.** Nelson 1975; **6–7.** DeKorte and DeVries 1978; **8.** Diamond 1975b; **9–10.** Nelson 1975; **11.** Diamond and Schreiber 2002; **12.** Osorno 1999; **13.** Howell 1994; **14.** James 2004; **15.** Valle et al. 2006; **16.** Diamond 1975b; **17.** Howell 1994; **18–19.** Valle et al. 2006; **20.** Diamond 1972; **21.** Osorno 1999; **22.** Trivelpiece and Ferraris 1987; **23.** Diamond 1975b; **24.** Nelson 1975; **25.** Lande 1980; **26.** Valle et al. 2006; **27.** Howell 1994; **28.** James 2004; **29.** Valle et al. 2006; **30.** Osorno 1999; **31.** Paulson 1983.

ARDEIDAE
Herons
(CBS, SAS?; 12 species)

These long-necked and long-legged wading birds occur worldwide except in cold high latitudes, and they are well represented in North America, from the snow-white Snowy Egret and cryptic American Bittern to the crazily dashing Reddish Egret and the nocturnally stealthy night-herons. All species are migratory or nomadic to some extent, although many southern populations are mostly resident. The sexes look alike except in the Least Bittern, and only the Cattle Egret has seasonal changes in the coloration of its plumage. Ages can look alike in plumage aspect, as in the white egrets, or different, as in the night-herons. Adult plumage aspect of all species is attained within 2 to 3 years, which coincides with the usual age of first breeding. The altricial young hatch with a sparse downy covering, and most species fledge in 1 to 2 months.

Most if not all North American herons exhibit the complex basic molt strategy, and stepwise patterns of wing molt can develop in several species. Prebasic molts occur mainly from late summer into winter and can be suspended for migration. Preformative molts are partial and can be protracted from fall into spring; they mainly involve head and body feathers and sometimes upperwing coverts. Depending on molt timing and on individual hormone levels, variable plumes can be acquired in the first cycle (see pages 17–19). Prealternate molts may

The long "breeding plumes" of herons and egrets, such as those on this adult Great Egret, are actually basic feathers that grow slowly over the course of the fall and winter, and not alternate feathers as has often been assumed. *Marin County, CA, 15 Mar. 2008. Steve N. G. Howell.*

The Cattle Egret is a common and conspicuous bird, but whether adults have an alternate plumage has yet to be satisfactorily resolved. Another puzzle is how its ornamental plumes are colored—does the orange color come from within and get deposited on the growing feathers, or is it stained on cosmetically? *Quintana Roo, Mexico, 4 Dec. 2007. Steve N. G. Howell.*

occur in the Cattle Egret but apparently not in other North American species.

PREBASIC MOLTS. Prebasic molts in herons are of interest in several ways, not least because they produce the long ornamental plumes traditionally associated with "breeding" (and, by analogy, alternate) plumages. A 2004 review of molts in herons and egrets[1] showed that the ornamental plumes in species such as the Great Blue Heron and Little Blue Heron are basic feathers that grow very slowly through the winter. These feathers are not simply produced quickly in a spring molt, but instead start growing in fall and may take 5 months or longer to develop; this long growth period may be related to their ornamental "feathery" structure.

Wing molts in herons may be sequential, synchronous, or stepwise (see pages 34–36). Most species have sequential molt, which may be punctuated by time constraints to produce up to three waves that run in stepwise patterns, such as those occurring in some populations of Black-crowned Night-Herons.[2] The secretive Least Bittern apparently undergoes a synchronous wing molt[3] like the rails, with which it shares its dense reed-bed habitat. Presumably this strategy is the most time-efficient way for it to fit molt between breeding and fall migration.

PREALTERNATE MOLTS. With the finding that the ornamental plumes of herons and egrets are slow-growing basic feathers, the question naturally arises: do any North American herons have alternate plumages? Certainly there is no evidence that bitterns and night-herons do, and neither do most herons and egrets. One possibility lies with the Cattle Egret, adults of which in spring have bright orange plumes on the crown, back, and chest, unlike the mostly white appearance they show in winter. The color of these plumage areas apparently intensifies as the breeding season approaches, suggesting that the coloration is not produced by molt (once feathers are grown, they are dead tissue and do not receive "internal" pigmentation). Instead, they may be colored by topical application of pigment,[4,5] a process that has also been implicated in the pale yellowish coloration shown by the Whistling Heron (*Syrigma sibilatrix*) of South America.[6] For color to be applied so precisely suggests that certain feathers may be specially structured to take up pigment that does not adhere to other feathers around them—or are Cattle Egrets just incredibly good at painting themselves?

Prealternate molts do occur in some herons, such as the Old World pond herons of the genus *Ardeola*, most of which molt from a dark-streaked basic plumage into an unstreaked alternate plumage. The extensive prealternate molts of these species include feathers of the head, neck, back, and some upperwing coverts and tertials. Thus there is a precedent in herons for the Cattle Egret to have a prealternate molt. That such an easily answered unknown surrounds such a common and conspicuous bird reflects how little critical attention has been paid to molt.

REFERENCES

1. Pyle and Howell 2004; **2.** Shugart and Rohwer 1996; **3.** Pyle 2008; **4.** Siegfried 1971; **5.** Pyle and Howell 2004; **6.** Humphrey and Parkes 1963a.

THRESKIORNITHIDAE
Ibises and Spoonbills
(SAS, CBS? SBS?; 4 species)

This family of striking and often attractively colored, long-legged wading birds occurs locally worldwide, mainly in warm climates and with the greatest diversity in tropical regions. All North American species are migratory to some extent, although none is truly a long-distance migrant, and all are prone to disperse in response to changing water levels. The sexes look alike but males average larger, and some species have distinct seasonal changes in their appearance. Ages differ, and adult plumage aspect is attained in 2 to 3 years. The semialtricial young hatch with a downy covering and fledge in 1 to 2 months. The age of first breeding is usually 2 to 4 years.

Ibises and spoonbills apparently exhibit the simple alternate molt strategy, although details of molting in this family are not well known. At least in White-faced and Glossy ibises (the genus *Plegadis*), there is a variably extensive molt of head and body feathers in the first cycle and a prealternate molt of head and body feathers in subsequent cycles. The same pattern may hold for Roseate Spoonbills, although their "alternate" plumes may prove to be basic feathers that grow in slowly, as occurs with the plumes of herons. In the White Ibis, it has been reported that the juvenile plumage is held through the winter and a protracted complete molt occurs from late winter into fall, starting in February with back and other body feathers and ending in October–November with head and neck feathers;[1] the flight feathers are replaced in fall. The molts of adult White Ibises have not been described critically, but if the head and neck feathers are always the last to be replaced in a prebasic molt, then would they be replaced again in a prealternate molt? The buff wash acquired by some White Ibises on their crown and chest, and suggested by some to be alternate plumage,[2] is apparently acquired by staining rather than by molt.[3] Thus, might the White Ibis have a simple basic molt strategy?

Prebasic wing molts in all species start in late summer or fall and may suspend for migration to finish on the nonbreeding grounds. Stepwise wing molts have not been documented but seem possible, given that they occur in comparably sized herons. Or is there something about the life cycles of ibises and spoonbills that differs from those of herons, such that wing molts in ibises and spoonbills can be completed without the development of stepwise patterns?

The Problem of First-cycle Molts

Naming the first-cycle molts in *Plegadis* ibises and in spoonbills exemplifies a difficulty that crops up in other groups of birds, such as cormorants and large

The head and neck feathers of this young White Ibis are clearly worn and faded, as are the brown feathers of the upperparts. But are the new and incoming white feathers part of a preformative molt or the start of a protracted and complete second prebasic molt? *Yucatán, Mexico, 28 Nov. 2007. Steve N. G. Howell.*

The Roseate Spoonbill is a stunning and unmistakable bird, even in its immature plumages. The uniform-generation flight feathers and lack of wing molt indicate that this individual is in its first plumage cycle. *Nayarit, Mexico, 14 Jan. 2007. Steve N. G. Howell.*

gulls. In first-cycle ibises it appears that only a single molt, involving a variable amount of head and body feathering, occurs from fall into spring. When compared with other families such as herons, or even hawks, this molt resembles a preformative molt in its timing and extent. Relative to the prealternate molts of adult ibises, however, the first-cycle molt differs in timing but not necessarily in extent. So is it a prealternate molt or a preformative molt?

In species that do not breed in their first cycle, and thus start their prebasic molt earlier than do breeding adults, it makes sense that a protracted preformative molt may be all that is needed to keep the plumage functional through the first cycle. In subsequent cycles the basic plumage could become worn enough for an additional molt to evolve, and hence alternate plumages could have developed that are not present in the first cycle. Or perhaps a prealternate molt developed at some time in all cycles and since has merged with the preformative molt, such that effectively only one molt now occurs in the first cycle?

Although there are presumably answers to such intriguing questions, for now they appear to be unknowable. If nothing else, this dilemma highlights that the frameworks we build are simply imperfect attempts to house natural systems, which do not necessarily lend themselves to intellectual domestication.

REFERENCES

1. Beebe 1914; **2.** Palmer 1962; **3.** Kushlan and Bildstein 1992.

CICONIIDAE
Storks
(SBS, CBS?; 1 species)

Storks are an ancient and well-defined family of large, long-legged wading birds found almost worldwide in warm climates, with the greatest diversity in Africa and tropical Asia. They are of relatively marginal occurrence in North America, where one species—the Wood Stork—nests in the southern U.S. Storks breeding at higher latitudes tend to be migratory. The sexes of Wood Stork look alike, and ages differ slightly in appearance, with adult plumage aspect attained in the second or third cycle. Young storks are altricial and hatch with a

This immature Wood Stork has apparently interrupted its primary molt at a point typical of stepwise patterns. Note the contrast between the newer and glossier inner and middle primaries (p1–p8) and the older, faded, and frayed outer two primaries (p9–p10). For a stepwise pattern to develop, the next episode of molt would be expected to continue with p9 and start again at p1 (see pages 36–45). *Nayarit, Mexico, 14 Jan. 2008. Steve N. G. Howell.*

Like several species of long-winged birds, the Wood Stork probably develops stepwise waves of wing molt so that all of its flight feathers can be replaced between breeding attempts. Note the elongated undertail coverts projecting between the legs on this adult—are these basic or alternate feathers? *Nayarit, Mexico, 17 Jan. 2007. Steve N. G. Howell.*

sparse downy covering; they soon grow a thicker woolly down and fledge in 2 to 3 months. The age of first breeding is probably 4 years or older.

The molting strategies of American storks are surprisingly poorly known and may be complex basic, simple basic, or even simple alternate! A limited preformative molt of some head and body feathers may occur, and the first wing molt might also be a preformative molt if stepwise waves develop on an accelerated schedule (see pages 38–45). It has also been suggested that the long plumelike undertail coverts of breeding adult Wood Storks represent an alternate plumage.[1] However, these specialized feathers may simply be basic feathers that grow in slowly, like the plumes of herons and egrets.

An interesting aspect of age development in Wood Storks is that juveniles have soft "woolly" feathering on the head and neck, whereas adults have a naked head and neck; second- and third-cycle birds have some neck feathering. Is neck feathering renewed in each prebasic molt and then not grown in adults? Or is the juvenile feathering slowly lost by abrasion?

The first wing molt may be incomplete, with some juvenile outer primaries retained, but in subsequent cycles all primaries are usually replaced, perhaps via two waves and possibly in a stepwise pattern. Given that storks soar frequently, either while commuting to feeding areas or while migrating, it makes sense that they would develop a stepwise molt to renew their flight feathers as economically as possible. But when does the wing molt occur? Does it overlap with the breeding season, or does it occur mainly after breeding? Clearly, a careful study of molt in Wood Storks could answer a lot of questions.

REFERENCES

1. Palmer 1962.

PHOENICOPTERIDAE
Flamingos
(SAS, CBS?; 1 species)

Synonymous with pink, flamingos are a small family of unmistakable birds found mainly from the Caribbean south into South America and from the Mediterranean region south into Africa. Their unique appearance has spawned many ideas about what their closest relatives might be, from storks and shorebirds to grebes and geese, and the jury is still out. Several species are nomadic or partly migratory, and birds from Caribbean populations wander in winter to the southern U.S. The sexes look alike, and the adult plumage aspect is attained at 2 to 3 years of age. Young flamingos are semi-precocial and downy, usually leaving the nest within a few days to form crèches with other young; they fledge in about 10 weeks. The age of first breeding is 3 to 6 years.

Despite the popularity of flamingos in captivity, their molts have avoided critical description, and wild birds are difficult to study in the extensive salt lagoons and lakes they typically inhabit. There may be a preformative molt of head and body feathers in the first 6 months or so of life, although one study of captive birds reported "little change in body contour feathers" between the acquisition of juvenile plumage and the complete second prebasic molt at about a year of age.[1] Another study[2] reported highly variable "juvenile" plumages that might instead refer to variable formative plumages. Alternate plumages likely occur (see below), at least in second and subsequent cycles. Thus the molt strategy of flamingos is probably simple alternate, although it could be complex basic.

PREBASIC MOLTS. Prebasic molts in flamingos, as in cranes, show a dichotomy in the patterns of flight-feather molt. Most birds apparently undergo a synchronous wing molt and become flightless for a few weeks, which may occur during or after the breeding season;[3] protection presumably comes from living in vast and relatively inaccessible salt lagoons inhospitable to predators. In populations where birds need to commute to and from feeding areas, molt of the flight feathers can be sequential, and birds can still fly with three primaries in molt; other birds in these populations, however, undergo synchronous wing molt.[4,5] It is not known whether flamingos, like cranes, invest all of their molting resources into the synchronous wing molt, to regain their power of flight as soon as possible, and then follow up with the head and body molt. Among one group of captive American (or Caribbean) Flamingos, molt of head and body feathers was found to be mostly separate in timing from wing molt, but wing coverts were renewed at the same time as the primaries.[6]

What determines whether an individual flamingo has sequential or synchronous wing molt? Does it relate to age or breeding status or simply to environmental conditions in a given year? Can a bird

Are the deep pink head and neck feathers of these adult American (or Caribbean) Flamingos basic or alternate feathers? Critical study could answer this and several other unknowns in the world of flamingo molt. *Yucatán, Mexico, 29 Nov. 2007. Steve N. G. Howell.*

How many molts and plumage cycles would be required for this young American Flamingo to attain its fully bright pink adult plumage aspect? As is so often the case, nobody really knows. The feathers on this individual's head and neck, plus its shorter scapulars, may well be formative plumage. *Yucatán, Mexico, 29 Nov. 2007. Steve N. G. Howell.*

switch from one strategy to another and back again? There are many questions that could be answered by studying molt in a population of wild flamingos.

PREALTERNATE MOLTS. Prealternate molts have yet to be documented unambiguously in flamingos, and distinguishing them from potentially protracted prebasic molts could be difficult. It is quite possible, however, that some head and neck feathers are renewed in prealternate molts, as suggested by several observers.[7] It has also been suggested that plumage may become brighter prior to the breeding season from pigmentation added by preening.[8] However, while yellow pigment can be added through preening (as in herons and pelicans), there is no evidence that any birds add pink pigment in this way,[9,10] although this has often been claimed. The pink coloration of flamingos is synthesized from their food, and pigment is deposited onto the growing feathers; captive adults fed an inappropriate diet molt into a weakly pigmented or white plumage. Once a feather grows, its color should not change, although older feathers fade over time.

REFERENCES

1. Shannon 2000; **2.** Johnson et al. 1993; **3.** Kear and Duplaix-Hall 1975; **4.** de Boer 1979; **5.** Johnson et al. 1993; **6–7.** Shannon 2000; **8.** Pyle 2008; **9.** Hardy 2003; **10.** Hudon and Brush 1990.

CATHARTIDAE
New World Vultures
(SBS, CBS; 3 species)

Changing fortunes mark the taxonomic history of the New World vultures (which include condors) and illustrate well the futility of trying to keep up with taxonomic decisions in the sequences we use in guidebooks. Taxonomy aims to reflect evolutionary relationships, which cannot be portrayed in linear sequences such as those in books. Field guides have, or should have, a different purpose: to help in field identification, for which taxonomy can be helpful—or not. Traditional views placed vultures near hawks, but genetic and anatomical studies indicated that New World vultures were closely related to storks, so they were shifted forward in the sequence. Further genetic evidence, however, suggests that this earlier conclusion may have been in error, so vultures have been put back to where they were placed before. Next week, who knows where they might be. Meanwhile, their molts remain unchanged.

New World vultures make up a small family found from southern Canada south to Tierra del Fuego. Although northern populations of the Turkey Vulture are migratory, most species are resident. The sexes look alike. Ages are alike in most vultures but are different in condors, which achieve their adult plumage aspect in 6 to 8 years. Young are altricial and downy and fledge in 2 to 6 months. The age of first breeding is not well known but may be about 6 years of age. The molting strategies of New World vultures are simple basic and complex basic, with stepwise patterns of wing molt that reflect their long wings and very long outer primaries.

PREBASIC MOLTS. Prebasic molts in both vultures and condors are protracted and overlap extensively with breeding. This presumably reflects the need to have unimpaired flying abilities during winter, when periods of bad weather can constrain foraging. It also indicates that resources from spring through fall are sufficient to support molting and breeding at the same time. Breeding individuals, however, typically have more-protracted molts than do nonbreeders. Thus, in breeding California Condors, primary shedding starts in February, slows or suspends during spring, and picks up again in late summer, spanning at least 5 to 6 months. This compares with about a 4-month span in nonbreeders, which shed primaries starting in April. The schedule of breeders corresponds to relatively rapid primary molt during incubation (eggs are laid in February–March) and slower molt during the early chick stage in April–May.[1]

Turkey and Black vultures usually replace all of their primaries in a single cycle, and during their second calendar year of life they replace the inner primaries twice—in effect molting 12 to 14 primaries in one season.[2] For example, among Turkey Vultures in central California, the first primary molt starts with p1 in late winter and continues through to p10 by late fall, with a second wave starting again at p1 in early fall and usually reaching p2–p4 before it suspends for the winter.[3] In subsequent years, primary molt starts up again in spring where it left off (usually at p3–p5) and continues out to p10; the inner few primaries molt again when molt of the outer primaries has reached the point at about when p8 is shed. Thus, each wing molt subsequent to the first wing molt starts before the preceding one has finished. This pattern appears to represent stepwise molt on an accelerated schedule (see pages 38–45), with all primaries consistently replaced in two waves; in the second calendar year, it recalls the molting pattern of Northern Gannets, another temperate-breeding species that suspends wing molt in midwinter but has an accelerated schedule of stepwise molt. The start of successive waves in the vultures apparently is delayed until birds somehow "know" they can complete molt of their outer primaries before the onset of winter or migration.

Did the need for having the outer primaries in good condition cause the development of this strategy? Or might it be the consequence of stepwise

Turkey Vultures have an unusual pattern of primary molt. In their first summer they molt all of the primaries once and then start a second wave of wing molt by renewing the inner primaries again. Each year thereafter, the primary molt starts at around p4, and when it has reached out to about p8 or p9 the inner primaries are molted again, as this individual is doing. *Marin County, CA, 23 Sept. 2006. Steve N. G. Howell.*

molt in larger ancestors being inherited by smaller species? If vultures simply had a standard, sequential p1–p10 molt each year, and if adverse conditions curtailed molt, it would be the outer primaries that did not get replaced. However, the strategy of Turkey and Black vultures means that the outer primaries should always get replaced before winter; any cutbacks in molt will be limited to the inner primaries, which are less important for flight maneuverability and less prone to get worn as birds land and take off in vegetation. In this regard, it would be interesting to know how the wing-molt patterns of North American populations compare with those of resident tropical populations, and with other tropical species of vultures, which likely experience different seasonal constraints on foraging.

The huge wing area of condors and the great length of their outer primaries have led to stepwise wing molt developing on a normal schedule (see pages 36–38), as detailed by Noel Snyder and colleagues.[4] The second prebasic molt, when the bird is about a year old, renews three to six inner primaries during spring through fall before molt is interrupted for winter. The extent of this first molt correlates with hatching date, and later-hatched birds replace fewer primaries. In the next season, molt continues from where it left off (at p4–p7) and starts again at p1, such that two waves are running out through the primaries. Successive waves start in successive cycles such that by the time birds reach breeding age there may be four or five waves of molt. Detailed studies of individual condors followed over 3 or 4 years suggest that the theoretical stepwise model is not adhered to strictly, although Snyder et al. had no data on p1–p2, which might have affected their conclusions. As in other large birds, sooner or later the primary molt waves in condors can go out of sync between right and left wings, making it even more difficult to trace their development.

The longest outer primaries of a California Condor are very long, and each requires 3 to 4 months to grow fully; on average, only four or five primaries are molted in a given cycle. Curiously, although the longest and heaviest primaries are p5–p7, the somewhat shorter and lighter outermost primary (p10) grows the most slowly.[5] Might this be because of some specialized structural modification? Or might it reflect the fact that these feathers are typically grown over the winter, rather than during summer and fall? Either way, it is another piece of evidence that the growth rates of primary feathers vary and that studies employing uniform growth rates for modeling purposes may be compromised.

REFERENCES

1. Snyder et al. 1987; **2.** Pyle 2008; **3.** S. N. G. Howell, pers. obs.; **4–5.** Snyder et al. 1987.

ACCIPITRIDAE
Hawks
(CBS, SBS; 24 species)

Hawks (including kites, harriers, eagles, and the Osprey) have long associations in human culture and are among the most popular birds in North America, where hawk-watching stations attract thousands of visitors during spring and fall migration. Although hawks and falcons are often grouped together as birds of prey, it has long been recognized that they may not be closely related.[1] Among several morphological, biochemical, and genetic differences, the two families have quite different wing-molt patterns: hawks have the standard sequential molt from innermost primary to outermost, whereas falcons molt inward and outward from a fixed starting point in the middle primaries.

Many small species of hawks, such as this Red-shouldered Hawk, are able to completely replace all of their flight feathers in the course of a single molting period, and thus do not develop stepwise waves (see Figure 89). This adult is of the brightly marked California subspecies *elegans. Monterey County, CA, 19 Oct. 2006. Brian L. Sullivan.*

In many species of hawks, such as this Snail Kite, stepwise waves of wing molt develop in some individuals but not in others. *Nayarit, Mexico, 15 Jan. 2007. Steve N. G. Howell.*

Hawks occur worldwide in almost all habitats, and several species breeding in temperate regions are short- to long-distance migrants. The sexes look alike in most species but strikingly different in a few, with females averaging larger, and there are no seasonal changes in appearance. Adult plumage aspect is attained in 1 to 5 years. Young hawks are altricial and downy, with North American species fledging in 1 to 3 months. The age of first breeding generally correlates with size: smaller species start to breed at 1 to 2 years of age, the large eagles at 4 to 7 years.

The molting strategies of hawks were not critically studied until recently, and it appears that most North American species have the complex basic molt strategy.[2] Preformative and prebasic molts can be protracted, and in northern-wintering species they are mostly completed before winter; among longer-distance migrants, these molts can be suspended for migration and completed on the nonbreeding grounds. Stepwise patterns of wing molt often develop in the larger and longer-winged species (see pages 36–45 and below, Variable Evolution of Stepwise Wing Molt).

PREBASIC MOLTS. Many hawks are large and long-winged birds with protracted breeding seasons, and fitting a complete molt into their annual cycle can be difficult. Thus, molt in hawks often overlaps with breeding. Females often molt one or more inner primaries in spring or early summer, while incubating their eggs, at which time males provide most of the food for the pair. Males often do not start wing molt until after the young have hatched, when both parents are provisioning food. Males are smaller than females and molt at a faster rate,[3] so they can afford to start wing molt later. Both sexes, but more often females, may also suspend primary molt in summer, when feeding young,[4,5] and in resident and short-distance migrant populations of North American hawks the prebasic wing molt is typically completed by fall or early winter.

Long-distance migrants start wing molt on the breeding grounds but may suspend it to complete on the nonbreeding grounds. The complete second prebasic molt of Mississippi and Swallow-tailed kites and of Broad-winged Hawks occurs mostly on or near the breeding grounds during the birds' first summer when they are not breeding, but especially in the kites it can be suspended to complete on the nonbreeding grounds.

In some species or individuals there is insufficient time for a complete wing molt in the annual cycle, and stepwise patterns of molt have developed. A 2005 review[6] found that in only 5 species

A spectacular kettle of migrating Broad-winged Hawks, with some Swainson's Hawks and Turkey Vultures mixed in, circles on thermals as the birds head south for the winter. Fewer than 0.1 percent of these migrants show wing molt, which would compromise their ability to soar. *Veracruz, Mexico, 6 Oct. 2007. Steve N. G. Howell.*

of North American hawks does stepwise molt occur in all individuals; in 10 other species its occurrence varies from rare to common, being more frequent in females, which are larger than males. Species exhibiting stepwise molt share one or more of the following characteristics: relatively long wings and large mass (thus, high wing-loading), longer migration distance, and relatively open habitat. High wing-loading constrains the number of feathers that can be molted at one time while maintaining flight capabilities, particularly during migration (when wing molt is usually suspended altogether). Species inhabiting more open areas presumably rely more on sustained flight than do species of wooded habitats and thus are more likely to develop stepwise patterns.

PREFORMATIVE MOLTS. Preformative molts reportedly occur in most species of North American raptors, but perhaps not in all individuals of all species.[7] They can start at any time from fall into early spring, before the start of the complete or near-complete second prebasic molt that occurs at about 1 year of age. In most species the preformative molt involves scattered head and body feathers, but in some it can be extensive. For example, White-tailed Kites molt most of their body plumage and often some tail feathers, and Mississippi Kites molt most or all of their head and body plumage on the nonbreeding grounds in South America.

There are no clean-cut differences among species in the timing and extent of preformative molts, but several trends are apparent. Birds migrating farther and those living in southern areas or in association with water, all factors that can contribute to greater feather wear, tend to have more-extensive preformative molts. Preformative molt is least extensive in species of northerly distribution and in species inhabiting forest interiors—that is, birds with feathers less exposed to strong sunlight. The two extremes in preformative molt are represented by the Northern Goshawk, an inhabitant of cold temperate forests (and a species that may lack a preformative molt), and the White-tailed Kite, an inhabitant of sunny open country in southern latitudes (and a species that has an extensive molt, as noted above).[8]

An interesting aspect of the preformative molt, and one that relates to field identification, is that body feathers acquired from the first fall through spring may look variably intermediate in pattern between those of the juvenile and the adult.[9] This may partly explain some of the confusing immature plumages associated with hawks, such as the notably varied appearance of first-cycle Hook-billed Kites.

Variable Evolution of Stepwise Wing Molt

The phenomenon of stepwise molt, in which two or more waves of molt occur in the primaries at the same time, is well documented as a means of replacing more feathers in a cycle than would be possible in a straightforward sequential molt. The stepwise molting patterns of hawks include both normal and accelerated schedules, as well as two types of strategies that may be called "obligate" and "opportunistic."

OBLIGATE STEPWISE MOLT. This occurs in species unable to replace all of their primaries sequentially in a single cycle. It can develop via a normal or an accelerated schedule. The normal schedule is exemplified by the Golden Eagle, which lives in temperate regions with well-defined seasons. Its first wave of primary molt starts about 7 to 8 months after fledging, in March–April, and ends in fall, before the onset of winter; during this period, three to six inner primaries are usually molted. Each year thereafter, the primary molt starts where it left off and again at p1, and all of the primaries are renewed over 2 or 3 or more cycles of prebasic molt[10] (Figure 89).

The Osprey, a long-distance migrant that passes its first year or two in tropical regions without strong seasonal constraints, develops stepwise molt on an accelerated schedule. For example, the first wave of primary molt starts about 5 months after fledging (in January), the second wave at 15 months, and the third at 23 months.[11] After this point, molt typically suspends for northward migration and breeding, and then waves resume on an annual cycle whenever time is available, mainly on the nonbreeding grounds (Figure 89).

In species such as the Red-shouldered Hawk, which replaces all of its primaries sequentially in each molt cycle, the complete primary molts

Figure 89. Diagrammatic representation of the development of successive accelerated waves (W1, W2, etc.) of stepwise primary molt in an Osprey (Prevost 1983) relative to the development of normal stepwise waves in a Golden Eagle (Bloom and Clark 2001) and the complete, standard sequential primary molts (M1, M2, etc.) of a Red-shouldered Hawk. In the Osprey and eagle, note how each new stepwise wave starts before the preceding wave has ended. Note, too, that the number of cycles of molt in the eagle and hawk are comparable; in contrast, by 23 months of age the Osprey has "gained" a primary molt relative to the other two species. In terms of its cyclic pattern, the Osprey's third wave of primary molt appears to be the "odd one out" by starting in July, whereas the other waves start in November–January.

F represents fledging; **p1- - -p6** and such indicate the number of primaries molted per molting period; ······ indicates an interruption of molt (as for migration, breeding, and winter).

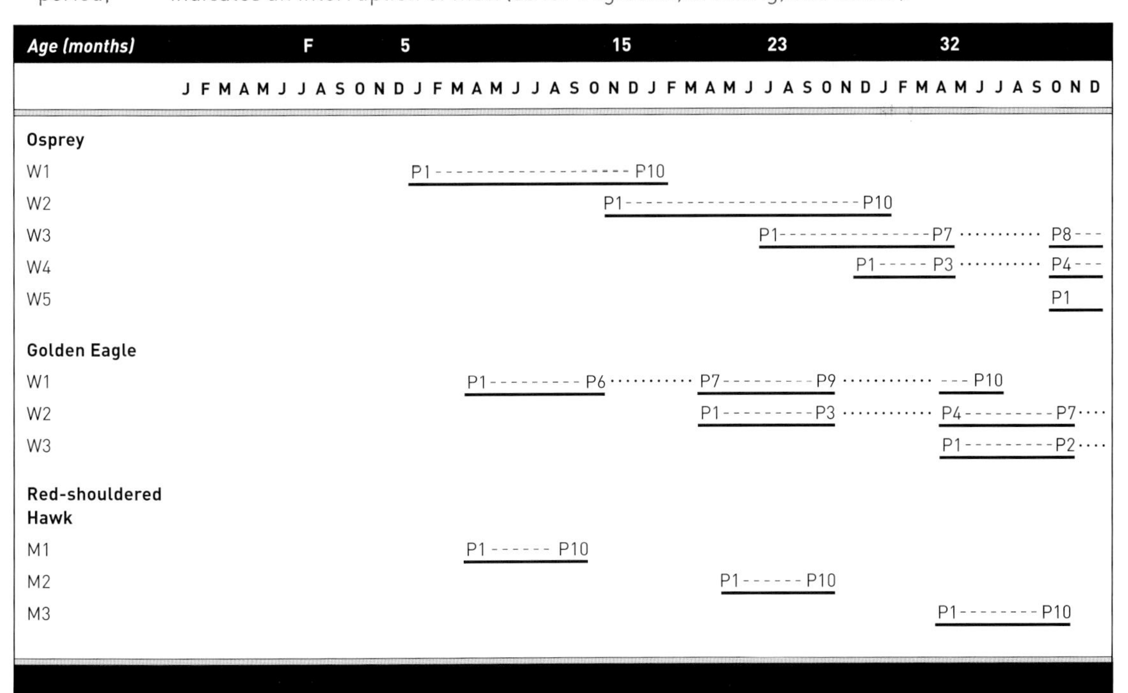

start about 9 months after fledging, and then at 23 months, 35 months, and so on every 12 months. Relative to the Red-shouldered Hawk, the first two waves of primary molt in an Osprey are pulled forward 4 months and 8 months, respectively. What might have caused the waves to be pulled forward? Another way to look at this is that by 23 months after fledging, the Osprey has initiated an extra wave of primary molt relative to the Red-shouldered Hawk (three versus two waves; Figure 89). Given this discrepancy, the initial wave can be considered to represent a preformative wing molt.

Stepwise molts pose a puzzle for nomenclature: given that one complete wave of molt may span three cycles, how should each wave of primary molt be named? Thus, is a bird in its fourth cycle undergoing its fourth prebasic body molt as well as different stages of its second, third, and fourth prebasic wing molts? In a 2006 review of stepwise molt, Peter Pyle reasonably suggested that incomplete stepwise waves should be considered part of the molt cycle to which the annual body molt pertains.[12] Thus, a fourth-cycle hawk would be undergoing its fourth prebasic molt, which includes head and body feathers along with up to three waves of primary renewal.

Applying this rule to accelerated schedules, without recognizing the first wing molt as a preformative molt, would mean that an Osprey in the fall of its second calendar year undergoes its second prebasic body molt at the same time as its second *and* third waves of prebasic primary molt (see Figure 89). This discrepancy in cycles is further evidence that it is most parsimonious to view the first wave of primary molt in Ospreys and other species with accelerated schedules as a preformative molt rather than as the second prebasic molt.

OPPORTUNISTIC STEPWISE MOLT. The finding that stepwise patterns occur to varying degrees both among and within hawk species[13] suggests that stepwise molt could develop in breeding adults of a species that did not set up stepwise patterns at an early age. That is, any individual hawk unable to complete its wing molt in a given season (perhaps because of delayed breeding, adverse weather, food shortages, or sickness) may undergo an incomplete molt that year, and then in the next year continue its primary molt where it left off, as well as starting anew at p1. Once the molt has fallen short of complete it seems unlikely to catch up, as there would be little or no benefit to replacing any primary twice in a single molt cycle. Thus, stepwise molt could be introduced as a novelty in the wings of a hawk 10 years old after it had undergone complete, sequential primary molts in the preceding 9 years.

Stepwise patterns likely develop in a similar manner among some tropical birds, such as chachalacas. Species in tropical regions can encounter suitable breeding conditions at varying times of year, and, if the timings of wing molt and potential breeding overlap, these species may interrupt their molt to breed. In such cases, if molt resumes at a time corresponding to the start of the next molt cycle, a new wave of primary molt may start while the interrupted wave may continue where it left off. Thus, an opportunistic stepwise molt pattern would be set up.

The largest and longest-winged hawks are unable to replace all of their flight feathers during a single cycle and are obliged to set up stepwise waves of molt. In the Golden Eagle, stepwise waves develop on a normal schedule, and the second prebasic molt usually involves three to six inner primaries before molt is interrupted for the winter (see Figure 89). On this individual, the inner primaries are new and p4 has been shed on the top wing, p3 on the bottom wing; the other flight feathers, including the ragged tail, are juvenile feathers. *Tooele County, UT, 29 June 2008. Steve N. G. Howell.*

REFERENCES

1. Olson 1985; **2.** Pyle 2005b; **3.** Schmutz and Schmutz 1975; **4.** Henny et al. 1985; **5–6.** Pyle 2005c; **7–9.** Pyle 2005b; **10.** Bloom and Clark 2001; **11.** Prevost 1983; **12.** Pyle 2006; **13.** Pyle 2005c.

FALCONIDAE
Falcons
(CBS, SBS; 7 species)

Falcons (including caracaras and the so-called typical falcons of the genus *Falco*) are a worldwide family of raptors with its greatest diversity in South America. Falcons differ from hawks in several features, including their molt patterns, and the two families apparently are not closely related despite frequently being associated together as birds of prey.[1,2] Falcons live in almost all habitats, and species breeding in temperate regions are short- to long-distance migrants. The sexes look alike in most species but different in others, including the familiar American Kestrel, and females average larger. No species show distinct seasonal changes in their appearance, and adult plumage aspect is mostly attained in 1 to 2 years. Young falcons are altricial and downy, with North American species fledging in 1 to 2 months. The age of first breeding is usually 1 to 3 years.

Most North American falcons apparently exhibit the complex basic molt strategy.[3] Molting in typical falcons occurs mainly from summer through fall, often overlapping with breeding, and is mostly completed before migration or the onset of winter. In longer-distance migrant populations of the Peregrine, molt can suspend for migration and complete on the nonbreeding grounds in winter.[4] Low-level preformative molt also may continue through winter into spring.

PREBASIC MOLTS. Prebasic molts in typical falcons often overlap extensively with breeding. Once the eggs are laid, both sexes may begin wing molt, which can be as early as late March or April in some populations of the Prairie Falcon.[5] As in hawks, female falcons often start primary molt while incubating, but most males wait until the young have hatched. Molt may progress slowly and without interruption through the nesting season, or it may be suspended as needed. Despite their smaller size, males more often suspend primary molt, presumably because they tend to provision most of the food for the female and young[6,7] and need to be in the best shape for hunting.

One long-known oddity of molt in falcons is that primary molt moves inward and outward from its starting point at p4. The exact sequence after p4 has dropped may vary between species or even individually, but differences are slight and p9, p10, and p1 usually drop within a few days of each other.[8]

Primary molt in the Crested Caracara, as in other falcons, starts with p4 and progresses outward to p10 and inward to p1. This explains why the middle primaries (especially p4–p5) are the most distinctly worn feathers on the wing of this individual. *Yucatán, Mexico, 5 Dec. 2007. Steve N. G. Howell.*

One difference between hawks and falcons, such as this male American Kestrel, is their pattern of wing molt. Falcons start their wing molt at a midpoint in the primaries rather than with the more conventional innermost primary as do hawks. *Morgan County, CO, 22 Dec. 2007. Christopher L. Wood.*

Most studies have been on captive birds, and whether this might affect the exact molt sequence is not known. The molt sequence in the secondaries is also unusual among birds and proceeds in both directions from s5, as well as outward from the tertials. No explanation for these patterns is apparent, and any thoughts on the subject are highly speculative. For birds that hunt in flight, like typical falcons, perhaps the resultant paddle-shaped surface area that this sequence produces in the primaries is aerodynamically more efficient than that produced by a sequential molt. However, if the ancestors of falcons were like the omnivorous caracaras of South America, aerial hunting would not have been so important. Another advantage to the falcons' sequence of wing molt is that it creates two to four small gaps in the wing area rather than one or two larger ones, which is a better aerodynamic arrangement for birds that rely on aerial hunting. Unlike in hawks, no examples of stepwise molt have been found among falcons.[9]

PREFORMATIVE MOLTS. Preformative molts reportedly occur in most North American falcons but perhaps not in all individuals of a species.[10] It has been argued that in some cases a limited preformative molt in late winter or spring has been mistaken for the start of the second prebasic molt,[11] but without following individual birds it is difficult to resolve this point. In most falcons, the preformative molt involves scattered head and body feathers, but in kestrels it is extensive, often involving all head and body plumage but no flight feathers. As with hawks, it may be that birds of southern or longer-distance migrant populations have more-extensive molts, but this appears not to have been investigated.

REFERENCES

1. Olson 1985; **2.** Hackett et al. 2008; **3.** Pyle 2008; **4.** Cramp and Simmons 1980; **5.** Steenhof and McKinley 2006; **6.** Holthuijzen 1990; **7.** Espie et al. 1996; **8.** Willoughby 1966; **9.** Pyle 2005c; **10.** Pyle 2005b; **11.** Pyle 2008.

RALLIDAE
Rails and Coots
(CBS, CAS?; 9 species)

The North American members of this almost worldwide family of ground-living birds run the gamut from the sought-after Yellow Rail to the unavoidable American Coot. Most species are associated with wetland habitats, ranging from coastal salt marshes and prairie swales to large lakes and even estuaries. Despite their frequently furtive nature and a seeming reluctance to fly, many rails are short- to moderate-distance migrants, and many species have the ability to disperse considerable distances if their wetland habitats dry out. The sexes look alike with little or no seasonal variation in appearance, but juveniles usually look somewhat different from adults. Adult plumage aspect is attained with the formative plumage. Young rails are semi-precocial and downy, and they usually leave the nest within a few days of hatching. Most species breed at 1 year of age.

The molt strategies of rails and coots are complex basic and perhaps complex alternate, although details are poorly known. Prebasic molts occur mainly in late summer and fall, before the onset of migration or winter. Prealternate molts, typically limited to head and neck feathers, are presumed to occur in several species. However, as in other groups of birds such as ducks and quail, it is uncertain whether late-winter and spring molts in rails and coots are really prealternate or simply a continuation of suspended prebasic molts.[1,2]

PREBASIC MOLTS. A feature of the prebasic molt in

Like most rails and gallinules, the Purple Gallinule apparently has the complex basic molt strategy. Following the preformative molt, young birds resemble adults but retain the juvenile flight feathers; the brightly blue-edged primary coverts of this individual suggest it is an adult. *Nayarit, Mexico, 14 Jan. 2008. Steve N. G. Howell.*

The protracted preformative molt of coots, moorhens, and gallinules often produces birds with a somewhat patchy-looking appearance, like this first-cycle Purple Gallinule. By the end of this molt, these birds will look much like adults. *Nayarit, Mexico, 15 Jan. 2007. Steve N. G. Howell.*

The skulking habits of most rails, such as this Clapper (or King?) Rail, are well suited to their synchronous wing molts, when birds become flightless for a few weeks. This individual is undergoing its preformative molt into a plumage that will resemble that of the adult; note the barred flank feathers starting to appear on the mostly plain underparts. *Baja California, Mexico, 9 Sept. 2006. Steve N. G. Howell.*

rails and coots (except for a few species of tropical rails) is that the flight feathers are molted synchronously. Birds are thus flightless for a few weeks in fall, which appears to be little problem for marsh-haunting species that are reclusive to begin with, or for aquatic species such as coots, which can swim and dive to avoid predators. This strategy may have developed in rails as the best compromise to having heavy bodies and relatively small wings, but in some species it is also a good way to undergo wing molt quickly, before migration (see page 36).

PREFORMATIVE MOLTS. These molts involve mainly head and body feathers and sometimes wing coverts. In resident and short-distance migrants they occur mostly in fall, although they may be suspended or protracted through the winter and into spring. In longer-distance migrants, such as the Sora and Purple Gallinule, some molting can start in fall, but many birds migrate south in juvenile plumage and undergo a protracted preformative molt on the nonbreeding grounds.

REFERENCES

1. McKnight and Hepp 1999; **2.** Pyle 2008.

ARAMIDAE
Limpkin
(CBS; 1 species)

The Limpkin is the sole living member of a New World family of wading birds whose closest relatives are not agreed upon, although the consensus is that this curious-looking bird is not too distantly related to cranes and rails. Limpkins inhabit freshwater marshes with woodland or trees nearby, and they are ostensibly resident. If marshes dry up, however, Limpkins can disperse hundreds of miles in search of suitable habitat, a trait they share with other marsh-living birds such as rails. The ages and sexes look similar, and the downy precocial young are able to run and swim shortly after hatching. Limpkins first breed at 1 year of age.

Molt in the Limpkin has not been well studied, but it appears that the species exhibits the complex basic molt strategy, with a partial preformative molt of head and body feathers. Unlike cranes and rails, which have synchronous wing molts, Limpkins molt their primaries inward from the outermost feather (p10). Molt proceeds in a sequence that has been considered as irregular, or transilient,[1] and also as stepwise.[2] Curiously, the trumpeters, a small South American family of odd-looking birds believed to be related to the Limpkin, also may have an inward stepwise molt.[3] Worldwide, extremely few species of birds molt their primaries from the outermost feather inward, and it is hard to envision what benefit this has, or may have had to an ancestral species. The outermost primary is unusually curved on adult Limpkins but not on juveniles,[4] so perhaps it is important to replace this feather early in the molt if its shape has something to do with display.

REFERENCES

1. Stresemann and Stresemann 1966; **2.** Pyle 2008; **3.** Stresemann and Stresemann 1966; **4.** Pyle 2008.

The Limpkin is unique among North American birds in having a "backward" primary molt, which starts with the outermost primary and proceeds inward. The strongly curved outer primary of this individual indicates that it is an adult. *Nayarit, Mexico, 15 Jan. 2007. Steve N. G. Howell.*

GRUIDAE
Cranes
(CBS; 2 species)

Few families of birds evoke the notion of wilderness better than cranes—whether it is a migrant flock calling and circling high over the Bering Straits or a winter group trumpeting and dancing in misty wetlands. These tall-standing, elegant birds compose a small family found locally in open habitats throughout the world except in South America. Most species are medium- to fairly long-distance migrants, although some U.S. populations are resident in the South. The sexes of cranes look alike, but juveniles of North American species look different from adults. Adult plumage aspect is attained in 1 to 2 years. Young cranes are precocial and downy, able to run and swim within a day of hatching, and the age of first breeding is 3 to 6 years.

Cranes have the complex basic molt strategy, although most studies relate to captive birds and the molting patterns of free-living birds are not well known. Prebasic and preformative molts occur mostly on the breeding grounds, but body molt may suspend for migration and continue into the winter. Preformative molts are partial and protracted, much like those of swans and geese, with which cranes share many life-history traits.

One intriguing aspect of Sandhill Crane plumages is the frequent occurrence of rusty coloration in the plumage. In spring and summer, many birds stain their plumage with soil,[1] which accounts for extensively brownish- and rusty-toned birds in the breeding season. Fall and winter birds are grayer overall but often show faded brown feathers that have not been molted, and sometimes also a few darker and brighter rusty feathers. These darker rusty feathers are heavily frayed at the tips, and clearly have been retained from a previous plumage cycle, but how they relate to the faded brown feathers is unclear. Are they of the same generation as the faded brown feathers? Might they have been stained by different pigments and in different geographic regions? Might some types of cosmetic staining intensify with age, or perhaps cause feathers to fray more quickly? A better understanding of feather staining might prove useful in evaluating variation in the extent of molt between years and among populations, and perhaps even in identifying locations of molt.

PREBASIC MOLTS. Prebasic molts in cranes show an interesting dichotomy in the patterns of flight-

The upperwings of these Sandhill Cranes have some faded brown feathers and some darker, more contrasting brown feathers. Both types are distinct from the fresh gray feathers of the new basic plumage, but what do the different types of brown feathers indicate? *San Joaquin County, CA, 17 Nov. 2007. Steve N. G. Howell.*

Few birds evoke the image of wilderness better than cranes, such as this Sandhill Crane flying over the winter plains. The primaries of this individual appear uniformly fresh, suggesting it may have undergone a synchronous wing molt. *Socorro County, NM, 22 Nov. 2005. Bill Schmoker.*

feather molt. At least some individuals of several species undergo a synchronous wing molt much like that of geese and swans, such that birds are flightless for a few weeks.[2,3] In North America, synchronous wing molt may be most frequent among migratory western- and northern-breeding populations that inhabit remote regions relatively free from predators. In other populations, such as Sandhill Cranes resident in the southern U.S., a complete renewal of primaries occurs gradually and requires at least 2 years.[4]

Among migrant and resident Sandhill Cranes in Florida,[5] the second prebasic molt involves only head and body feathers plus tertials; thus, the juvenile primaries and outer secondaries are retained through the second cycle, as has been reported for the Common Crane[6] and for the Whooping Crane.[7] No carefully articulated study appears to have been undertaken for the subsequent wing molts of Florida Sandhills, which involve some but not all primaries.[8] Thus, the details of wing molt in southern Sandhill Cranes remain to be elucidated. Do they have stepwise molt patterns, or perhaps some other, novel sequence of primary replacement?

Wing molt in Whooping Cranes is synchronous, but how frequently wild birds molt is uncertain and perhaps variable. A study of Whooping Cranes translocated to Florida found that wing molt occurred only every 2 to 4 years in a resident population.[9] Thus, an individual might retain its wing feathers for more than 3 years before molting them! In some populations of Common Cranes in Europe, the flight feathers are molted synchronously but apparently only every 2 years,[10] which suggests that nonannual wing molts may be characteristic of some species of large cranes. But are the wing molts of resident, translocated Whooping Cranes typical of migratory populations? And did all native resident populations of Whoopers have synchronous wing molt, or might some have developed other strategies, such as gradual wing molts like those of southern Sandhill Cranes?

The wing molt of breeding cranes starts in spring or early summer, before or shortly after the young have hatched, and is mostly completed before the molt of head and body feathers begins in late summer. A nesting pair of Sandhill Cranes in Oregon involved a full-winged female and a flightless male,[11] which raises the question: might the timing of flightlessness be offset between the sexes of breeding pairs, as occurs in some swans and geese?

The molt of head and body feathers occurs annually, regardless of how frequently the wings are molted. In Whooping Cranes, the normally snow-white birds become stained dirty grayish for a short period before body molt.[12] Whether staining is more pronounced in years when birds undergo wing molt is unclear. Might this staining, and that of Sandhill Cranes, be a form of camouflage to help molting birds become less conspicuous? Effectively, cranes could be gaining an "eclipse plumage" analogous to that of ducks, but without having to molt.

REFERENCES

1. Nesbitt 1975; **2.** Walkinshaw 1949; **3.** Littlefield 1970; **4–5.** Nesbitt and Schwickert 2005; **6.** Stresemann and Stresemann 1966; **7.** Erickson and Derrickson 1981; **8.** Nesbitt and Schwickert 2005; **9.** Folk et al. 2008; **10.** Cramp and Simmons 1980; **11.** Littlefield 1970; **12.** Folk et al. 2008.

CHARADRIIDAE
Plovers
(CAS, SAS? CBS?; 11 species)

This worldwide family of shorebirds is well represented in open habitats throughout North America, from noisy Killdeers in parking lots and handsome golden-plovers on Arctic tundra, to Snowy and Piping plovers, two unobtrusive "beach ghosts" that human recreation is pushing toward extinction. All North American species are migratory, although southern populations of some are resident, and several plovers are long-distance migrants that travel all the way to southern South America and New Zealand. There are marked seasonal differences in the appearance of several species, and the sexes of most species look somewhat different in the breeding season. Adult plumage aspect is usually attained with the formative plumage. Young plovers are precocial and downy, able to run within a day of hatching, and North American species fledge in 3 to 5 weeks. The age of first breeding is 1 to 3 years.

The Killdeer is common, widespread, noisy, and conspicuous—but not well known in terms of its molt. Does it have a prealternate molt like other North American plovers? And might the extent of its prealternate molt vary geographically, being more extensive in northern-breeding populations and less extensive, or even absent, in southern-breeding populations? *Marin County, CA, 2 Jan. 2008. Steve N. G. Howell.*

The molting patterns of plovers, like those of sandpipers, largely reflect the latitudes where birds spend their nonbreeding season. This subject is discussed more fully in the family account for sandpipers. Most plovers, at least in North America, have the complex alternate molt strategy, although Black-bellied Plovers may have the simple alternate strategy and Killdeer may have the complex basic strategy. The location of prebasic and preformative molts varies with breeding latitude: high-latitude breeders have short summer seasons and mostly migrate before molting, whereas lower-latitude breeders often start molting on or near the breeding grounds. Prebasic wing molts can be suspended over fall migration in long-distance migrants and occasionally are interrupted over the winter in northerly-wintering birds, which then molt their outer primaries in spring. Prealternate molts occur mainly on the nonbreeding grounds. Presupplemental molts have been reported in golden-plovers but require confirmation (see below).

PREBASIC MOLTS. As noted above, the timing and duration of prebasic molts reflect migration distance and wintering latitude. For example, adult Black-bellied Plovers wintering in northern mid-latitudes (around 50° N) usually molt during August–October and undergo a relatively rapid prebasic molt in fall, before winter sets in.[1] Adults that migrate to southern latitudes, however, molt later and tend to have a slower molt over the "winter." For example, birds wintering in northwestern Australia (at 18–20° S) molt during September–January, and those even farther south, in southeastern Australia (at 38–39° S), molt during October–February.[2] Interestingly, it has been proposed that a slower growth rate may allow more keratin (the protein that strengthens feathers) to be deposited on a feather; thus, more-durable primaries may be produced in a slower molt, which is obviously advantageous for long-distance migrants wintering in latitudes with strong sunlight.[3]

Wintering latitude also has a bearing on how much prebasic molt occurs on the breeding grounds, before migration.[4] American Golden-Plovers are long-distance migrants that do not molt their primaries until they reach the nonbreeding grounds in South America. Conversely, European Golden-Plovers are short-distance migrants that usually start molting their primaries during incubation and replace up to six inner primaries, and sometimes all ten, before migration. The intermediate-distance migrants, the Black-bellied Plover

Some Wilson's Plovers apparently have an eccentric primary molt in their first winter, something typically associated with long-distance migrants. However, this species is not known to be a long-distance migrant, and the first-cycle wing molt may be driven by fading and wear experienced in the sunny latitudes it inhabits. These birds are of the Pacific Coast subspecies *beldingi*, which is darker above than the Atlantic Coast *wilsonia. Nayarit, Mexico, 15 Jan. 2009. Steve N. G. Howell.*

and Pacific Golden-Plover, replace up to three and five primaries, respectively, before suspending wing molt for fall migration. Thus, if you see a golden-plover in fall migration in North America with suspended wing molt (fresh inner primaries and faded outers) it's likely to be a Pacific.

FIRST-CYCLE MOLTS. First-cycle molts of birds wintering in northern mid-latitudes usually involve only head and body feathers, as well as tertials and sometimes upperwing coverts. These preformative molts occur mainly in fall. Birds wintering in the Southern Hemisphere, however, can have complete or almost-complete preformative molts which, like prebasic molts, can be protracted through the winter. In general, the preformative molt of head and body feathers occurs in fall and early winter, and any wing molt occurs from midwinter into spring; the later stages of wing molt can thus overlap with the prealternate molt of head and body feathers.

Among golden-plovers, the Pacific generally does not replace primaries in its first winter, whereas the American does. It has been suggested that this difference reflects the food available to each species: Pacifics mainly inhabit relatively food-poor tropical beach strands, whereas Americans inhabit extensive grasslands where food may be plentiful.[5]

Incomplete preformative wing molts show eccentric patterns (see page 35) and are typical of long-distance migrants. Some Wilson's Plovers, however, also have eccentric first-cycle wing molts,[6,7] yet this is not a long-distance migrant. Has strong sunlight, perhaps combined with abrasion from wind-blown sand, caused this molt to develop in some individuals? If so, we might expect such a molt to also occur in populations of Snowy Plovers with similar geographic ranges.

For the Black-bellied Plover, it is unresolved whether one or two first-cycle molts occur.[8] Some birds have a fairly extensive molt in fall, whereas others retain their relatively unworn juvenile plumage into midwinter. In spring, some birds are heavily worn and have a gray nonbreeding aspect, whereas others have a fresh, black-bellied plumage much like that of breeding adults. Do these two types reflect wintering latitude, or sex? Did the black-bellied birds retain their juvenile plumage into the winter and then molt straight into bright alternate plumage? Or did they have a preceding preformative molt? Do two molts occur in the fittest birds, which then migrate north to the breeding grounds, while less-fit individuals can manage only one? This last possibility mirrors the situation among Bar-tailed Godwits undergoing presupplemental molt at migration stopovers—only the fittest birds may have this molt.[9] Such questions are difficult to answer in highly migratory species with non-breeding ranges that extend over wide spans of latitude.

PREALTERNATE MOLTS. Prealternate molts in plovers vary from extensive (among the Black-bellied Plover and golden-plovers) to perhaps lacking altogether, as in the Killdeer.[10] It has been suggested that the extent of prealternate molt among some species varies with their wintering latitude, the idea being that birds wintering farther south require more time for migration and so have less time for molt. Thus, in the Black-bellied Plover, females usually winter farther south than males and have less-extensive prealternate molts.[11] However, species such as the American Golden-Plover, which molts in spring at migration staging sites in North America, shows similar sex-related differences in the extent of its prealternate molt,[12] suggesting that more than wintering latitude is behind such differences.

Whether large plovers have one or two molts added into their first cycle remains unresolved. This first-cycle Pacific Golden-Plover has replaced much of its juvenile plumage with golden-spangled feathers that contrast with the frayed whitish edgings to its retained juvenile feathers. But will it undergo another molt before its complete second prebasic molt? *Truk, Micronesia, 15 Apr. 2007. Steve N. G. Howell.*

In songbirds, it has been found that northern-breeding individuals of a species, which nest later, have more-extensive prealternate molts than do southern-breeding birds, which start to nest earlier. This difference has been attributed to later breeders having more time available for molt.[13,14] This trend may also occur among plovers, with the tundra-breeding species having more-extensive molts than species breeding at mid-latitudes. Might this pattern even be found within a species, such as the Killdeer, whose prealternate molt remains enigmatic? Thus, might northern-breeding Killdeers have prealternate molts while southern-breeding populations lack them?

PRESUPPLEMENTAL MOLTS. Presupplemental molts of some chest and belly feathers have been reported for breeding adult golden-plovers,[15,16,17] although other authors have considered this as the start of the prebasic molt.[18] This molt coincides with the incubation and chick period, and it produces pale feathers that somewhat break up the solidly black underparts. But why would golden-plovers do this and not the Black-bellied Plover? In general, golden-plovers have a somewhat passive nest defense strategy, whereas the Black-bellied Plover is more aggressive, actively chasing off potential predators. Being cryptic may thus be of more benefit to golden-plovers,[19] although this still doesn't resolve whether the molt is prebasic or presupplemental. The feathers acquired at this time could be basic feathers that fade and look much like other basic feathers by winter, and this may be the most likely explanation. However, presupplemental molts of body feathers help other tundra-breeding birds, such as ptarmigans, become less conspicuous. A wide-open, seasonally changing habitat with numerous predators but ample food seems like a good environment in which to have molts that can change your appearance. How any "extra" (presupplemental) molts might have developed would be an interesting subject for study.

REFERENCES

1. Cramp and Simmons 1983; **2.** Minton and Serra 2001; **3.** Serra 2001; **4.** Byrkjedal and Thompson 1998; **5.** Johnson 1985; **6.** Pyle 2008; **7.** P. Pyle, pers. comm. (Museum of Vertebrate Zoology specimens 84191 from Florida, 59301 from Baja California, both collected in June); **8.** Howell and Pyle 2002; **9.** Piersma and Jukema 1993; **10.** Pyle 2008; **11.** Cramp and Simmons 1983; **12.** Pyle 2008; **13.** Foster 1967; **14.** Willoughby 1991; **15.** Jukema and Piersma 1987; **16.** Byrkjedal and Thompson 1998; **17.** Jukema et al. 2003; **18.** Johnson & Connors 1996; **19.** Jukema et al. 2003.

HAEMATOPODIDAE
Oystercatchers
(SAS, CBS?; 2 species)

Oystercatchers are a small family of big, boldly patterned shorebirds found almost worldwide. In North America they occur from Florida beaches to rocky shorelines in Alaska. Oystercatchers are mainly resident, but many birds breeding at higher latitudes move south in winter. The sexes look similar year-round, and juveniles look slightly different from adults; only the Eurasian Oystercatcher (*Haematopus ostralegus*) shows obvious seasonal differences in appearance. Like other shorebirds, young oystercatchers are precocial and downy, able to run within a day of hatching. Unlike most other shorebirds, oystercatchers often remain with their parents into the winter, when they can learn foraging techniques. The main food of oystercatchers is bivalve mollusks such as clams, which take some practice to open; Black Oystercatchers require more than 3 years to develop their repertoire of feeding skills.[1] Adult plumage aspect is attained in 1 year, but the age of first breeding is 3 to 5 years.

The molt strategy of oystercatchers is simple alternate or complex basic. Prebasic molts occur from summer into early winter, whereas preformative molts occur mainly from winter into spring and can include tertials and central tail feathers. In some cases the prebasic molt may start during incubation and be suspended when adults are feeding chicks.[2] In the first cycle there appears to be simply a single added molt, as in ibises and large gulls. A variable prealternate molt has been described for adult Black Oystercatchers,[3] although other authors discount this information,[4] and no data are available for the American Oystercatcher. The Eurasian Oystercatcher, which is the longest-distance migrant in the family, has a prealternate molt of head and neck feathers, but might a prealternate molt be lacking among mostly resident species of oystercatchers?

REFERENCES

1. Andres and Falxa 1995; **2.** Pyle 2008; **3.** Webster 1942; **4.** Pyle 2008.

A boldly patterned mollusk-eater, the Black Oystercatcher may or may not have a prealternate molt. As is so often the case, even with big and conspicuous species, more study is needed. *Sonoma County, CA, 26 Feb. 2008. Steve N. G. Howell.*

RECURVIROSTRIDAE
Stilts and Avocets
(CAS; 2 species)

"Like a barking dog on a farm road, a stilt colony compromises the peacefulness of nature." So wrote shorebird expert Dennis Paulson[1] of the Black-necked Stilt—and avocets are little better. Stilts and avocets compensate for their "marsh-poodle voices" with elegant shapes and handsome plumages. This small family of long-legged shorebirds occurs worldwide in warm climates. They are migrants in the mid-latitude portions of their range, but southern populations of stilts are resident. Among stilts and avocets as a whole, plumage differences between the ages, sexes, and seasons are slight. The American Avocet is an exception in that it exhibits distinct seasonal changes in its appearance. Adult plumage aspect is attained within 1 year. Young stilts and avocets are precocial and downy, and juveniles can fly at about a month of age. The age of first breeding is 1 to 2 years.

North American stilts and avocets apparently have the complex alternate molt strategy. Prebasic and preformative molts occur from summer into fall on the breeding grounds, at staging sites, or on the nonbreeding grounds. Preformative molts often include the tertials and sometimes tail feathers, but no primaries. In the Black-winged Stilt (*Himantopus himantopus*), the Old World counterpart to our Black-necked Stilt, some birds molt primaries in their first cycle.[2] The prealternate molt in avocets occurs on the nonbreeding grounds and involves at least the head and neck feathers, but its extent in North American stilts is not well known. In Black-winged Stilts, a prealternate molt of head and some body feathers is well documented in both first and subsequent cycles.[3,4] This may also be true of the Black-necked Stilt, although it has also been stated that a prealternate molt is lacking in this species.[5]

REFERENCES

1. Paulson 1993:135; **2–3.** Marchant and Higgins 1993; **4.** Cramp and Simmons 1983; **5.** Pyle 2008.

Like many large birds, American Avocets start their prebasic wing molt while breeding. They may then suspend it to finish at staging sites on or near the nonbreeding grounds. This bird was attending small chicks and had renewed p1, with p2 growing. *Sonoma County, CA, 3 July 2007. Steve N. G. Howell.*

Is this Northern Jacana in juvenile plumage or in a formative plumage that resembles the juvenile plumage? Contrasts in wear among the scapulars and between fresher-looking scapulars and faded upperwing coverts indicate the latter option. *Nayarit, Mexico, 11 Jan. 2007. Steve N. G. Howell.*

JACANIDAE
Northern Jacana
(CBS; 1 species)

Jacanas are a small family of distinctive, rather rail-like shorebirds found worldwide in freshwater marshes of wet tropical regions. The popular name "lily-trotter" reflects the extremely long toes of jacanas, which allow them to walk on floating vegetation. Like many tropical waterbirds, jacanas are mostly resident, but they can disperse considerable distances in response to changing water levels. The sexes look alike, but females are larger. New World species do not undergo seasonal changes of appearance. Like other shorebirds, young jacanas are precocial and downy, able to walk and swim shortly after hatching, and they fledge in about 2 months. Adult plumage aspect is attained in about 1 year. The age of first breeding is unknown but may be around 2 years.

In their molting, resident species of jacanas apparently have the complex basic strategy. The migratory Pheasant-tailed Jacana (*Hydrophasianus chirurgus*) of Asia, however, has distinct alternate and basic plumages. Prebasic molts are reportedly protracted, with molt of flight feathers in the Northern Jacana being a sequential process. Interestingly, some Old World species of jacanas undergo synchronous wing molt, and one species apparently molts "backward" from p10 inward to p1.[1]

The plumage sequences of immature Northern Jacanas are described differently by different authors, and tracing this species' plumage development may, in certain areas, be confounded by protracted breeding seasons. Careful field and museum studies reveal that the soft juvenile plumage is replaced by a similar-looking formative plumage, which is worn from fall into spring, or at least for several months; this is then replaced by an adultlike plumage via a complete second prebasic molt.[2,3] Despite this, some authors report that a protracted and usually complete preformative molt begins at about 8 weeks of age and gradually results in an adultlike plumage being acquired at age 10 to 12 months.[4,5]

It has been suggested that gradual wing molt in Northern Jacanas might be an adaptation for breeding at unpredictable times, whenever water levels are suitable.[6] The synchronous wing molt of other jacana species is perhaps an even better way of being available to breed for most of the year, as it gets molt done in a short period. In Northern Jacanas,

Potentially year-round breeding by the Northern Jacana has confounded an understanding of its molt, which still needs to be studied critically. The adult plumage aspect, as on this individual, is attained via the second prebasic molt. *Veracruz, Mexico, 7 Oct. 2007. Steve N. G. Howell.*

neither an interruption of wing molt for breeding nor an overlap of wing molt and breeding have been described, and the species might simply have a typical complete prebasic molt following the breeding season.

Clearly, a careful study of the Northern Jacana could answer several questions. For example, do jacanas actually breed throughout the year in any area, as has been claimed?[7] If so, do all individuals breed in all months, or is breeding staggered evenly through the year, or is it concentrated into distinct periods? And how do molts relate to breeding cycles? Are these two activities overlapping or separate?

REFERENCES

1. Stresemann and Stresemann 1966; **2.** Dickey and Van Rossem 1938; **3.** Howell and Webb 1995; **4.** Jenni and Mace 1999; **5.** Pyle 2008; **6.** Jenni 1996; **7.** Jenni and Mace 1999.

SCOLOPACIDAE
Sandpipers
(CAS, SAS, CBS; 39 species)

From cryptic woodcocks crouched motionless in the leaf-litter of eastern winter woodlands to silvery phalaropes sparkling on the open ocean, members of the sandpiper family occur throughout North America, mainly in open and marshy habitats. This large shorebird family occurs worldwide, although most species breed in the Northern Hemisphere. In North America, most sandpipers breed in the far north, often on the tundra, and are seen only as nonbreeding migrants in the Lower 48 states. All North American breeding species are migratory, although some southern populations may be resident. Like plovers, several species of sandpipers are long-distance migrants that move to southern South America and even New Zealand for the nonbreeding season. Among the larger species, first-cycle birds, and sometimes second-cycle birds, do not migrate north to nest but remain on the nonbreeding grounds. Most species, except those of grasslands and grassy marshes (such as curlews and snipes), show marked seasonal and age-related differences in their appearance. The sexes of a few species differ in the breeding season, with males being either brighter (as in godwits) or duller (as in phalaropes) than females. Adult plumage aspect is usually attained in the first cycle, with the formative plumage. Young sandpipers are precocial and downy, able to run within a day of hatching, and most fledge in 3 to 7 weeks. The age of first breeding is usually 1 to 3 years, which correlates somewhat with size, the larger species breeding when they are older.

The molting strategies of sandpipers are complex alternate (in most smaller species), simple alternate (in some larger species), and complex basic (in the American Woodcock, at least). These differences reflect body size, age of first breeding, migration distance, and ancestry. Sandpipers can be subdivided into several groups, such as the yellowlegs and allies (including the Willet and Spotted Sandpiper; complex alternate), the curlews (including the Upland Sandpiper; simple alternate), the godwits (complex alternate, and perhaps simple alternate), the typical sandpipers (such as the Sanderling and Least Sandpiper; complex alternate), the snipes (including the American Woodcock; complex alternate and complex basic), and the phalaropes (complex alternate). Most molting in sandpipers occurs on the nonbreeding grounds and at staging sites, but a handful of species that winter in northern latitudes undergo their prebasic and preformative molts on or near the breeding grounds. Prebasic wing molts can be suspended over fall migration in some spe-

Many adult phalaropes start their prebasic body molt during late summer or fall, but this adult female Wilson's Phalarope on southward migration shows little sign of losing its handsome alternate plumage. *Marin County, CA, 7 July 2008. Steve N. G. Howell.*

cies. Preformative molts vary in extent with nonbreeding latitude (see below, Molt and Migration), and first-cycle wing molts are usually eccentric (see page 35). Prealternate molts are usually partial and often include upperwing coverts and tertials, and sometimes also tail feathers, especially the central pair. The prealternate molt of some species, such as the Little Stint, may also include three or four outer primaries,[1] but this has not been critically documented. Presupplemental molts occur in a few species.

PREBASIC MOLTS. The timing and location of adult prebasic molts are summarized below, under Molt and Migration and in Table 2. In several species, nonbreeding birds can remain south of the breeding grounds for their first year of life, occasionally even for 2 or 3 years. In such cases, the second prebasic molt tends to start earlier than the adult prebasic molt, since the young birds aren't breeding and may not be migrating. This wing molt can't be too early, however, because the newly acquired feathers can't become too worn to support the bird through the two migrations that follow. In some species, the start of the second prebasic wing molt can overlap in timing with the end of the preforma-

Variation in molt timing is common among many species of migratory shorebirds, reflecting things such as a bird's age, its breeding status (failed breeders and nonbreeders usually molt earlier than successful breeders), and particularly its migration distance. Adult Surfbirds wintering in California typically complete their prebasic molt by October, before the onset of winter. At this time, however, adults wintering in Chile can still be in heavily worn alternate plumage and only just starting their prebasic molt, like this individual. *Region I, Chile, 7 Nov. 2006. Steve N. G. Howell.*

First-cycle Surfbirds wintering in South America experience a warmer climate than do birds in California, and their preformative molt occurs later and may be more protracted overall. The righthand bird is mostly in worn and faded juvenile plumage, with only a few new gray feathers apparent on the head, back, and chest. The lefthand bird is at a stage of primary molt shown by California birds in August–September. *Region I, Chile, 7 Nov. 2006. Steve N. G. Howell.*

tive wing molt, as has been found in the Red Knot and the Bristle-thighed Curlew.[2,3]

As in most birds, the prebasic wing molts of sandpipers are sequential, and birds retain the ability to fly, such as to escape predators. Among Bristle-thighed Curlews, however, some adults wintering on remote, predator-free Pacific islands shed so many primaries at one time that they become flightless for up to 2 weeks during the height of their molt.[4] Only some individuals become flightless, which is a good thing given that non-native predators such as rats and cats could potentially decimate molting populations.

FIRST-CYCLE MOLTS. First-cycle molts in sandpipers show considerable variation in extent, as discussed below, under Molt and Migration. Many of the smaller species, which can breed in their first year, have preformative and prealternate molts in their first cycle. However, it is unclear how many molts occur in some larger species, which usually do not breed in their first year. Among curlews there appears to be only a single molt, and thus these species have the simple alternate strategy. In godwits there may be one or two molts—and perhaps some individuals (nonbreeding) have one molt and others (that return to the breeding grounds) have two? This same question exists for the Black-bellied Plover, and it is discussed more fully in the family account for plovers.

PREFORMATIVE MOLTS. The preformative wing molts of most shorebirds (such as yellowlegs and allies, typical sandpipers, and phalaropes) can be complete but more often show eccentric patterns (see Figure 42). Thus, the exposed juvenile outer primaries are replaced but the protected inner ones are retained. This makes sense for long-distance migrants that return to the breeding grounds in their first spring; without a preformative wing molt, the juvenile outer primaries would have to last through one northbound as well as two southbound migrations before being renewed in the second prebasic molt. It also makes sense for birds remaining on sunny nonbreeding grounds for their first summer. Strong sunlight could wear down the juvenile outer primaries so much that they would not last until being renewed late in the second prebasic molt (see below, Why Oversummer?).

In Whimbrels and Bristle-thighed Curlews, however, the first-summer wing molt apparently follows the "typical" sequence, starting at p1 and including a variable number of primaries,[5] although eccentric patterns also occur in Whimbrels.[6] Why would these species be different, and are these "typical" primary molts really incomplete? Might they simply be interrupted complete molts? In larger shorebirds such as these two curlews, replacing the long outer primaries twice in a relatively short period may be too costly, especially for birds wintering in fairly food-poor environments such as tropical beach strands. Because it is important to have the outer primaries in good condition for migration, however, might the inner primaries be molted in spring and summer, with molt then suspended until winter when fresh outer primaries could be grown?

A study of Bristle-thighed Curlews suggests this isn't the case, and that the complete preformative wing molt may be followed by an incomplete second prebasic molt.[7] The preformative wing molt spans March to October, sometimes being complete and less commonly ending with one to three juvenile outer primaries unmolted. Before the preformative molt has finished, in July–August the second prebasic wing molt starts at p1, and by early winter it either completes or apparently arrests with two to eight formative outer primaries not renewed. Presumably the relatively fresh formative primaries are adequate to carry the bird to its next molt. But why are these molts incomplete?

In Ruddy Turnstones, first-cycle birds wintering in Australia replace from none to all of their primaries in the preformative molt, and incomplete molts also follow the "typical" sequence.[8] The authors of this study conclude that birds will molt their primaries if they can, because of the benefits to having wings in good condition, such as for avoiding predators. Thus, birds that fail to complete their wing molt are suspected to do so because they are not in good enough condition. Turnstones with complete preformative wing molts have fresh primaries in summer and, not surprisingly, undergo their second prebasic wing molt later than birds that have retained their juvenile primaries, which are heavily worn by summer. This means that the former birds will have slightly fresher primaries for forthcoming long-distance migrations, demonstrating a follow-on benefit to having first-cycle wing molts.

So, might the extent of preformative and second prebasic wing molts in Bristle-thighed Curlews and Whimbrels reflect individual fitness? If so, we might predict that birds retaining formative primaries are those that stay on the nonbreeding grounds for their second summer, while those with complete second prebasic molts are fitter and can migrate north to the breeding grounds. It would also be interesting to know what determines whether an individual shorebird has an eccentric or typical first-cycle primary molt, both patterns having been reported in Whimbrels and Ruddy Turnstones.

PREALTERNATE MOLTS. Prealternate molts are well documented in many sandpipers. In species that don't have appreciable seasonal changes in their appearance, however, it can be difficult to see if such molts occur or not. A careful study of the American Woodcock found no evidence for a prealternate molt,[9] but snipes have prealternate molts even though their appearance doesn't change much. Buff-breasted Sandpipers may lack a prealternate molt,[10] although their long-distance migrations suggest they should have one. Prealternate molts in curlews bring about no striking changes in appearance and are variable in extent, and in some Whimbrels they may even be lacking.

In general, prealternate molts are most extensive and bring about the most pronounced changes in appearance among several high-latitude breeders. These molts may have evolved originally as a consequence of long-distance migrations, but they have enabled birds to undergo changes in their appearance. The drab winter colors of several species, such as Western Sandpiper and Dunlin, are well suited to make birds less conspicuous on mudflats. Prealternate molts produce plumage patterns that balance the needs of display and camouflage for birds nesting on the open tundra. In some high-latitude breeders, such as Pectoral and Baird's sandpipers, relatively little change in appearance is caused by prealternate molts. These are species that inhabit marshy and grassy habitats year-round where their generic "cryptic brown" appearance serves them well.

In other species, striking reddish underparts have developed that may serve in display and courtship, such as in males of the Bar-tailed Godwit, Red Knot, and Curlew Sandpiper, and in female Red Phalaropes, although how much these may represent supplemental plumages is unclear (see below, Presupplemental Molts). Either way, why reddish

First-cycle Surfbirds wintering in California typically undergo an extensive preformative molt of head and body feathers before the northern winter sets in. This individual has molted most of its head and body feathers, including the scapulars, and the new feathers resemble adult basic feathers in pattern. The small, neatly white-edged upperwing coverts and the flight feathers are still juvenile feathers. *San Diego County, CA, 5 Nov. 2006. Lyann A. Comrack.*

feathers are so popular is unknown. Perhaps they reflect an ancestral trait or a physiological disposition among sandpipers?

PRESUPPLEMENTAL MOLTS. Presupplemental molts are difficult to confirm because of the problems involved in establishing that some feathers are molted three times in a cycle. Another possibility is that fluctuating hormone levels simply result in differently pigmented feathers being produced depending on the season (see pages 17–18). Evidence of presupplemental molts has been presented for three sandpipers: the Ruff,[11] Bar-tailed Godwit,[12] and Great Knot.[13] In these species, wintering adults undergo a fairly extensive head and body molt in late winter before leaving the nonbreeding grounds, and then at least some birds undergo another period of body molt at staging sites on their northward spring migration. Because the late-winter molts appear most like the prealternate molts of other sandpipers, the spring molt at staging sites is considered as a novel, presupplemental molt.[14]

In Ruffs, males acquire extravagant neck feathers (which form the eponymous "ruff") and also renew some head and back feathers in the presupplemental molt. The resultant plumages are strikingly different from basic and alternate plumages, and males gather at traditional display grounds, known as leks, to attract females and mate with them. A less extensive but homologous molt occurs in females but does not produce a ruff. Females implanted with testosterone, however, can develop ruffs, whereas castrated males fail to develop ruffs and instead molt into a plumage similar to that of a normal female.[15] Among breeding male Ruffs there are two well-known types: independent males (whose ruffs can be various colors from black to rufous, plain or barred) and satellite males (whose ruffs are usually white). Independent males defend small mating courts within the leks, whereas satellite males don't defend courts but rely on stealing copulations when independent males aren't looking or are otherwise distracted.

The Bar-tailed Godwit, here a male of the eastern subspecies *baueri* fattened up for migration, is one of three sandpipers in which there is evidence for a presupplemental molt. This intense brick red alternate body plumage may be partially renewed via a presupplemental molt at staging sites on the northward migration. *North Island, New Zealand, 25 Mar. 2007. Steve N. G. Howell.*

A further twist to the molts and breeding strategies of Ruffs came recently, with the discovery of a third type of breeding male Ruff. About 1 percent of male Ruffs are intermediate in size between males and females and acquire a femalelike "breeding" plumage, although whether this is alternate or supplemental is unclear.[16] It has been suggested that these "feminine males," which have been termed faeders, may represent an ancestral state from which the supplemental plumages developed. These birds retain their femalelike plumage in successive years and have not been found to develop ruffs. Faeders appear to us as female mimics, but their sexual identity may be known to other males. Although observations at leks found that 24 of 25 attempted copulations involving faeders were with males (both independents and satellites), in half of these cases the faeder mounted the other male.[17] Might faeders be a novel manifestation of homosexuality among birds?

In Bar-tailed Godwits, supposedly only the fittest birds undergo the presupplemental molt, which in males produces red feathers that may replace barred alternate feathers grown only a month or two previously. Because shorebirds that breed in the short Arctic summer have tight schedules, it has been suggested that godwits arriving at the staging grounds with low body weight invest their time in simply maintaining their metabolism and not in molting. Thus, the bright red supplemental plumage is viewed as an honest indicator of fitness exhibited by breeding males,[18] although it may simply be a signal of age, with second-cycle males having duller plumage than adults.[19]

In Great Knots, both sexes acquire brighter and more extensively red back feathers at the staging sites than the alternate feathers grown a few months earlier before starting northward migration. Intense sexual selection is easily implicated in the supplemental plumages acquired by Ruffs and perhaps Bar-tailed Godwits, whereas in Great Knots the plumage changes may reflect bi-directional sexual

Phalaropes often start their preformative body molt on southbound migration, although this first-cycle Red-necked Phalarope, intent on hunting mating flies, appears to be still in almost full juvenile plumage. *Marin County, CA, 19 Aug. 2008. Steve N. G. Howell.*

selection[20] or perhaps selection for cryptic plumage while nesting.

A conclusion of many molt studies tends to be that evolving a molt reflects the benefits of having colorful plumage. This implies that birds can somehow "choose" to have a molt in order to change their appearance. From an evolutionary standpoint it may be more likely that the molts came first because some feathers needed to be replaced, plumage color changes resulted incidentally, and selection followed. In the Ruff, Bar-tailed Godwit, and Great Knot, however, selection may actually have driven the evolution of presupplemental molts.

So how could a third molt evolve in some sandpipers but not others? Surely, feathers renewed only a month or two earlier would not be so worn that they needed to be molted again. One possible pathway for the development of these presupplemental molts is via suspended prealternate molts. That is, the prealternate molts are rarely completed prior to migration and can be completed at staging sites (for example, 80 percent of molt may occur on the nonbreeding grounds, 20 percent on the staging grounds). Fluctuating hormone levels as the breeding season approaches may result in differently pigmented alternate feathers being produced at staging sites. And if subsequent selection favors those birds with the greatest number of second-stage alternate feathers, then perhaps the molts at staging sites may become more extensive, such that a supplemental plumage develops. For example, if 80 percent of molt occurs on the nonbreeding grounds, and 30 percent or more on the staging grounds, then two molts are occurring.

Might other sandpipers have supplemental plumages that have been overlooked? Shared features of the three species having such plumages are long-distance migration and staging sites where fueling a molt is possible. Two other candidates that fit this job description are Red Knots wintering in southern South America and staging in Delaware Bay, and Surfbirds wintering in Chile and staging in the Gulf of California. In both cases these species also have northern-wintering populations in the U.S., which might offer interesting "control" groups to see whether presupplemental molts also occur in short-distance migrants.

Molt and Migration

The timing and location of prebasic and preformative molts in sandpipers vary with the wintering latitudes of species and individuals, as does the extent of the preformative molt. These molting patterns are perhaps most easily viewed simply as Northern Hemisphere and Southern Hemisphere strategies.[21,22,23] Some species with wide nonbreeding ranges exhibit both strategies, depending on where individuals spend the winter, such as Red Knots wintering in the southeastern U.S. versus those wintering in southern South America[24] (Table 2).

The Northern Hemisphere strategy includes two groups of birds, with the second group effectively grading into the Southern Hemisphere strategy. The first group comprises species that molt on or near their breeding grounds (mainly during July–September), before migrating south the relatively short distance to their mainly mid-latitude, Northern Hemisphere wintering grounds. In these species the preformative molt is limited to head and body feathers, often tertials, and sometimes upperwing coverts and central tail feathers. Breeding and molt can even overlap extensively at high latitudes with short summer seasons. For example, Dunlin nesting in Alaska at around 71° N start wing molt when the eggs are laid and have mostly completed

The Buff-breasted Sandpiper is a long-distance migrant that winters in southern South America, where this juvenile would undergo a complete preformative molt. Whether a prealternate molt occurs in any age of Buff-breasted Sandpiper remains unclear. *Marin County, CA, 3 Oct. 2006. Steve N. G. Howell.*

TABLE 2

MOLT STRATEGIES OF NORTH AMERICAN SANDPIPERS

Molting strategies of North American sandpipers relative to wintering latitude (Northern Hemisphere versus Southern Hemisphere; see text). **Northern^{br}** indicates molt on or near breeding grounds. **Northern^{nbr}** indicates molt on nonbreeding grounds. **Preformative wing molts** of species wintering in South America are not well known; a question mark indicates that such molts may occur but have not been reported.

	NORTHERN^{br}	NORTHERN^{nbr}	SOUTHERN	PREFORMATIVE WING MOLT
Willet		X		
Greater Yellowlegs		X	X	?
Lesser Yellowlegs		X	X	X
Solitary Sandpiper		X	X	X
Wandering Tattler		X	X	X
Spotted Sandpiper		X	X	X
Upland Sandpiper			X	?
Whimbrel		X	X	X
Bristle-thighed Curlew		X[1]	X	X[1]
Long-billed Curlew		X		
Hudsonian Godwit			X	?
Bar-tailed Godwit		X	X	X

Marbled Godwit		X		
Ruddy Turnstone		X	X	X
Black Turnstone		X		
Surfbird		X	X	?
Red Knot		X	X	X
Sanderling		X	X	X
Semipalmated Sandpiper		X	X	X
Western Sandpiper		X		
Least Sandpiper		X	X	X
White-rumped Sandpiper			X	X
Baird's Sandpiper			X	X
Pectoral Sandpiper			X	X
Sharp-tailed Sandpiper			X	X
Purple Sandpiper	X			
Rock Sandpiper	X			
Dunlin	X			
Curlew Sandpiper			X	X
Stilt Sandpiper		X	X	X
Buff-breasted Sandpiper			X	X
Ruff		X	X	X
Short-billed Dowitcher		X	X	X
Long-billed Dowitcher		X		
Wilson's Snipe	X	X		
American Woodcock	X			
Wilson's Phalarope		X	X	X
Red-necked Phalarope		X	X	X
Red Phalarope		X	X	X

[1] Adults molt on a northern schedule, but first-cycle birds have a preformative wing molt.

molt when the young fledge. Dunlin nesting at 61° N in western Alaska, however, have time to molt after breeding, and their molt is also less intense: a complete molt at 71° N requires about 10 weeks, whereas at 61° N it requires about 14 weeks.[25]

The second group of Northern Hemisphere molters are birds that winter from mid-latitudes of the Northern Hemisphere south into the tropics and undergo most or all of their molt at staging sites or on the wintering grounds (mainly during August–November). Several species may start head and body molt, and rarely primary molt, on the breeding grounds and then suspend molt for migration. The partial preformative molt involves head and body feathers, usually tertials, and sometimes upperwing coverts and tail feathers, and may be protracted into the winter. Prealternate molts of both Northern Hemisphere groups average later (mainly March–May) and less extensive than those of birds with the Southern Hemisphere strategy. This may be because northern birds have less time available in warm weather for molt prior to migration, which is countered by their feathers not receiving so much sun damage in winter.

In the Southern Hemisphere strategy, prebasic and preformative molts occur later, usually after birds arrive on their nonbreeding grounds in the tropics and the Southern Hemisphere, and they tend to be more protracted. Some birds molt a few head and body feathers near the breeding grounds; others, perhaps mainly nonbreeding birds or failed breeders, may replace a few inner primaries before suspending wing molt for migration. Some molt can occur during migration, as in Wilson's Phalaropes at staging sites,[26] but most species molt after arrival on the nonbreeding grounds, mainly from September–October to January–February. The preformative molt varies from partial to complete, with head and body feathers renewed mainly during September–December and any wing molt following, mainly in December–April. Prealternate molts are not constrained by cold winter weather, can start in January or February, and can overlap in timing with the prebasic or preformative wing molts.

In species whose wintering range is concentrated in temperate South America (such as Baird's, Pectoral, and Buff-breasted sandpipers), the preformative molt is usually complete. In other species (such as the Semipalmated Sandpiper, Sanderling, and phalaropes) it varies from complete to partial. In most cases, incomplete molts involve eccentric

This Surfbird undergoing its prebasic molt (primary molt has reached p5 being shed) offers some food for thought. Note how its outer primaries are contrastingly darker and less worn than the adjacent middle primaries. Usually, however, the outer primaries of a shorebird are the most heavily worn because they are the most exposed. The pattern of fresher outer primaries is typical of an eccentric preformative molt (see page 35), but might also result from a prebasic molt that was interrupted, with the outer primaries being renewed sometime later than the middle primaries. Neither preformative wing molt nor interrupted outer primary molt has been documented in Surfbirds. If the molt contrast resulted from a preformative molt, this would suggest the bird spent its previous winter in South America (preformative wing molts in sandpipers do not occur in North America), and now it appears to be wintering in North America (most sandpipers undergo prebasic primary molt on or near their nonbreeding grounds). *Sonoma County, CA, 8 Aug. 2008. Steve N. G. Howell.*

The Dunlin is a relatively short-distance migrant that winters in the Northern Hemisphere. Unlike most other sandpipers in North America, both adults and first-cycle birds molt before migrating south, which helps explain why the Dunlin is a notably late fall migrant. The buffy white edgings to the upperwing coverts and tertials indicate that this is a first-cycle individual. *Marin County, CA, 21 Oct. 2008. Steve N. G. Howell*

patterns of primary renewal, with the more exposed juvenile outer primaries being replaced and the protected inner ones retained (as discussed above, under First-cycle Molts). The differences in the extent of first-cycle molts presumably reflect a balance between a bird's fitness and ability to molt with the need to counter increased feather bleaching and wear caused by stronger sunlight in tropical and southern latitudes.

While the principles of these patterns are easily understood and make sense, a few species don't fit the patterns. For example, many Western Sandpipers winter in the tropics yet apparently do not renew primaries in their preformative molt.[27] While this may be an artifact of small sample sizes from South America, might it also reflect an ancestral northern-wintering range? If a southward extension of the Western Sandpiper's wintering range into the tropics is sufficiently recent, in evolutionary terms, then perhaps wing molt strategies haven't caught up.

More surprising is the Upland Sandpiper, which winters wholly in South America and apparently does not replace primaries in the preformative molt.[28] Yet the Buff-breasted Sandpipers and American Golden-Plovers wintering alongside Upland Sandpipers have complete preformative molts. Why the Upland would be different is unclear, as it has to undergo three long-distance migrations with its juvenile primaries before they are renewed.

Why Oversummer?

Why do so many first-cycle shorebirds remain on the nonbreeding grounds in their first summer, whereas landbirds typically migrate north to the breeding grounds? A simple question, perhaps, but there's no simple answer. Several explanations have been proposed for oversummering in shorebirds, and the reasons for any given species probably involve a combination of factors. Relative to landbirds, most shorebirds winter in open habitats where their feathers can become faded and worn, and they may be longer-lived so that foregoing the first season is not so critical over their lifetime.

Larger species generally live longer[29] and may mature more slowly, so it is not surprising that oversummering is more common among larger shorebirds, such as curlews, godwits, Black-bellied Plovers, and Pacific Golden-Plovers, in which some birds even stay on the nonbreeding grounds through their second summer. Another factor may be nonbreeding habitat. As a rule, shorebirds inhabiting coastal habitats delay their return to the breeding grounds considerably longer than do species of freshwater and other inland habitats.[30] This may be due to the challenges of feeding in habitats that are available only half of the time (being covered at high tide) combined with the costs of finding safe roost sites. Another factor that has been proposed

Similar but different in appearance, the first-cycle Semipalmated Sandpiper (left) and Western Sandpiper (right) also differ in the extent of their preformative molt. Semipalmated Sandpipers have a more southerly winter distribution and can replace some to all of their flight feathers in the preformative molt. The more northerly wintering Western Sandpiper is not known to replace any primaries in its preformative molt. *Marin County, CA, 13 Aug. 2007. Steve N. G. Howell.*

is that young birds may have poorer parasite resistance and thus may be physiologically compromised from undertaking a successful migration.[31] However, parasite loads are lower in coastal habitats flushed regularly by the tide than in freshwater habitats,[32] and first-cycle individuals of most larger species, perhaps the fittest individuals, can return to the breeding grounds.

The "cost of migration" is an obvious factor, but how exactly is this defined? Two components of this are feeding and flight. Most sandpipers breed at high latitudes, where summer is short, so they have to keep to a tight migration schedule if they are to arrive in time to breed. Migration involves stopovers to feed and refuel. Young shorebirds forage less efficiently than do adults, which means they may be unable to refuel quickly enough on northward migration—and thus may not get to the breeding grounds on time.[33]

Migration also involves sustained flight, for which having the wings in good condition is important. Linked to this are the distance between the breeding and nonbreeding grounds—and molt. Birds wintering nearer to the breeding grounds tend to migrate north in summer more commonly than those wintering farther south. For example, first-cycle Western Sandpipers winter in large numbers in Panama and northwestern Mexico, but those in Panama typically oversummer whereas those in Mexico typically migrate north.[34] Birds in Panama experience stronger sunlight, and their flight feathers likely get more faded and worn than those of Mexican birds, yet they would need to migrate farther. As noted earlier, Western Sandpipers are unusual among tropical-wintering sandpipers in not having a first-cycle wing molt, so their worn juvenile primaries would have to carry them through a northward and southward migration before being molted. Thus, wing-molt considerations may also play into whether Western Sandpipers oversummer.[35] Birds remaining in Panama can undergo their second prebasic wing molt in the absence of competing adults and largely in the absence of avian predators (mainly wintering Peregrine Falcons and Merlins).

It has often been assumed that shorebirds do not migrate north in their first summer simply because they are inexperienced and may die on migration. But the migration needs to be justified, and there is little benefit to surviving the journey only to arrive too late to breed—when you could have stayed in Panama and molted. In support of the latter, it has been found that individual survival between the first and second winters is higher for Western Sandpipers in Panama than in northwestern Mexico.[36] How might this compare to species with a southern hemisphere strategy, in which some individuals have first-cycle wing molts and others don't? Do birds having wing molts migrate north while those lacking wing molts remain to oversummer? And is survival of migrants linked to the extent of wing molt, with birds that have the most extensive molts having a better chance of surviving their migrations?

REFERENCES

1. Pearson 1984; **2.** Higgins and Davies 1996; **3.** Marks 1993; **4.** Marks et al. 1990; **5.** Pyle 2008; **6.** Higgins and Davies 1996; **7.** Marks 1993; **8.** Skewes et al. 2004; **9.** Keppie and Whiting 1994; **10.** Pyle 2008; **11.** Jukema and Piersma 2000; **12.** Piersma and Jukema 1993; **13.** Battley et al. 2006; **14–15.** Jukema and Piersma 2000; **16–17.** Jukema and Piersma 2006; **18.** Piersma and Jukema 1993; **19.** Battley 2007; **20.** Battley et al. 2006; **21.** Holmes 1971; **22.** Marks 1993; **23.** Pyle 2008; **24.** Harrington et al. 2007; **25.** Holmes 1971; **26.** Jehl 1987; **27.** O'Hara et al. 2002; **28.** Pyle 2008; **29.** Summers et al. 1995; **30.** Rogers 2006; **31.** McNeil et al. 1994; **32.** Rogers 2006; **33.** Hockey et al. 1998; **34.** Fernández et al. 2004; **35.** O'Hara et al. 2002; **36.** Fernández et al. 2004.

LARIDAE
Gulls
(CAS, SAS, SBS; 25 species)

Gulls, or seagulls as they are often known, are familiar birds in many parts of North America, from beaches to shopping malls and inland lakes. They are a cosmopolitan family of about 50 species that are sometimes treated in the same family as skuas, terns, and skimmers. Gulls can be divided into two well-defined groups that generally differ in their molt strategies.[1] One group is the generally smaller "ternlike" gulls, with more slender bills and raspy and shrieky calls. These are considered closer to the common ancestor with terns, and examples include Bonaparte's, Sabine's, and Ivory gulls. The second group comprises the generally larger "typical" gulls, with stouter bills and crowing or laughing calls, and which reflect a more recent offshoot. Examples include the Laughing, Ring-billed, and Western gulls. All North American gulls are migratory, from the short-distance migrant Ivory Gull, which remains year-round near the Arctic ice edge, to the trans-equatorial Sabine's and Franklin's gulls. The sexes of gulls look alike, but in most species there are distinct seasonal and age-related differences in appearance. Adult plumage aspect is usually attained in 1 to 4 years. The semi-precocial downy young are fed by their parents and fledge at 3 to 8 weeks. The age of first breeding is usually 2 to 10 years.

Gulls as a group exhibit three of the four fundamental molt strategies. Most small gulls have the complex alternate strategy, but large gulls, like this adult Western Gull, have the simple alternate strategy. The most recent revision of molt terminology was precipitated by simple field observations of molt in the Western Gull. *Monterey County, CA, 29 Dec. 2008. Steve N. G. Howell.*

Like most small gulls, Sabine's Gull has the complex alternate molt strategy, but it is unusual in that the preformative molt is complete. Juveniles, like this individual, usually migrate south in juvenile plumage before starting their molt, but occasional birds start their body molt during fall migration. This species also offers a good example of how different molts can overlap in timing. Both the preformative and prebasic wing molts start in mid- to late winter and in their later stages overlap with the prealternate molt of head and body feathers. *Monterey County, CA, 15 Sept. 2008. Steve N. G. Howell.*

Gulls apparently exhibit three of the four fundamental molt strategies. The strategy of most smaller gulls is complex alternate, whereas large gulls appear to have simple alternate strategies, and the unique Ivory Gull has a simple basic strategy.[2] The differences among species reflect ancestry, body size, breeding latitude, and migration distance. Prebasic molts often start on the breeding grounds and suspend to complete on the nonbreeding grounds. Prealternate molts occur mainly on the nonbreeding grounds.

Challenges in divining the molt strategies of gulls lie with the protracted nature of the molts and with the varied appearance of a generation of feathers, such as the scapulars, that can occur in a bird's first year or two of life. However, when it is recognized that the colors and patterns of feathers do not necessarily correlate with molts (see pages 17–18), and that molts can overlap in timing, patterns are revealed. In fact, it was a study of molt in the Western Gull[3,4] that precipitated the revised molt terminology[5] that forms the basis for this book—an example of how simple observations of a common species can offer insights into the bigger picture.

PREBASIC MOLTS. A pattern found in almost all gulls is for some inner primaries to be shed during incubation; molt then suspends during chick-feeding and is resumed later. Two high-Arctic breeders are exceptions to this pattern and exhibit extremes in prebasic wing-molt timing. Sabine's Gull suspends its wing molt until reaching the nonbreeding grounds, although birds often molt some head and body feathers during southward migration. Thus, the protracted prebasic wing molt (during December–March) overlaps with the prealternate molt, such that some authors have considered that Sabine's Gull has a complete "prebreeding" molt. The strategy of Sabine's Gull is much like that of the Long-tailed Jaeger and Arctic Tern, two species with similar life-history traits.

At the other extreme, in Ivory Gulls the primary molt is mostly completed *before* the breeding season.[6] In spring there is enough food in the Arctic to fuel molt before conditions become suitable for breeding (wing molt starts as early as March, when it's still damn cold in the far north!). And there isn't time for a complete molt after breeding, before winter sets in. This timing recalls that of some loons and puffins (which also molt their wings in spring) and allows molt and breeding to fit into the short Arctic summer. Breeding Ivory Gulls suspend their wing molt during egg-laying and incubation and then finish molt of the outer primaries in August–September, before the onset of winter. Some birds do not complete molt before winter, however, and they retain one or two old outer primaries through to the next spring. Whether such birds set up opportunistic stepwise waves of primary molt (as described in the family account for hawks) or might have their next breeding attempt compromised is not known. Perhaps as another concession to their high-Arctic environment, Ivory Gulls appear to molt simply from one ivory white basic plumage to another, and thus lack an alternate plumage.

Delayed prebasic molts can also occur in kittiwakes, which winter in an unforgiving environment—at sea in the North Pacific. In some years (perhaps mainly those with food crashes related to El Niño events), many kittiwakes do not complete their primary molt by spring.[7] It seems likely that these birds would need to skip a year of breeding to get their molt back on track, but this hasn't been established.

FIRST-CYCLE MOLTS. Most small gulls have a partial preformative molt in fall and a partial first prealternate molt in spring. The head patterns of small gulls (such as Black-headed) in first alternate plumage vary in appearance from breeding to nonbreeding aspect, presumably a reflection of the differing sexual maturity of individuals. The molt is a constant, but the color depends on other cues.

Large gulls appear to have only a single molt

The first-cycle molt of head and body feathers in gulls varies greatly in timing and extent among species and even among individuals of the same species. This molt averages less extensive in species breeding in high latitudes, such as this Thayer's Gull, which is still largely in juvenile plumage in late winter. Compare this with a first-cycle California Gull on the same date. *Sonoma Co., CA, 12 Mar. 2008. Steve N. G. Howell.*

The color and pattern of feathers grown by first-cycle gulls can vary greatly over the course of a single molt. In general, feathers grown earlier are more immature-like in appearance, whereas those grown later are more adultlike. On this first-cycle California Gull, the postjuvenile scapulars vary from dark brown with broad whitish tips to fairly plain gray overall, yet all are of the same generation and were produced by the same molt. The greater coverts and tertials are worn and faded juvenile feathers, whereas the whitish head is produced by fading rather than by the molt of new white feathers. Contrast the extent of molt and plumage wear with a first-cycle Thayer's Gull on the same date. *Sonoma Co., CA, 12 Mar. 2008. Steve N. G. Howell.*

added into their first cycle. Compared with the adult molts, this added molt resembles a prealternate molt in its extent and timing. But it also resembles a preformative molt when compared with the molts of small gulls with a complex alternate strategy. The puzzles of naming this molt, and how it may have evolved, are discussed in the family account for ibises and spoonbills.

The first-cycle molt of large gulls typically is partial and averages more extensive in southern-breeding species (such as California and Yellow-footed gulls) than in northern-breeding species (such as Thayer's and Glaucous-winged gulls). Some individuals of high-latitude species, such as the Glaucous Gull, may even skip this molt altogether and in their first summer molt directly from juvenile (first basic) into second basic plumage. At the other extreme, in the southernmost breeding populations of Heermann's Gull, the first-cycle molt can include all of the primaries, and sometimes it may even be complete.[8] It seems likely that these extremes reflect latitude, sunlight intensity, time of fledging, and food. Thus, the largely unpigmented plumage of a juvenile Glaucous Gull, grown in late summer from round-the-clock feeding of rich summer food, is not exposed to intense sunlight at high latitude, and it doesn't get worn or need to be replaced. But sometimes even the all-dark plumage of a juvenile Heermann's Gull, grown in early summer from an unpredictable food supply in waters prone to El Niño events, may not be enough to protect it under an intense tropical sun.

Two small ternlike gulls, the Red-legged and Black-legged kittiwakes, also exhibit the simple alternate molt strategy. These are high-latitude breeders that grow a strong juvenile plumage and winter at sea in cold northern latitudes, where plumage doesn't get degraded by strong sunlight; thus they get by with a single first-cycle molt.

An exception to the molts of other gulls is again found in the Ivory Gull.[9] These beautiful birds breed relatively late in the year because they nest in the

In their first winter, Heermann's Gulls typically molt only some head and body feathers, and sometimes upperwing coverts. Some individuals of the southernmost populations, however, have an extensive molt that might even be complete on occasion. This individual has undergone an eccentric primary molt: the outer two primaries are still growing, and only the relatively protected innermost primary is a juvenile feather. *Nayarit, Mexico, 22 Jan. 2007. Steve N. G. Howell.*

high Arctic, and juveniles fledge in September with a strong plumage. This juvenile plumage is retained through the winter, and then in spring–summer birds undergo a complete second prebasic molt into the ivory white adult plumage, without any intervening first-cycle molt.

PREALTERNATE MOLTS. The prealternate molts of most gulls are partial, involving head and body feathers, scapulars, and often some upperwing coverts. In smaller gulls these molts occur mostly during late winter and spring and are discrete in timing from the prebasic molts. In large gulls the prealternate molt often starts in fall, before the prebasic molt has completed, and at this time it can include upperwing coverts and scapulars; these prealternate molts probably suspend over midwinter and resume in late winter and spring.

An exception to the typical patterns is found in Franklin's Gull, all ages of which undergo a complete or near-complete prealternate molt in the food-rich Humboldt Current before they head back north for the summer. The prealternate molt is more often complete in adults than in first-cycle birds, and some northerly-wintering first-cycle birds may replace no primaries. In poor food years even adults may not always manage a complete molt, and the ratio of complete to incomplete prealternate molts observed among spring migrants or birds on the breeding grounds might be an index of food supply the previous winter. But why should Franklin's Gull have a complete prealternate molt? Is the more extensive molt triggered by a Southern Hemisphere daylight regime, which mimics summer in the Northern Hemisphere? Do the costs of migration and strong sunlight take their toll on the wings, which then need to be replaced?

An interesting trend among large gulls is that the second and sometimes third prealternate molts average more extensive than do adult prealternate molts. This situation is similar to that found with prealternate primary molts of terns, which average more extensive on prebreeding immatures than on adults. In general, these extensive molts involve upperwing coverts, but sometimes even primaries are replaced, as in the Yellow-footed Gull,[10] and they result in a more adultlike appearance. Is this because some prebreeding birds have time and energy for more extensive molts? And do such molts produce plumage signals that accrue benefits in social interactions such as feeding and mate selection?

REFERENCES

1. Howell and Dunn 2007; **2.** Howell 2001a; **3.** Howell and Corben 2000a; **4.** Howell and Corben 2000b; **5.** Howell et al. 2003; **6.** Howell 2001c; **7.** Howell and Corben 2000c; **8.** Howell and Wood 2004; **9.** Howell 2001c; **10.** Howell and Dunn 2007.

STERNIDAE
Terns
(CAS, CBS, SAS; 15 species)

Looking like gulls that got a makeover, terns are a family of about 45 species that occur around the world in marine and freshwater habitats. They have variously been treated as their own family and as part of an expanded family, Laridae, along with skuas, gulls, and skimmers, and are usually considered most closely related to gulls. A 2005 review of tern taxonomy[1] proposed the recognition of eight genera for North American species, including *Onychoprion* for dark-backed terns (Bridled, Sooty, and Aleutian), *Sternula* for little terns (including Least), and *Thalasseus* for crested terns (Royal, Elegant, and Sandwich). Thus, for North American species, the genus *Sterna* now includes only Roseate, Arctic, Common, and Forster's terns.

All North American species are migratory, with several species being mostly or wholly pelagic in winter. The sexes of terns look alike, but in most species there are distinct seasonal and age-related differences in appearance. Adult plumage aspect is attained in 1 to 2 years. Young terns are semiprecocial and downy, are fed by their parents, and fledge in 3 to 6 weeks. The age of first breeding in most terns is 2 to 6 years.

The molt strategies of terns are not well known, largely because of the difficulty in following birds through their first year or two of life. Most North American species appear to have the complex alternate molt strategy, although some species have the simple alternate, and Brown Noddy appears to have the complex basic. Details of molt strategy relate both to life-history traits (especially migration distance) and to ancestry. Most molting occurs away from the breeding grounds, but several species start their prebasic molt while nesting.

PREBASIC MOLTS. Prebasic molts often start on the breeding grounds and suspend to complete on the nonbreeding grounds, or they can occur wholly away from the breeding grounds. Patterns vary among and within genera. For example, crested terns (genus *Thalasseus*) typically start prebasic head molt while nesting, and have white-spotted or fully white forecrowns as early as May–June, whereas "typical" terns (genus *Sterna*) retain full black caps until after breeding or even until after migration, and little terns (genus *Sternula*) can molt their wings throughout much of the breeding season, starting when the eggs are laid.[2] White crowns early in the season among the dense-packed colonies of *Thalasseus* may convey social information that is less important, or conveyed by other means, in colonies of typical terns. The small size of *Sternula* terns, the potential for abrasion of their pri-

The Black Tern winters in large numbers off the Pacific Coast of Mexico and Central America, where sea turtles provide opportunistic perches. This individual still has three unmolted basic outer primaries and has not yet started its prealternate molt of inner and middle primaries. *Oaxaca, Mexico, 16 Dec. 2008. Steve N. G. Howell.*

maries by sand, and perhaps a differing food supply between summer and winter may all contribute to molting through the breeding season.

Wing-molt timing can be independent of genus, instead reflecting life-history traits such as migration distance. Thus, species in different genera (such as Bridled, Least, Gull-billed, Common, Black, and Sandwich terns) can replace some inner primaries while nesting and then suspend wing molt until they reach the nonbreeding grounds. In such species, molt of the outer primaries completes in mid- to late winter and may overlap in timing with prealternate molt of the inner primaries in late winter to spring. In Elegant Terns, prebasic primary molt of adults starts on the breeding grounds, suspends for migration north to the food-rich California Current where molt resumes but rarely completes, and then suspends for migration south to the food-rich Humboldt Current where it is completed.

Extreme differences in prebasic wing-molt timing are manifest by two typical terns, Forster's and Arctic. Forster's Tern winters in temperate mid-latitudes and completes its prebasic primary molt during late July–November, before the onset of winter. Arctic Tern is a long-distance migrant that migrates to Antarctica, where adults undergo a rapid prebasic primary molt at the food-rich ice edge, mainly during late December–late February, when they may be almost flightless at times.[3,4,5] Thus, Forster's Tern may have completed molting before the Arctic Tern starts.

Among North American terns, the Brown Noddy is an exception, although its molt strategy may be more representative of ancestral terns. Among tropical terns such as noddies, the breeding and molt cycle in aseasonal equatorial regions may be less than annual, with wing molt being a continuous, sequential process that is interrupted for breeding. There are no prealternate primary molts, and wing molt simply continues from where it left off after any interruptions, restarting at p1 when p10 has been renewed. Brown Noddies nesting in Florida molt their primaries year-round, starting with the inner primaries in April–June and completing with the outer primaries in February–March. Wing molt is interrupted in spring and summer for breeding, and apparently also in winter, such that the outer three or four primaries are contrastingly fresh in spring. Having the outer primaries in good condition at this time may be important to reduce the effects of wear incurred from the scrubby bushes in which the noddies build their stick nests.

FIRST-CYCLE MOLTS. Most North American terns have a complete preformative molt, which occurs mainly on the nonbreeding grounds. A few species molt some head and body feathers on the breeding grounds or during migration, and Least, Roseate, and the three crested terns (*Thalasseus*) can also molt a few inner primaries on migration. The preformative primary molts of most species (including the Brown Noddy) tend to start in mid- to late winter and complete from spring (in smaller species) through fall (in larger species), when they can overlap with the start of the second prebasic wing molt.

An overlap in timing between prebasic and prealternate wing molts is frequent among terns. On this adult Royal Tern the outermost primary is just starting to grow, marking the end of the prebasic wing molt. At the same time, the prealternate molt of inner primaries has reached p3 growing. *Nayarit, Mexico, 22 Jan. 2007. Steve N. G. Howell.*

At least in the Gull-billed Tern and the crested terns, there can be a variable molt suspension in winter (spanning about 3 months) after the inner primaries have been molted, with primary molt resuming in February–March. Could this relate to limited food resources? Or might it reflect the period when young are first independent of their parents and fending for themselves, which could conveniently minimize resource competition with adults that are molting?

As with first-cycle jaegers, a first prealternate molt of some head and body feathers is presumed to occur in first-cycle terns, but data are few. If no second molt occurs, as may be the case in larger species such as the Caspian and Gull-billed terns, then these species would have the simple alternate strategy, like large gulls. At the other end of the scale, the first prealternate molt of some small and medium-sized terns can include inner primaries and tail feathers. These primary molts occur from late winter to spring in Least, Black, and Roseate terns, and in spring or summer in the larger Common, Sandwich, and Elegant terns. The short-winged Least, Black, and Roseate terns might also even fit in a first presupplemental molt of one or more inner primaries, thus undergoing four molts of their inner primaries within a year!

Two exceptions to this general pattern of first-cycle molts are Forster's and Bridled terns. These species winter relatively far north and in relatively unproductive marine waters, respectively. Their preformative molts occur from fall into spring (likely being suspended in midwinter) and involve only head and body feathers plus some to all tail feathers, but no wing feathers. There is no evidence for a first prealternate molt before the complete second prebasic molt starts in April–May. The first-cycle molts of these two species thus recall those of most gulls.

PREALTERNATE MOLTS. Prealternate molts of terns occur on the nonbreeding grounds and may complete during northward migration. The Brown Noddy is an exception in that it appears to lack a prealternate molt; hence its molt strategy is complex basic. The prealternate molts of most terns include head and body feathers, some to all tail feathers, and often a variable number of inner primaries. The prealternate molt commonly overlaps in timing with completion of the prebasic molt of outer primaries, such that inner (alternate) and outer (basic) primaries can be growing simultaneously.

The Bridled Tern, an adult shown here, shares with the similar-looking Sooty Tern a common ancestor and a year-round pelagic habitat, and both species rarely renew inner primaries in their prealternate molt. In its first cycle, however, the Bridled Tern's molt is more like that of Forster's Tern—neither species replaces primaries in its preformative molt (in most other terns, including Sooty, the preformative molt is typically complete). Why might this be? *Dare County, NC, 4 Aug. 2007. Steve N. G. Howell.*

Some populations of the Sooty Tern breed on a 9- to 10-month cycle.[6] In such cases, the prealternate molt may be compressed in extent compared with that of birds on a 12-month cycle. For example, the protracted prebasic tail molt can finish with the next-to-outermost feather at the same time as the prealternate molt of adjacent tail feathers, and thus a tail that appears uniform in wear may actually include feathers from two molts. In the most compressed cycles, it is even conceivable that the basic outermost tail feather could grow long and white on birds in breeding condition, but short and dark on birds in nonbreeding condition.

The number of inner primaries acquired in prealternate molts reflects wing length, migration distance, and perhaps ancestry.[7] The following data for North American populations are based on specimens of adults[8] that likely included breeding and nonbreeding birds, but they still reveal patterns. In the genus *Onychoprion*, Sooty and Bridled terns winter in food-poor tropical oceans and sometimes molt one to three inner primaries (and often none), whereas the Aleutian Tern molts four or five inner primaries, suggesting that it winters, or at least molts, in areas with richer food resources than its congeners. The Least Tern molts seven to nine primaries, a high score that is likely due at least in part to its relatively short primaries being less costly to renew than, say, those of the giant Caspian Tern, which sometimes molts one or two primaries (and often none). The Gull-billed Tern molts one to four primaries, a low to average score for its size. Forster's and Arctic terns molt none to four and none to three primaries, respectively (and most often none), presumably because of time constraints on molting caused by their opposite extremes of migration distance (discussed above under Prebasic Molts). Common and Black terns both molt three to seven primaries and are medium- to long-distance migrants that often winter together and thus experience similar conditions. The relatively high score of five to eight primaries in the Roseate Tern is likely due in part to its short wings. The crested terns (Sandwich, Elegant, and Royal) all have fairly high scores and replace three to seven primaries. This is a relatively extensive molt for the Royal Tern, given its large size, so perhaps it reflects some shared life-history or ancestral trait in the genus *Thalasseus*.

Before they reach breeding age, immature terns (mainly those aged 2 to 3 years) often have more-extensive prealternate molts of the inner primaries and outer secondaries than do older age classes. These extensive molts by immatures probably occur because they have more time available to molt than do adults engaged in breeding. For example, in Forster's Tern, which winters relatively far north, there may be insufficient time for a prealternate wing molt in spring; hence, any such molt occurs mostly in fall and early winter. Because their second prebasic molt is relatively early and finishes in August–September, second-cycle Forster's have time in September–November for a relatively extensive prealternate molt of inner primaries, whereas breeding adults have little if any time for a prealternate wing molt.

PRESUPPLEMENTAL MOLTS. Presupplemental molts of one to six inner primaries and some adjacent outer secondaries have been found in Least, Black, Common, Roseate, and Sandwich terns and may occur in other species. These molts presumably occur mainly during February–March, following the prealternate primary molts. Presupplemental wing molts are most frequent in those species that replace the most primaries in their prealternate molts. As with prealternate molts, the presupplemental wing molts appear to be most common, and most extensive, in prebreeding immatures, and exceptionally there may be two waves of presupplemental molt (see Figure 34). The existence of a prealternate wing molt is puzzling enough, but presupplemental molts seem even more esoteric—so what is going on? These "extra" wing molts of terns are discussed below.

Extra Wing Molts in Terns

The "extra" wing molts of terns (that is, their prealternate and presupplemental molts) have long attracted attention, but their evolution has avoided satisfactory explanation. A 2007 review of this phenomenon[9] noted that shorter-winged species usually replace more primaries than do longer-winged species (although the age and breeding status of birds included in the study were not specified) and that extra wing molts are typical of migratory species.

The tropical noddies and allies are generally regarded as ancestral to the migratory tern species breeding at higher latitudes.[10] Among these tropical terns, stepwise wing molt and continuous sequential wing molt (discussed above, under Prebasic Molts) are strategies that allow birds to maintain functional wings and take advantage of aseasonal environments for nesting.[11,12,13,14]

A striking difference between stepwise or interrupted continuous molts and the extra wing molts of North American terns is that the extra molts do not start up again where the preceding molt ended, but instead restart with the innermost primary. Assuming the ancestral tern was predisposed to continuous or stepwise wing molt, how has this been punctuated into the repeating wing molts we see today in so many tern species? One possibility is that interruptions of molt for migration may have somehow caused the ancestral molt clock to

The distinct molt contrast on the upperwing of this adult Common Tern suggests that a moderate period has elapsed between the end of the complete prebasic molt and the prealternate molt of the inner six primaries. *Dare County, NC, 22 May 2007. Steve N. G. Howell.*

The degree of molt contrast between alternate and basic primaries varies with the relative age of the two feather generations. Terns have a silvery bloom on the dorsal surface of fresh primaries, and as this wears off a blackish "undercoat" is revealed. On this adult Common Tern, it is likely that relatively little time has elapsed between the end of the complete prebasic molt and the prealternate molt of the inner six primaries. The molt contrast is thus subtle, but it will become more noticeable as p7 becomes more worn. *Dare County, NC, 26 May 2007. Steve N. G. Howell.*

be reset, such that each wave of molt restarts with the innermost primary.[15] A study of migratory and resident subspecies of Gull-billed Terns in Australia suggests that the reverse may also be possible—that stepwise molts could develop from a migratory ancestor with distinct prebasic and prealternate wing molts. In Australian Gull-billed Terns the prebasic and prealternate wing molts often overlap in resident populations, and both molts are sometimes interrupted, leading to patterns that might be mistaken for (or might be developing into?) waves of stepwise molt.[16]

But what might select for and reinforce a change from stepwise molts to repeating wing molts? In most terns with extra wing molts, the uppersides of the primaries have a frosting in fresh plumage, caused by elongated barbules on the outer sides of the feather barbs;[17] when these barbules wear off, the black basal portions of the feathers become conspicuous. The contrast between fresh silvery inner primaries and worn black outer primaries can be striking and is an obvious sign of how extensive the prealternate molt has been. The extent of extra wing molts may represent an honest signal of fitness—basically, the more feathers you can replace, the better mate you may be. In Common Terns, individuals with more-extensive extra primary molts tend to mate with each other, rather than with birds having less-extensive molts,[18] which lends support to this idea.

Could the benefits of social signaling have helped cause wing molts to restart with the innermost primary? Hence, the "badge of fitness" worn on the outer wing is kept for as long as possible. And if advantages gained by birds with a greater number of new inner primaries extend late into the breeding season (such as in group feeding interactions), this would be further reason to maintain a point of contrast in the outer primaries rather than resume molt at that point.

The inner primaries of terns that have prealternate wing molts are often notably soft and weak relative to the outer primaries. Might selective pressure to have more-extensive wing molts have led birds to "cheat" and produce more, but weaker, feathers in prealternate molts, and did this then lead to basic inner primaries also becoming weaker? Because the alternate inner primaries are weak, they need to be replaced fairly quickly, which could reinforce the need to restart the prebasic wing molt with the innermost primary.

REFERENCES

1. Bridge et al. 2005; **2.** Thompson and Slack 1983; **3.** Higgins and Davies 1996; **4.** D. I. Rogers, pers. comm.; **5.** Voelker 1997; **6.** Ashmole 1963; **7.** Bridge et al. 2007; **8.** S. N. G. Howell, unpub. data; **9.** Bridge et al. 2007; **10.** Bridge et al. 2005; **11.** Ashmole 1963; **12.** Ashmole 1965; **13.** Ashmole 1968; **14.** Dorward and Ashmole 1963; **15.** Bridge et al. 2007; **16.** Rogers et al. 2005; **17.** Dwight 1901; **18.** Bridge and Nisbet 2004.

Forster's Tern winters relatively far north and has to finish its prebasic molt before winter sets in. Second-cycle birds complete this molt up to 2 months earlier than do breeding adults and thus have time for a prealternate wing molt in late fall. First-cycle birds, in common with Bridled Tern, do not molt any primaries in their first winter. This trio of birds comprises a first-cycle bird at front, with uniformly worn and dark juvenile primaries; a second-cycle bird in the middle, with somewhat worn (basic) outer primaries contrasting with relatively fresh silvery (alternate) middle primaries; and an adult at the back, with relatively fresh, uniformly silvery (basic) primaries. *Jalisco, Mexico, 24 Jan. 2007. Steve N. G. Howell.*

RYNCHOPIDAE
Black Skimmer
(CAS, SAS?; 1 species)

Skimmers are a small family of angular, tern-like birds that breed and range in warm latitudes, with one species occurring in North America. They have variously been treated as their own family and as part of the family Laridae, along with gulls, terns, and skuas,[1] and are usually considered most closely related to terns. Skimmers are short- to medium-distance migrants that mainly inhabit large rivers and estuaries. They feed in flight by "skimming" their elongated lower mandible through the water surface and snapping the bill shut when it makes contact with food. The sexes of skimmers look alike, although males are appreciably larger and longer-billed than females, and there are distinct seasonal and age-related differences in appearance. Adult plumage aspect is attained at 1 year. The semi-precocial downy young are fed by their parents and fledge in 4 to 5 weeks. Age of first breeding is usually 2 to 4 years.

The molting strategy of the Black Skimmer is probably complex alternate but could be simple alternate (see below). Most molting occurs on the nonbreeding grounds, although prebasic molts can start on the breeding grounds.

PREBASIC MOLTS. The prebasic molt of skimmers starts mainly in August–October, either on the breeding grounds (when a few inner primaries may be molted) or after arrival on the nonbreeding grounds. Wing molt often suspends during January–March, and the outer primaries grow in April–May, perhaps so they are relatively fresh to help withstand sand abrasion on the nesting grounds.

FIRST-CYCLE MOLTS. Juvenile plumage is generally retained through fall migration, followed by a protracted and apparently complete molt. By midwinter, most birds have molted feathers of the head, back, tail, and often some upperwing coverts; wing molt starts in January–April and ends in July–September, with this variation in timing presumably reflecting food supply, individual fitness, and perhaps wintering latitude. A second molt of some head and neck feathers may occur during April–June, but this requires confirmation. If only a single first-cycle molt occurs, the molt strategy would be simple alternate. Starting with their second prebasic molt, skimmers enter the definitive molt cycle.

As an aerial feeder, the Black Skimmer has a gradual primary molt that spans much of the fall and winter. But unlike most migratory terns, skimmers lack a prealternate molt of their inner primaries. The inner primaries on this individual are notably worn relative to the outers, suggesting that its last prebasic wing molt was suspended between late summer (when the inners were renewed) and mid- to late winter (when the outers were renewed). *Imperial County, CA, 19 July 2008. Steve N. G. Howell.*

The juvenile Black Skimmer undergoes a protracted and complete preformative molt over the course of its first year, after which it resembles an adult and loses this attractive scaly patterning. *Baja California, Mexico, 10 Sept. 2006. Steve N. G. Howell.*

Given that skimmers are not long-distance migrants, why would they have a complete preformative molt? It would be interesting to know whether all species of skimmers have a complete preformative molt (as in skuas and jaegers) or only some species (as in terns). If the former, then perhaps the skimmer ancestor was a long-distance migrant, although this would not be my guess based on the present-day pantropical distribution of skimmers. A more likely explanation may lie with skimmers being aerial feeders that need relatively unimpaired flight powers to skim over the water with precision. Thus, their wing molt needs to be gradual. In the second prebasic molt, the outer primaries aren't replaced until late winter. Thus, without an intervening preformative wing molt, the juvenile outer primaries would be more than 18 months old before being renewed, and by that time they might well be too worn to function properly. This compromise of aerial feeding and protracted wing molt balanced by a complete preformative molt occurs in other groups such as hummingbirds, swallows, and some terns.

Like many species of gulls, terns, and jaegers, the Black Skimmer has a somewhat variable juvenile plumage aspect. This individual is relatively dark and its plumage is fairly worn, with a few new formative upperwing coverts already starting to be acquired. *Baja California, Mexico, 10 Sept. 2006. Steve N. G. Howell.*

PREALTERNATE MOLTS. Prealternate molts involve at least some head and neck feathers and occur mainly during March–May. At this time they can overlap with completion of the prebasic wing molt, as with Sabine's Gull, jaegers, and some terns.

REFERENCES

1. Zusi 1996.

STERCORARIIDAE
Skuas and Jaegers
(CAS, SAS? CBS?; 5 species)

Skuas and jaegers (the latter also known as skuas in the Old World) are a small family of gull-like birds that breed at high latitudes and winter at sea. They have variously been treated as their own family and as part of the family Laridae, along with gulls, terns, and skimmers, and they are usually considered most closely related to gulls. Skuas are large, bulky, and dark brown overall, whereas jaegers are smaller, often polymorphic and boldly patterned, and have elongated central tail feathers (most pronounced on adults). The diet of jaegers on the breeding grounds ranges from lemmings to berries, but at sea they often feed by pirating other bird species. All species are short- to long-distance migrants, and the sexes look mostly alike. There are distinct seasonal and age-related differences in appearance among jaegers, and less-distinct age-related differences in skuas. Adult plumage aspect is attained in about 3 years. The semi-precocial downy young are fed by their parents, and most fledge at 4 to 7 weeks. Age of first breeding is usually 3 to 8 years.

The South Polar Skua (left) has been recorded, appropriately, at the South Pole, but in order to molt this species migrates north across the equator into the North Pacific and North Atlantic. Adults undergo wing molt during April–September, scavenging at fishing boats and tracking gulls and shearwaters that they pirate for food. This bird is completing its primary molt, with the outer two primaries growing concurrently. While it is conceivable that this bird could fly back across the Pacific to reach the Antarctic breeding grounds by the end of October, it may be a nonbreeding subadult (skuas don't breed until they are 6–10 years of age). This bold attempt to pirate a juvenile Laysan Albatross was unsuccessful, but the skua did force the albatross to settle on the water! *Marin County, CA, 30 Sept. 2007. Steve N. G. Howell.*

Jaegers and skuas probably exhibit the complex alternate molt strategy, with complete prebasic and preformative molts in fall and winter and partial to incomplete prealternate molts in late winter and spring. It is possible, however, that their molt strategy is simple alternate (especially in skuas), or that skuas may even have the complex basic strategy.[1,2,3,4,5] The molt strategies of all species appear to be broadly similar, with differences in timing related to migration distances and breeding hemispheres. Almost all molting occurs at sea, away from the breeding grounds, and the preformative molt is complete (see below, Life History versus Ancestry).

PREBASIC MOLTS. The prebasic molt of jaegers occurs mostly at sea during fall and winter, but light molt of body feathers can start on the breeding grounds.[6] Molt of head and body feathers generally proceeds quickly from fall to early winter or midwinter, often during southward migration; wing molt, however, tends to await a bird's arrival on the nonbreeding grounds. The elongated central tail feathers are often dropped early in the adult prebasic molt, sometimes during fall migration (especially in the Long-tailed Jaeger). The wintering ranges, and thus wing-molt timings, of the three jaegers are correlated with the species they pirate for food. Pomarine Jaeger is the largest species and winters farthest north, where it pirates mainly gulls and shearwaters. It starts wing molt earlier, in August–November, and completes it in December–April. The smaller Parasitic and Long-tailed jaegers pirate mainly smaller species such as terns, which are longer-distance migrants. Thus, they have later wing molts, starting mainly in October–December on the nonbreeding grounds and completing in February–April.

Among adults of the shorter-distance migrant skuas, which winter in less-hospitable temperate waters, some inner primaries may be molted on the breeding grounds in late summer,[7,8] and the wing molt of the Great Skua may be spread over the winter (August–April). In the transequatorial migrant South Polar Skua, prebasic wing molt occurs relatively rapidly during April–September in pro-

The late wing-molt timing of this South Polar Skua, which still has four unmolted outer primaries, indicates a first-cycle bird; also note the extensively pale legs, a feature of first-cycle skuas. The complete preformative wing molt of South Polar Skuas starts during July–September and finishes sometime during the northern winter (or southern summer, if birds migrate back south across the equator). The late-summer wing molt of young skuas often corresponds in timing with the wing molt of Northern Hemisphere gulls, perhaps increasing the odds of a successful chase by a relatively inexperienced immature skua. *Monterey County, CA, 30 Sept. 2006. Steve N. G. Howell.*

ductive Northern Hemisphere waters during the southern winter. The reciprocal wing-molt timings of these two species reflect their summer breeding schedules in different hemispheres, even though both molt in the Northern Hemisphere.

FIRST-CYCLE MOLTS. In all species of skuas and jaegers, juvenile plumage is generally retained through fall migration and even into winter, when it is replaced by a protracted and complete molt. A partial second molt of some head and body feathers presumably occurs in first-cycle jaegers, and perhaps also in skuas, but this requires confirmation. This is made challenging by factors such as the difficulty in following individual birds at sea, and by protracted molts that span periods of hormonal development. In jaegers, at least, developing hormones may affect both feather pigmentation and even the length and shape of the elongated central tail feathers. Thus, within a single molt, earlier-molted feathers may look more immature-like, later-molted feathers more adultlike!

The complete preformative wing molt of jaegers starts in December–February and completes in April–October (usually by June in the two smaller species). Wing-molt timing in first-cycle birds shows much individual variation, likely linked

Jaegers and skuas molt almost exclusively at sea, away from the breeding grounds, and the long central tail feathers of jaegers continue to grow during northward migration. Thus, the relatively short streamers on this Long-tailed Jaeger may yet attain the full length of an adult, or this might be a subadult bird, perhaps in its third cycle. *Dare County, NC, 27 May 2008. Steve N. G. Howell.*

Like immature large gulls, first- and second-cycle jaegers can be quite similar in appearance. Molt timing can be a useful clue to determining the age of immature Pomarine Jaegers in spring. First-cycle birds usually do not complete their preformative primary molt until July, and sometimes not until October. This individual, with wing molt virtually completed in late May (p10 looks to be not quite full-grown) is thus presumably in its second cycle. As well as its molt timing, note the plain greater underwing coverts (which tend to barred and juvenile-like on first-cycle birds). *Dare County, NC, 22 May 2007. Steve N. G. Howell.*

to migration distance and food resources. For example, in Pomarine Jaegers the complete first-cycle molt usually finishes between July and October and can overlap with the second prebasic wing molt, which starts during August–October. Starting with their second prebasic molt, skuas and jaegers enter the definitive molt cycle.

In the South Polar Skua, the complete preformative wing molt starts in July–September, after migration to the Northern Hemisphere, and presumably completes in November–February, which is broadly similar in timing to the other species of southern skuas.[9] This timing is reciprocal with that of the preformative wing molts in the small jaegers, which are also long-distance migrants. In the Great Skua, which winters mainly in the North Atlantic, the complete preformative wing molt starts in February–April and completes in August–October, thus being much like that of the Pomarine Jaeger, another shorter-distance migrant that shares similar wintering latitudes.

PREALTERNATE MOLTS. Prealternate molts in adult jaegers involve head and body feathers plus at least

Adult Long-tailed Jaegers often start their prebasic molt on fall migration, although they do not usually commence wing molt before reaching the nonbreeding grounds. This individual has shed its elongated central tail feathers and has also started to acquire a dusky chest collar. *Marin County, CA , 7 Sept. 2007. Steve N. G. Howell.*

This immature Pomarine Jaeger still has two old juvenile outer primaries remaining, p7 is almost grown, and p8 appears to be shed. Thus it is in its first cycle, and its preformative wing molt could complete in July, much later than the wing molt of a second-cycle bird. *Dare County, NC, 31 May 2008. Steve N. G. Howell.*

the central pair of tail feathers, and occur mainly during March–May when they may overlap in timing with completion of the prebasic molt. Prealternate molts of head and neck feathers are presumed to occur in adult skuas in late winter or spring, but it is also possible that molt of these feathers at this season represents the last stages of a protracted prebasic molt, in which case the molt strategy of skuas would be complex basic.

Life History versus Ancestry

Complete preformative molts occur in representatives of diverse families that share long-distance (especially transequatorial) migrations, such as Wilson's Storm-Petrel, American Golden-Plover, Pectoral Sandpiper, Sabine's Gull, Common Tern, and Eastern Kingbird. In all of these cases, though, complete preformative molts occur only in some species within a family, or even within a genus. So, based simply on life-history traits, why don't Great Skuas and Brown Skuas (*Catharacta antarctica*) have partial preformative molts like the Herring Gulls and Kelp Gulls (*Larus dominicanus*) alongside which they nest and with which they share similar migration distances?

One possibility may be that skuas and jaegers need the best flight powers possible to give them an edge in aerial chases at sea. Thus, as in other aerial feeders such as swallows and skimmers, the complete preformative wing molts of jaegers and skuas need to be gradual. Without an intervening preformative wing molt, the juvenile outer primaries could become too worn to function properly by the time they would otherwise be replaced, in the second prebasic molt at about 18 to 20 months of age.

A 1999 study of the relationships among skuas and jaegers[10] found that the Pomarine Jaeger is more closely related to the skuas than it is to the more similar-looking small jaegers. It may be that a common ancestor gave rise to two lineages, one producing the small jaegers and the other producing the Pomarine Jaeger and the skuas. So might it be that ancestral, jaeger-like molt traits have been retained in the skuas? Over time, might the complete preformative molts of Great and Brown skuas become partial or incomplete? As is so often the case, thinking about molt strategies can offer more questions than answers.

REFERENCES

1. Cramp and Simmons 1983; **2.** Howell 2004; **3.** Howell 2005; **4.** Howell 2007; **5.** Howell 2008; **6.** Parmelee et al. 1967; **7.** Jiguet 2007; **8.** S. N. G. Howell, pers. obs. for Brown Skua; **9.** Howell 2008; **10.** Andersson 1999.

ALCIDAE
Auklets, Murres, and Puffins
(CAS, SAS, CBS?; 20 species)

Auklets, murres, and puffins make up a fairly small, specialized family of marine birds that breed and range in cold waters of the Northern Hemisphere, with most species found in the North Pacific. They are often collectively termed "alcids," an abbreviation of their family name, and are usually considered an offshoot of the assemblage that includes gulls, terns, and skuas. Alcids are short- to medium-distance migrants that winter at sea, sometimes hundreds of miles from land. They have stocky, streamlined bodies that are well suited for underwater pursuit of their prey, during which they use their relatively small, almost flipperlike wings to "fly" underwater, a little like penguins. Conversely, their flight through the air is labored, and they need to flap fast and constantly to keep their relatively heavy bodies aloft.

The sexes of alcids appear similar in plumage aspect, and in some species there are distinct seasonal and age-related differences in appearance. Adult plumage aspect is attained in 1 to 2 years. Young alcids are downy and semi-precocial, fed by their parents, and usually fledge in 1 to 2 months. In most species the young remain at the nest site until fledging, but some murrelets follow their parents out to sea within a day or two of hatching, and young murres leave their nest ledges before they are fully winged. The age of first breeding in alcids is usually 3 to 6 years, and nest sites range from exposed cliff ledges to burrows dug into soil.

The molt strategies of alcids are not well known, in large part because of the difficulty in following birds at sea over their first year of life. Most if not all species appear to have the complex alternate or simple alternate molt strategy, with prealternate molts being partial. Most molting occurs at sea, away from the breeding grounds, although adult prebasic molts in the small auklets can start during the breeding season.

PREBASIC MOLTS. The alcids represent an interesting case of a family in which many, but not all, species become flightless during their prebasic molt. As a rule, the larger and heavier-bodied species, which have proportionately smaller wings and thus a higher wing-loading (such as puffins and murres), become flightless, while the smaller, lighter-bodied species (such as small auklets) do not.

Synchronous or near-synchronous wing molt is not simply a reflection of the balance between heavy bodies and small wings, however. For example, the heavy-bodied Parakeet Auklet (weighing 250 to 350 grams) undergoes a sequential wing molt, whereas the lighter-bodied Xantus's Murrelet (at 160 to 175 grams) has a synchronous wing molt. And the Dovekie, a very small alcid, undergoes synchronous wing molt. Why would this be?

It has been argued that Dovekies undergo synchronous wing molt in order to molt quickly before their long-distance migration in October,[1] and this may be the reason. If Dovekies were to overlap molt and breeding as auklets do, however, then they might have time to complete molt before migration. Perhaps a synchronous wing molt is simply the best fit with the seasonal distribution and abundance of food resources. In fall, Dovekies move to areas with pack ice, which is a very food-rich habitat that can fuel an intense, short-duration molt.[2] Another explanation may lie with ancestry: the Parakeet Auklet shares a common ancestor with the smaller auklets, all of which undergo sequential primary molt, whereas the Dovekie is related to a group of larger alcids,[3,4,5] all of which undergo synchronous wing molt.

Although traditionally considered "synchronous," the prebasic wing molt of many alcids may have an underlying sequence in which the middle primaries molt before the inner and outer primaries.[6,7] Why this should be, nobody knows. Perhaps the paddle-shaped surface area produced by regrowing in this sequence is a more efficient wing shape for underwater propulsion than would result from regrowth in a standard sequential order. A study of

Dovekies are typical of most alcids in having a distinct alternate plumage, but they are unusual among the smaller alcids in having a synchronous wing molt. *Spitsbergen, Norway, 22 June 2008. Lyann A. Comrack.*

In so-called synchronous wing molt, not all feathers are necessarily shed at exactly the same time, although they are usually shed within a day or two. In some alcids, such as this adult Tufted Puffin, the middle primaries can be shed before the outermost and innermost ones. *Marin County, CA, 24 Sept. 2006. Steve N. G. Howell.*

captive Tufted Puffins found that some birds shed their primaries sequentially whereas others started at a midpoint in the primaries; these captive birds also took about 2 weeks to shed all of their primaries.[8] Although interesting, the Tufted Puffin study is compromised by its use of captive birds, and corroboration from wild birds would be valuable. How long it takes to grow new flight feathers may vary with interannual variation in food resources. For example, in good food years the wing molt of Common Murres may take only 3 to 4 weeks, but in poor food years it may last as long as 10 weeks.[9]

The prebasic molt of most alcids occurs after the breeding season, at sea. An exception to this pattern occurs among the auklets, which start their molt when nesting. They may interrupt wing molt while feeding young and then complete it after the young have fledged. In Cassin's Auklet, the only alcid known to sometimes raise two broods in a season, molt may overlap extensively with breeding, and birds can adjust their molt schedule according to the energy demands of breeding in an unpredictable marine environment.[10]

Another exception to the general alcid pattern is found in Atlantic and Horned puffins, adults of which undergo their wing molt in spring, during February–April. The conventional interpretation of this strategy has been that the prebasic molt is interrupted over the winter and completed in spring. By tracing molts starting with the first cycle, however, it appears that the wings are molted *before*, not after, the breeding season.[11] This situation mirrors that of the larger loons and shows how different groups of birds manage to fit molt and breeding into the short northern summer.

FIRST-CYCLE MOLTS. All alcids have one or two partial molts added into their first cycle, but the difficulty lies in determining whether there are one or two—not easy with birds that spend most of their life at sea. Puffins appear to keep their fairly strong juvenile plumage through the winter and then undergo a limited first prealternate molt in spring, before their complete second prebasic molt in summer; thus they have the simple alternate molt strategy.[12]

The Dovekie, Razorbill, and murres, which form a related group, have a preformative molt of head and body feathers in fall and a partial prealternate molt at some point from winter to spring, and thus follow the complex alternate strategy. Young murres and Razorbills leave their cliff nests before their wings are grown, so they grow their juvenile flight feathers concurrent with their preformative molt of head and body feathers. The guillemots and Marbled Murrelet, which may all be fairly closely related, also appear to exhibit the complex alternate strategy. Among the murres, guillemots, and allies, the first alternate plumage can look similar to a typical adult basic plumage or alternate plumage, or may fall at some point in between, presumably reflecting individual hormone levels at the time of molt.

Among the auklets and other murrelets, it is harder to see what is going on, in large part because relatively few birds have been collected in their first winter. The Rhinoceros Auklet may have a single molt and thus exhibit the simple alternate strategy like puffins, whereas Cassin's Auklet follows the complex alternate strategy.[13] Data for the other small auklets and murrelets are insufficient for any conclusions to be reached.

PREALTERNATE MOLTS. All alcids, except perhaps the Whiskered Auklet,[14] apparently have a partial prealternate molt in their definitive cycles, which in most species is limited to feathers of the head and body and occurs from mid- or late winter into spring. In some southerly-breeding species exposed to relatively sunny environments, such as Cassin's Auklet, the prealternate molt can be extensive and include upperwing coverts; conversely, in northerly-breeding species such as Parakeet and Crested auklets, the prealternate molt may be quite limited in extent, mainly involving feathers on the breast and flanks, areas that are important for flotation.[15] It has also been suggested[16] that the ornamental head plumes of auklets are simply basic feathers that grow in slowly, as occurs with the ornamental plumes of many herons. In Atlantic and Horned puffins, the prealternate molt of head and body feathers overlaps in spring with the prebasic wing molt, which effectively produces a complete "prebreeding" molt.[17]

REFERENCES

1. Thompson et al. 1998; **2.** Bradstreet 1982; **3.** Friesen et al. 1996; **4.** Moum et al. 1994; **5.** Strauch 1985; **6.** Sealy 1977; **7.** Thompson et al. 1998; **8.** Thompson and Kitaysky 2004; **9.** Thompson et al. 1998; **10.** Emslie et al. 1990; **11–12.** Howell and Pyle 2005; **13–16.** Pyle in press; **17.** Howell and Pyle 2005.

COLUMBIDAE
Pigeons
(CBS; 11 species)

In North America, this family of familiar birds ranges from the truly familiar Feral Pigeon (or Rock Pigeon) of city parks to the elusive Key West Quail-Dove that occasionally visits the Florida Keys and the newly arrived Eurasian Collared-Dove that is taking North America by storm as its population explodes. Pigeons are a worldwide family of distinctive birds with no obvious close relatives. They are most diverse in Australasia, absent only from cold northern latitudes, and occur in a variety of forested, semiopen, and open habitats. Some northern species are migratory, but most species are ostensibly resident. The sexes look alike or differ slightly in North American species, juveniles of which resemble duller, scaly versions of the adults. Adult plumage aspect is attained in the first year. The young are semi-altricial, hatch with a sparse covering of down, and fledge in 2 to 4 weeks. Pigeons tend to be accomplished breeders that can raise several broods a year when conditions are favorable, and some species start to breed at less than 6 months of age.

As far as is known, the molting of all pigeons follows the complex basic strategy. Prebasic and preformative molts mostly occur between the end of the breeding season and the onset of winter, although migrants can suspend their molt over migration and complete it on the nonbreeding

As in several resident tropical landbirds, opportunistic stepwise molt may develop in the wings of the White-tipped Dove, presumably as a result of unpredictable conditions during the breeding season. *Starr County, TX, 13 May 2007. Larry Sansone.*

Most birders barely grace Feral Pigeons with a second glance, but the next time you go for a walk in the park, take a second look at those resilient birds. They're certainly tame and easy to watch, and they make good subjects on which to practice seeing molt contrasts. Are birds in your park molting year-round, or are there seasonal patterns to their molting? This individual has interrupted its primary molt, with an obvious contrast between the newer and blacker inner primaries and the faded brown middle and outer primaries. Some of the secondaries are also old feathers, as is one greater covert. *San Francisco County, CA, 30 Jan. 2007. Steve N. G. Howell.*

grounds. In addition, some city populations of Feral Pigeons appear to breed and molt year-round. Molt and breeding sometimes overlap, or the protracted complete molt may be interrupted for breeding such that opportunistic stepwise wing molts develop[1] (see the family account for hawks for discussion of this phenomenon).

Preformative molts in pigeons are unusual in being complete, although sometimes a few juvenile middle secondaries and even an outer primary or two may be retained until the second prebasic molt. It is not clear why pigeons should have complete preformative molts. In some species the middle or outer primaries of adults are structurally modified to produce rattling, whirring, or whistling sounds. Does the preformative molt allow birds to acquire specialized primaries that are undeveloped on the juvenile wings and that may be important for breeding displays? Or might the rapid reproduction rates of doves result in poor-quality flight feathers that won't last through a full year of crashing about in foliage and brush?

Among some of the smaller species such as Common Ground-Dove, young birds can be sexually active at only a few months of age, when the preformative wing molt has reached only out to p5 or p6.[2,3] Thus, birds fledged from early spring broods could easily be breeding by late summer, while they are still actively undergoing their preformative molt! For one population of Common Ground-Doves it has even been suggested that birds less than 6 months of age may represent as much as 15 to 21 percent of an annual reproductive population.[4]

REFERENCES

1. Pyle 2006; **2.** Johnston 1962; **3–4.** Passmore 1984.

PSITTACIDAE
Parrots
(CBS, SAS; 20+ species)

Over the past 50 years or so, numerous species of parrots have escaped from captivity and colonized urban and suburban areas of the U.S., especially the subtropical regions of Florida and southern California. This cornucopia of "screaming green" is poor compensation for the extinct Carolina Parakeet, North America's only native breeding parrot, which disappeared 100 years ago courtesy of human persecution. But like it or not, non-native parrots look like they're here to stay. The general lack of interest in parrots shown by birders, however, means we don't know how many species, or even which species, have well-established populations in North America, although it's likely to be in the neighborhood of 20 to 30 species.[1]

Parrots are native to tropical and subtropical regions worldwide. They are a distinctive group of landbirds with no obvious close relatives, although a 2008 study suggested that they share a common ancestor with falcons and songbirds.[2] Some species are resident in their native ranges, while others undergo seasonal migrations related to tracking food resources. The ages and sexes look alike in some species, different in others, and there are no seasonal changes in appearance. Adult plumage aspect is attained in 1 to 2 years. Young parrots are altricial and hatch virtually naked, with most species fledging in 3 to 15 weeks. The age of first breeding is usually 2 to 4 years.

The complex basic molt strategy, with a partial preformative molt, appears to be typical of parrots,[3,4,5] although the Carolina Parakeet appears to have had a simple alternate strategy.[6] As in many tropical birds, though, molt in parrots can be a protracted process, making it difficult to draw lines between different molts. For example, Budgerigars (*Melopsittacus undulatus*), which live in tropical arid regions of Australia, appear to molt almost continually on a cycle of about 8 months, even when breeding, and one wing molt may start before the preceding one has finished.[7] The Eastern Rosella (*Platycercus eximius*) of Australia, however, lives in areas with better-defined seasons, and it undergoes a complete prebasic molt after breeding.[8] If we had data for parrots that occur in the U.S., it would be interesting to see if non-native populations have developed shorter molting periods in response to northern climates.

One aspect of molt in parrots that has been described is their sequence of primary molt, which is

The Mitred Parakeet, a native of South America, is commonly established in a number of cities in the U.S., but like most New World parrots its molt has not been studied critically. *Orange County, CA, 14 Nov. 2007. Bowles/Erickson–amazornia.us.*

Red-crowned Parrots have become established in south Texas as well as in southern California. How might the molt timings of these northern outpost populations compare with those of birds in their native Mexican range? *Orange County, CA, 11 Jan. 2009. Bowles/Erickson–amazornia.us.*

distinctive. It typically starts at p6 or p5 and spreads outward to p10 and inward to p1.[9,10,11,12] The molt waves that spread from p6 are sequential in some species, such as parakeets of the genus *Aratinga*, while in other species, such as parrots of the genus *Amazona*, the primaries (especially the inner ones) may be replaced out of sequence.[13] No explanation for these odd patterns has been proposed, although they may allow the longest and most-exposed middle and outer primaries to be replaced sooner than might occur in a sequential p1–p10 progression. Wing molts can be protracted, and in some cases a second cycle starts before the preceding one has finished, such that divergent stepwise waves are set up among p7–p10 and p5–p1.[14]

It may be a coincidence that both parrots and falcons have divergent molt sequences among their primaries, typically starting with p6 or p5 and with p4, respectively. But if these two families really do share a common ancestor, then might their unusual molt patterns be ancestral residues?

REFERENCES

1. Garrett and Dunning 2001; **2.** Hackett et al. 2008; **3.** Rowley 1988; **4.** Wyndham 1981; **5.** Wyndham et al. 1983; **6.** Snyder and Russell 2002; **7.** Wyndham 1981; **8.** Wyndham et al. 1983; **9.** Stresemann and Stresemann 1966; **10.** Wyndham 1981; **11.** Wyndham et al. 1983; **12.** Rowley 1988; **13–14.** Stresemann and Stresemann 1966.

CUCULIDAE
Cuckoos
(CBS; 6 species)

Cuckoos, like pigeons, are a worldwide family of landbirds with no obvious close relatives. The great diversity within cuckoos, however, has led some authorities to suggest that the family actually comprises five or more families,[1] three of which occur in North America: the "typical" American cuckoos (such as Yellow-billed), the anis, and the roadrunners. Cuckoos occur in forested, scrubby, and sometimes fairly open habitats. The species breeding at higher latitudes are migratory, whereas tropical species are mostly resident or undergo local migrations. Among New World cuckoos, the sexes usually look alike and there are no seasonal changes in appearance. Young cuckoos are altricial and fledge within 3 weeks, with juveniles usually looking slightly to appreciably different from adults. Adult plumage aspect is attained within 1 year, and the age of first breeding is presumed to be 1 year in most species.

North American cuckoos, such as this adult Black-billed Cuckoo, have the complex basic molt strategy, but the sequences in which they molt their flight feathers are seemingly haphazard and poorly known. *St. Louis County, MN, 31 May 2007. Christopher L. Wood.*

It has been suggested that the Greater Roadrunner may undergo a limited prealternate molt. If this proves to be the case, might other cuckoos in the roadrunner subfamily (or family?) have prealternate molts? Or might the prealternate molt simply be an adaptation to living in abrasive scrubby and grassy habitats, as with some sparrows and wrens? *Willacy County, TX, 18 Feb. 2009. Steve N. G. Howell.*

Despite their varied appearances, from the comic roadrunners to the ungainly looking anis, all cuckoos apparently exhibit the complex basic molt strategy. Prebasic and preformative molts are often protracted and in migratory species may be suspended between the breeding and nonbreeding grounds, with molt of flight feathers occurring mainly or wholly on the nonbreeding grounds.

PREBASIC MOLTS. Saying that all cuckoos have the complex basic strategy might imply that they have typical molting patterns. However, while the number of molts does follow a common pattern, the sequence of wing molt in cuckoos seems almost haphazard and does not lend itself to any obvious interpretation of how or why it may have developed. The primary molt of cuckoos does not follow the typical inner-to-outer sequence, but instead starts from one or more points and proceeds forward or backward, typically with jumps across adjacent primaries and with up to four primaries growing at one time! Thus, in the well-studied Common Cuckoo of the Old World, a typical sequence of primary molt may be p7–p9–p4–p1–p2–p5–p10–p8–p3–p6.[2] Might critical studies documenting the exact sequences of primary molt in cuckoos enable some patterns to be discerned? And if sequences were found to vary among different groups of cuckoos, then might molt patterns offer clues as to how groups are related or how molting patterns developed? For example, are the molts of anis more like those of roadrunners than those of the "typical" American cuckoos? For now there does not appear to be enough data to address this intriguing speculation.

PREFORMATIVE MOLTS. Preformative molts are often complete, although not infrequently a few juvenile flight feathers may be retained, particularly among the secondaries and tail feathers.[3] The preformative primary molts apparently proceed in "irregular" sequences similar to those of adults, although detailed studies of most species are lacking.

PREALTERNATE MOLTS. Prealternate molts have not been certainly documented in cuckoos. A limited prealternate molt has been suggested for the Yellow-billed Cuckoo,[4] but may simply be part of the protracted prebasic molt, and some Greater Roadrunners may have a limited prealternate molt involving exposed feathers such as the tertials.[5]

REFERENCES

1. Sibley and Alquist 1990; **2.** Cramp 1985; **3.** Pyle 1995; **4.** Cramp 1985; **5.** Pyle 1997b.

Some authorities suggest that the cuckoos may comprise five or more different families, with the anis being distinct from other North American cuckoos. How the wing-molt patterns of the different groups of cuckoos might relate to each other remains unknown. Is the wing molt of this Groove-billed Ani more like that of a roadrunner than of a Yellow-billed Cuckoo? Or do all cuckoos have the same wing-molt patterns? *Nayarit, Mexico, 12 Jan. 2007. Steve N. G. Howell.*

Like most owls, the Barn Owl spends the daylight hours roosting in shady spots such as amid these palm fronds. Its plumage is thus protected from the damaging effects of sunlight, which may explain why some juvenile primaries can be retained until the third or even fourth prebasic molt. *Imperial County, CA, 19 July 2008. Steve N. G. Howell.*

TYTONIDAE
Barn Owl
(SBS; 1 species)

Typical encounters with barn owls are of a ghostly white image floating over grassland at dusk, or of a heart-stopping hissing shriek in the black of night, making it easy to see how superstitions built up around these nocturnal hunters. The barn owl family occurs in warm regions around the world and is most diverse in Australasia, with the widespread Barn Owl (or Common Barn Owl) being the only species found in North America. Features that set barn owls apart from typical owls (Strigidae) are their heart-shaped facial discs, relatively small eyes, and short, squared tails. Another difference is that young barn owls molt straight from their downy plumage into a juvenile plumage that looks like the adult basic plumage. In contrast, typical owls have a variable and generally weak juvenile plumage and undergo a partial preformative molt into adultlike plumage. Northern populations of barn owls are migratory, but many tropical populations around the world are resident. The sexes look alike, although females average darker, and there are no seasonal changes in appearance. Young fledge in 7 to 8 weeks, and the age of first breeding is 1 year.

Barn Owls have the simple basic molt strategy by virtue of acquiring a strong juvenile plumage that will last through the first year. In populations of temperate northern regions, most molting occurs from late summer into fall, before the onset of cold winter weather.

PREBASIC MOLTS. The wing molt of barn owls follows an interesting pattern, recalling that of falcons and of some typical owls. Primary molt starts in the middle primaries, with p6, from which point it progresses outward to p10 and inward to p1. In temperate-breeding populations, the complete primary molt requires two, and sometimes three, cycles. For example, the second prebasic molt may involve only p6, or sometimes p5–p7; the third prebasic molt then replaces adjacent primaries, including p8–p10 and usually p1–p4, but sometimes p1–p2 are retained for another cycle.[1] Among some tropical populations, the prebasic wing molt may be almost completed in one cycle,[2] although this finding

The uniform-looking flight feathers of this Barn Owl, with the narrow dark bars aligning between feathers, suggest it is a juvenile bird in fresh plumage. *Baja California, Mexico, 10 Sept. 2006. Steve N. G. Howell.*

was based on a study of captive birds, which may differ from wild birds. Complete molts may also occur in wild birds, however, given that more time is available for molt in a tropical climate versus in a colder climate where molt suspends during winter and perhaps also for migration.

But why start with p6? Given that prebasic molts may not always be complete, perhaps replacing the central (and outer) primaries is the best compromise available; these are the longest primaries and presumably the most important for maintaining the power of carefully controlled flight. If the protected innermost primaries are occasionally retained for an extra cycle, it is probably no big deal, but having old and worn outer primaries could compromise efficient hunting.

REFERENCES

1. Pyle 1997c; **2.** Lenton 1984.

STRIGIDAE
Typical Owls
(CBS; 19 species)

Owls are among the most popular of birds in North America, ranging from a stately Great Horned Owl atop a roadside utility pole at dusk, to a tiny elusive Elf Owl peeping from a cactus, to a majestic Snowy Owl hunting among a blizzard of auklets at an Arctic seabird colony. They have a worldwide distribution and occur in virtually all habitats, from deserts and rain forests to tundra. Most owls are birds of the night, although a few North American species, notably the Burrowing Owl and pygmy-owls, are also active in the daytime, as are Arctic-breeding species such as the Snowy Owl. Several northern-breeding owls are short- to medium-distance migrants, and southward incursions of some species are tied to cycles of their prey. When food is scarce in the far north, invasions of "northern owls" are a highlight of winter birding in many northern states.

The sexes of owls generally look alike, with females averaging larger, and there are no appreciable seasonal changes in appearance. Young owls are altricial, hatch with a sparse down, and fledge in 1 to 3 months. Most species look adultlike in their first winter, but Snowy Owls attain their adult plumage aspect in 1 to 2 years or later. The age of first breeding is typically 1 to 3 years.

Like most resident and short-distance migrant landbirds, owls exhibit the complex basic molt strategy. The prebasic and preformative molts generally occur between mid- to late summer and the start of winter. Molts in migratory species usually complete before birds head south, but in the Elf Owl molt can suspend and complete on the nonbreeding grounds.[1]

PREBASIC MOLTS. The prebasic molts of owls vary considerably in extent, which correlates mainly with size and habitat. The smaller species, such as the pygmy-owls, Elf Owl, Burrowing Owl, and screech-owls, typically have complete prebasic molts. Despite its small size, the Flammulated Owl often has an incomplete prebasic molt and retains some secondaries and, rarely, a few inner primaries until its next molt.[2] This may relate to it not having enough time to complete molt before migration, combined with a cool, high-elevation nonbreeding habitat where molt may not be feasible in winter. In the medium-sized Northern Hawk Owl, Short-eared Owl, and Long-eared Owl, prebasic molts can be complete, but often a few secondaries are retained until the next molt.

Molting patterns among the other owls are more complex. The relatively small Northern Saw-whet and Boreal owls rarely have complete prebasic molts. More often they replace their primaries over two or even three cycles; the outer primaries are molted in one year, middle and inner primaries in the alternating year(s). Perhaps the cold environments favored year-round by these two species, in combination with migration, limit the time available for molt. Thus, might prebasic molts be more extensive in resident western populations of saw-whets that experience less extreme conditions? As with all owls, molt might also be expected to vary in extent between years depending on the levels of prey abundance.

Wing molt of the medium-large owls in the genus *Strix* (that is, the Barred, Spotted, and Great Gray owls) has not been studied critically in wild birds. It appears that primary molt rarely may be complete, especially among birds not involved in breeding. However, in most years only some primaries are molted and two cycles, rarely three, are required to renew all of the flight feathers. Primary molt apparently starts among the middle primaries and proceeds both outward and inward.[3,4]

In the large Great Horned and Snowy owls, the primaries are molted over two or three cycles in a pattern similar to that of barn owls. Thus, the second prebasic molt includes up to three primaries among p6–p8, and the third prebasic molt includes up to seven primaries among those not replaced in the second prebasic molt, and so on. Thoughts on why these molting patterns may have developed are offered in the family account for barn owls.

One reason that many owls can take 2 or more years to renew their flight feathers is that the feathers don't get much exposure to sunlight and don't wear out quickly. After all, most owls hunt at night and spend the day in shady roosts. Although the Snowy Owl inhabits open country, it rarely ventures south to latitudes where strong sunlight could damage its feathers; in this respect it resembles other mostly white, high-Arctic birds such as the Ivory

The Burrowing Owl is unusual among North American owls in being mostly diurnal and in having a relatively strong juvenile plumage, as on these two birds. Like other owls, though, the juvenile head and body plumage of Burrowing Owls is soon replaced with a formative plumage that resembles the adult basic plumage; some barred formative feathers can be seen coming in on the lower bird's chest. *Imperial County, CA, 19 July 2008. Steve N. G. Howell.*

and Glaucous gulls. The specialized, soft-edged feathers of owls that are nighttime hunters also may require more time to grow than do "normal" primaries, which might weigh against having time for a complete molt, but this has not been investigated.

Patterns of tail molt in owls also show considerable variation, which again correlates mainly with size.[5,6] The smaller owls, such as the screech-owls, pygmy-owls, and Burrowing Owl, often molt their tail synchronously, with all feathers shed in a day or two. Among larger owls such as Snowy, Short-eared, and the genus *Strix*, tail molt can be synchronous, or it can be gradual with variations on an overall center–outward sequence. Why should synchronous tail molts be so frequent in owls? Perhaps for precision hunting the tail is important as a rudder, and having it in good shape (or not having it at all for a short period) is better than having an unevenly configured rudder over the course of a protracted molt.

PREFORMATIVE MOLTS. Owls have a relatively weak and soft juvenile plumage, and by fall they undergo a partial preformative molt into adultlike plumage; no flight feathers are molted until the second prebasic molt at about a year of age, when adultlike patterns of wing molt start (see above, under Prebasic Molts).

REFERENCES

1. Ligon 1968; **2.** Pyle 1997c; **3.** Cramp 1985; **4.** Pyle 1997c; **5.** Mayr and Mayr 1954; **6.** Cramp 1985.

CAPRIMULGIDAE
Nightjars
(CBS; 9 species)

Epitomizing the word "cryptic," the dead-leaf plumage patterns of nightjars allow them to "hide in plain sight" during the day while they wait for dusk, which is their "dawn." Nightjars and nighthawks (collectively termed nightjars) are nocturnal insect-eating birds that occur worldwide in warm climates. They live in a variety of habitats from forests to open grasslands. Like owls, most nightjars are best identified by their distinctive voices, which have lent themselves to names such as poorwill and Chuck-will's-widow. Nighthawks tend to be more aerial in their feeding, whereas nightjars mainly hunt from perches or from the ground. Nightjars have long rictal bristles, which are specialized, stiff, hairlike feathers that surround the gape and increase the "net" that can be effectively cast to catch flying insects; on nighthawks the rictal bristles are short. The more northerly breeding nightjars are short- to long-distance migrants, and during cold weather some species can even lower their body temperature and enter a state of torpor, or "short-term hibernation."

The sexes mostly look alike except for differences in tail pattern in many species, and there are no seasonal changes in appearance. Adult plumage aspect is typically attained by the second prebasic molt, and the age of first breeding for most species is presumed to be 1 to 2 years. The semi-precocial young are covered with a dense down and can move away from the nest within a day or two of hatching; they first fly at about 3 weeks of age.

Despite their unique nocturnal lifestyle, nightjars exhibit the complex basic molt strategy typical of most landbirds. The prebasic and preformative molts of residents and short-distance migrants occur on or near the breeding grounds, before the onset of winter, whereas the Common Nighthawk, a long-distance migrant, undergoes most of its molting on the nonbreeding grounds in South America. Stepwise patterns of wing molt may occur among some Common Poorwills of the Rocky Mountains,[1] suggesting that short summer nights and cold temperatures impose time constraints on molting.

PREBASIC MOLTS. As in many species of owls, the prebasic molt of nightjars is not always complete, and often some secondaries are not renewed in each cycle. Owls and nightjars are not closely related,[2] so it may be that these incomplete molts relate to their nocturnal lifestyles. Despite the refined adaptations that owls and nightjars have developed, feeding at night may simply be harder than daytime feeding because, well, it's dark. And since nightbirds don't expose themselves to as much sunlight as do "day-

The cryptic plumage patterns of this Pauraque, like that of most nightjars, allow it to pass the daytime in fairly open situations and even to nest on the ground where it will be overlooked easily by all but the most perceptive observer. *Hidalgo County, TX, 15 Feb. 2008. Christopher L. Wood.*

birds," there may be less need to replace some of their relatively protected secondaries.

An interesting study of the Chuck-will's-widow[3] examined how its molt relates to the rest of its annual cycle. Chucks breed early in the summer and have a single brood, then molt in late summer before southward migration. They usually start wing molt while nesting, with the inner six or seven primaries being replaced before molt starts in other areas. At about the time when p8 is shed, the tail and the rictal bristles appear to be shed almost synchronously. Chucks during this point in their life are an enigma, and it appears that they may be effectively flightless while the tail and rictal bristles are growing. Both of these parts of the plumage are important for capturing prey in flight, and Chucks may hunt for a while by hopping on the ground and picking up prey, for which long rictal bristles might actually be a handicap. Birds start to fly again once the tail and rictal bristles have mostly regrown. Chucks usually complete wing molt before migration, but some birds are growing their outer primaries while migrating. Thus, molt and breeding are compressed into the summer period.

A question posed in this study was: why don't Chuck-will's-widows raise two broods in summer and molt on the wintering grounds, as does the European Nightjar (*Caprimulgus europaeus*)? The European Nightjar, like the Common Nighthawk, is a transequatorial migrant whose nonbreeding season is spent in regions with warm weather and abundant food. The reason for the Chuck's molting pattern may simply be a reflection of food constraints and winter climate, which may also explain why all nightjars (indeed, almost all other bird species) in the Chuck's main wintering latitudes molt mainly in late summer and fall. However, a far more convoluted explanation was proposed. It was suggested that the Chuck's relatively small wintering range, combined with fierce competition for winter territories, limits the species' population as a whole, and that producing more young would be pointless—there is nowhere for them to go and they would only compete with adults.

PREFORMATIVE MOLTS. Nightjars have a soft juvenile plumage that is quickly replaced by adultlike plumage via a preformative molt of head and body feathers. In most species, no flight feathers are molted

The Lesser Nighthawk is a short-distance migrant that molts in late summer, whereas the Common Nighthawk is a long-distance migrant that molts on its nonbreeding grounds. This late-summer bird is in obvious wing molt, with old outer primaries, new inner primaries, and a big gap in the middle primaries. Therefore we can identify it safely as a Lesser Nighthawk, and a male, given its bold white wing band, which is here revealed across the full width of p8. *Imperial County, CA, 19 July 2008. Steve N. G. Howell.*

until the second prebasic molt at about a year of age. Common Nighthawks are an exception in that they molt the tail in their first winter on the nonbreeding grounds. Lesser Nighthawks, however, retain their juvenile tail through their first summer.

What might account for this difference? One possible reason may relate to the tail becoming worn by contact with branches and other surfaces on which nighthawks roost. Given that Common Nighthawks molt on the nonbreeding grounds, the juvenile tail (if not molted in the first winter) would need to last about 18 months before being renewed, whereas Lesser Nighthawks molt their juvenile tail in the summer, at about a year of age. The first-winter tail molt also allows male Commons to better signal their sex in the first summer by acquiring a white tail band like that of adult males (something that females and juveniles lack). Unlike Common Nighthawks, juvenile male Lessers have a white tail band similar to that of the adult male.

REFERENCES

1. Rohwer 1971; **2.** Hackett et al. 2008; **3.** Rohwer 1971.

APODIDAE
Swifts
(SBS, CBS; 4 species)

Swifts are supreme aerialists of worldwide distribution and are well named for their fast flight. Most swifts spend their time foraging high overhead and nest in remote caves or hollow trees, so they remain among the least known of all birds. A new species to science was described from Mexico as recently as the 1990s, and the nonbreeding grounds of the Black Swift, a summer resident in western North America, remain a mystery—although its molt strategy suggests an answer. The more northerly breeding species are short- to long-distance

The molts of swifts, such as this Chimney Swift, are poorly known, thanks to the birds' aerial and generally inaccessible (to humans) lifestyle. Not surprisingly for birds that feed in flight, the wing molt of swifts tends to be protracted. *McHenry County, ND, 16 June 2008. Steve N. G. Howell.*

migrants, and even when not migrating some species may range more than 100 miles a day in search of food. The sexes of swifts generally look alike, and there are no seasonal changes in appearance. Juveniles resemble adults, and the age of first breeding is presumed to be 1 year, at least in the small species. The altricial young are naked at hatching but soon grow a fuzzy down to help regulate their temperature. Swifts require 1 to 2 months to fledge, a relatively long period for their size. This may reflect the unpredictable nature of feeding conditions and food abundance faced by adults, and recalls the life-history strategy of storm-petrels, in many ways the marine counterparts to swifts.

Given their extreme lifestyle, it is not surprising that swifts remain a bit of an unknown when it comes to their molt strategies. Nobody has ever seen a Black Swift in wing molt, and a 2003 study of molt in the White-throated Swift[1] contradicted previous assumptions about its molt and documented that it follows the simple basic strategy. The smaller North American swifts apparently have the complex basic molt strategy.

PREBASIC MOLTS. Given the importance of flight in the life of swifts, it is not surprising that their complete prebasic molts are protracted, with 4 to 7 months needed to gradually replace all of the flight feathers. Among many resident and short-distance migrant swifts there is often overlap between molt and breeding, although in some cases wing molt may not start until the young are well grown.[2] Thus, White-throated Swifts undergo their complete molt in summer and fall, before migration. Some Vaux's Swifts may complete their wing molt in summer, but others may continue molting in fall migration, and others may interrupt wing molt and complete it on the nonbreeding grounds.

Among long-distance migrants in general, prebasic molts often occur on the nonbreeding grounds, or they may start on the breeding grounds and suspend to finish on the nonbreeding grounds. This is true of Old World swifts,[3] and given that Black Swifts do not molt in North America, we can predict that they are long-distance migrants, presumably wintering somewhere in South America.

Although the Chimney Swift is a long-distance migrant, its molt and breeding overlap, with some birds completing molt during migration or interrupting it for migration to finish on the nonbreeding grounds. This molt–breeding overlap may be an ancestral trait, since most species in the genus *Chaetura*, to which the Chimney and Vaux's swifts belong, are residents and short-distance migrants. In an evolutionary time frame, Chimney Swifts may not have been breeding in eastern North America for that long, given that the last ice sheets retreated "only" 12,000 to 15,000 years ago, so perhaps their summer molt is an ancestral trait at odds with their present-day life history.

Like swallows, the smaller swifts can migrate while molting, as illustrated by this Vaux's Swift. *Los Angeles County, CA, 15 Sept. 2007. Larry Sansone.*

PREFORMATIVE MOLTS. Partial preformative molts have been reported or been assumed to occur in fall, before migration, in Chimney, Vaux's, and White-throated swifts.[4,5,6] However, a careful study of molt in the White-throated Swift found no evidence of a preformative molt,[7] and it would be interesting to critically revisit the question in Chimney and Vaux's swifts. It has been suggested that Black Swifts have a complete preformative molt in their first winter,[8] but in reality the molt of Black Swift remains a mystery.

REFERENCES

1–2. Marín 2003; **3.** Cramp 1985; **4.** Bull and Collins 1993; **5.** Cink and Collins 2002; **6.** Pyle 1997b; **7.** Marín 2003; **8.** Pyle 1997b.

TROCHILIDAE
Hummingbirds
(CBS, CAS; 16 species)

These dazzling avian jewels, the smallest of all birds, are well known for their incredible powers of flight. As a family, hummingbirds are restricted to the New World and achieve their greatest diversity in the Andes of South America, from Colombia to Peru. They occur in almost all regions where there are flowers, and the more-northerly breeding species are short- to medium-distance migrants. However, if you compare body size with distance traveled, such as from Alaska to southern Mexico, then in proportion to their length hummingbirds are among the longest-distance migrants in the bird world. The males and females look alike in some species and strikingly different in others, while juveniles usually resemble females; there are no seasonal differences in appearance. The altricial young are virtually naked at hatching, and North American species develop their juvenile plumage over about 3 weeks, at which point they fledge. Adult plumage aspect is attained within the first cycle, and the age of first breeding is 1 year, at least in North American species.

The conventional view has long been that North American hummingbirds follow the complex basic molt strategy, like most landbirds. Although some of the larger species probably do this, recent careful observations have shown that Ruby-throated Hummingbirds, and perhaps other small gorgeted hummingbirds (those in the genera *Archilochus* and *Selasphorus,* at least), have an alternate plumage.[1] Moreover, some tropical hummingbirds apparently lack a preformative molt, and birds molt directly from juvenile into second basic (adult) plumage at about one year of age.[2] This last strategy has been noted so far only in hermits, which comprise a subfamily of relatively dull-colored hummingbirds in which the sexes usually look alike. Among migratory species, the prebasic and preformative molts often start on the breeding grounds and complete after migration; prealternate molts occur mainly on the nonbreeding grounds.

PREBASIC MOLTS. Because hummingbirds rely on having their wings in good condition, molt of the primaries is necessarily gradual. Wing molt may occur in pulses when food supplies are rich, then slow down or suspend in times when food is scarce or weather conditions adverse. It has been calculated that individuals of the small North American species require 3 to 7 months for a complete primary molt,[3,4,5] but such estimates have not been based on following individual birds. In a unique study, the molt of two individual Calliope Hummingbirds was tracked over four consecutive winters and a complete replacement of the primaries for these birds required 4 to 6 months.[6] Molt tended to start and end earlier as the birds got older, and interannual variation likely reflected environmental variables such as food availability and weather.

An interesting feature of wing molt in hummingbirds is the sequence in which the primaries are replaced: whereas many birds follow the typical p1 out to p10 sequence, this is modified in hummingbirds to end in the sequence p7–p8–p10–p9.[7,8] This variation on a theme has presumably evolved to maximize flight capabilities: because p9 is generally the biggest and heaviest primary, and because p10 is most heavily worn at the time of molt, a bird following the standard p9–p10 order would be missing its biggest primary and relying on a worn outer primary for support. Reversing the sequence means that a fresh p10 is there to support the loss of p9, and such a seemingly small difference may be important when your life depends on precision flying.

Effectively, the small gorgeted hummingbirds have a complete over-winter molt, but recent observations suggest this may actually be two molts overlapping: the later stages (mainly wing and tail feathers) of a complete prebasic molt (which started on the breeding grounds) and a late-winter or spring prealternate molt of head and body feathers.[9] Thus, while it has usually been considered that in Anna's Hummingbird the male's iridescent crown and gorget feathers are grown at the end of its complete prebasic molt, might this actually be a separate prealternate molt? It has also been noted that the male head molt in Anna's is delayed relative to that of the female, which lacks specialized feathering in this region.[10] This raises the question of whether females might have a less extensive prealternate molt. In contrast, among several tropical species of hummingbirds studied in Costa Rica, the molt of head feathers occurred at the same time in both males and females, even in species in which the males had iridescent crown or gorget patches that the females lacked. This may be attributed to a shorter time between breeding cycles in a less seasonal environment, such that males do not delay their head molt to accommodate seasonal breeding conditions.[11] It may also reflect the lack of a prealternate molt in these species.

PREFORMATIVE MOLTS. The preformative molts of North American hummingbirds are usually complete, at least in the small gorgeted species such as Ruby-throated and Rufous hummingbirds, although sometimes an outer primary or tail feather may be not be replaced,[12] presumably because of food stress or sickness over the winter. In such cases, retained feathers are presumably molted in the spring and summer, which would likely compromise any breeding attempt in a bird's first year.

As with other North American hummingbirds, the young male Allen's Hummingbird migrates south in late summer wearing a femalelike juvenile plumage. Over the winter it undergoes a complete preformative molt (and perhaps a partial prealternate molt), which produces a plumage indistinguishable from that of an adult male. Thus, this male Allen's Hummingbird might be 1 year old or even 5 years old. *Marin County, CA, 15 May 2007. Steve N. G. Howell.*

Among tropical hummingbirds, some of the larger species do not replace flight feathers in the preformative molt, which is thus limited to head and body feathers.[13]

PREALTERNATE MOLTS. One aspect of first-cycle hummingbird molts that has long been something of a puzzle is how many times immature males (and females?) replace some of their throat feathers. In species such as Ruby-throated and Rufous hummingbirds, juvenile males often have few or no iridescent throat feathers in early fall, but over the winter young males tend to sport variable patches or spots of adultlike iridescent color—yet the gorget region is completely molted in late winter to produce a solid, adultlike shield of color. Does the appearance of iridescent feathers in winter represent delayed activation of some juvenile follicles, or might it reflect an additional and limited preformative molt? Well, perhaps neither.

With the recent documentation of an alternate plumage in Ruby-throated Hummingbirds,[14] the acquisition of some colored gorget feathers over the first winter can simply be viewed as part of the (presumably) complete preformative molt, followed in late winter or spring by a prealternate molt of head and body feathers that is homologous to the adult prealternate molt.

In adult Ruby-throated Hummingbirds the prebasic molt starts with head (including male gorget) and body feathers in summer and completes with wing and tail feathers on the nonbreeding grounds, when it can overlap in timing with the prealternate molt of head and body feathers, including the male gorget.[15] Interestingly, this split-molt strategy (head and body first, wings and tail later) is much like that found in long-distance migrants such as Sabine's Gull, jaegers, and several sandpipers, in which the later stages of the preformative and prebasic wing molts overlap in timing with prealternate head and body molts. More than anything though, the fact that a prealternate molt was documented so recently in such a popular and common species shows how much there is to learn and how much can be contributed simply by careful observation.

REFERENCES

1. Dittmann and Cardiff 2009; **2.** Stiles and Wolf 1974; **3.** Williamson 1956; **4.** Baltosser 1995; **5.** Pyle et al. 1997; **6.** Dittman and Demcheck 2006; **7.** Williamson 1956; **8.** Stiles 1995; **9.** Dittmann and Cardiff 2009; **10.** Williamson 1956; **11.** Stiles 1995; **12.** Pyle et al. 1997; **13.** Stiles 1980; **14–15.** Dittmann and Cardiff 2009.

TROGONIDAE
Trogons
(CBS; 2 species)

Trogons are a family of highly colorful birds found in tropical forested and wooded habitats of the Americas, Africa, and Asia. They are a distinctive family of landbirds with no obvious close relatives. Most species are residents, although some populations of the two northernmost New World species are short-distance migrants. The sexes and ages look different, but there are no seasonal changes in appearance. Trogons nest in cavities and the altricial young hatch naked, quickly molting directly into their juvenile plumage without any appreciable downy stage. This naked-to-feathered progression is similar to that of kingfishers and woodpeckers and may be an ancestral character or perhaps simply a function of nesting in cavities where the temperature is fairly constant and down is not needed for protection or temperature regulation. Trogons fledge in about 3 weeks and attain adult plumage aspect with their second prebasic molt; the age of first breeding is likely 1 to 2 years.

Like most resident landbirds of forested habitats, trogons have the complex basic molt strategy, and their primary molt appears to follow the standard p1 out to p10 sequence. Thus, despite their flashy colors and exotic appeal to birders, trogons have the same molt strategy as the crows in your backyard.

Preformative molts are partial in extent, involving the head and body feathers plus a variable number of upperwing coverts and sometimes tertials. It would not be surprising if some individuals also replaced central tail feathers, although this has not been documented. At least in migrant populations, the preformative molt may be suspended over the winter, although most molt occurs in fall.[1]

REFERENCES

1. Pyle 1997b.

Like most resident landbirds of forested habitats, trogons have the complex basic molt strategy. This juvenile Elegant Trogon (left) will soon attain an adultlike plumage aspect, although typically the juvenile tail feathers are kept for the first year of life. The adult male (right) is undergoing wing molt, with the inner primaries having been shed. The details of molt in trogons are little studied, however. For example, whether adults complete their wing molt before migration or interrupt it until reaching the nonbreeding grounds in Mexico is not known. *Santa Cruz County, AZ, 1 Aug. 2006. Bill Schmoker.*

The preformative molt in Belted Kingfisher (here an adult male) varies from absent (in northerly-wintering populations) to incomplete (involving most if not all primaries) in tropical-wintering populations. *Nayarit, Mexico, 13 Jan. 2007. Steve N. G. Howell.*

ALCEDINIDAE
Kingfishers
(CBS; 3 species)

Kingfishers are a distinctive group of birds found worldwide, mainly in warm climates. Their greatest diversity is in Australasia, and only six or seven species occur in the New World. Some authorities[1] consider that kingfishers actually comprise three families and, interestingly, these three families differ slightly in their molting patterns. One group follows the standard p1 out through p10 sequence, whereas the other two start primary molt at different points among the middle and inner primaries.[2,3] Knowing which group of kingfishers is ancestral would offer food for thought on how different molting patterns might have arisen, but as yet there is no agreement on which group came first.

Kingfishers live in varied habitats from rain forests to temperate rocky seacoasts, and all species nest in burrows. The sexes and ages differ to varying degrees, but there are no seasonal changes in appearance. Young kingfishers are altricial and hatch naked, soon molting directly into juvenile plumage, and they fledge in about 4 weeks. Adult plumage aspect is attained within 1 year, and the age of first breeding is usually 1 year.

Despite differences in the sequence of wing molt among different groups of kingfishers, the underlying molt strategy is complex basic. Prebasic and preformative molts in resident populations occur between the end of the breeding season and the onset of winter, but migrants often suspend molt and complete it on the nonbreeding grounds.

PREBASIC MOLTS. New World kingfishers all belong to the subfamily Cerylinae (or family Cerylidae), which has outlying members in Africa and southern Asia. In the familiar Belted Kingfisher, primary molt progresses both outward and inward from its starting point at p7, with a second wave possibly moving outward from p1;[4] molt is assumed to be similar in other New World species, although these have not been studied.

PREFORMATIVE MOLTS. Preformative molts vary in extent depending on wintering latitude,[5] which is a pattern also found in many migratory populations of shorebirds and songbirds. Among northerly-wintering populations of the Belted Kingfisher, some individuals apparently retain all of their juvenile plumage through the first winter and summer and then undergo a complete second prebasic molt. While this effectively equates to the simple basic molt strategy, the assigning of molt strategies is based on the most extensive and typical patterns shown by a species. Because most Belted Kingfish-

The molts and plumage sequences of the Green Kingfisher are not well understood. This bird's mixture of malelike rufous chest feathers overlying a femalelike pattern of two green-spotted chest bands indicates it is a first-cycle male. *Nayarit, Mexico, 10 Jan. 2008. Steve N. G. Howell.*

ers have fairly extensive preformative molts (see below), the species has the complex basic strategy, with the preformative molt varying from absent (after all, zero is a number) to extensive.

Among first-cycle Belted Kingfishers wintering in tropical latitudes, all individuals replace head and body feathers, some replace tail feathers, and a few even replace primaries. The first-cycle primary molt recalls an eccentric sequence (see page 35), but given the sequence of prebasic primary molt, the first-cycle molt may simply be an arrested molt in typical adult sequence. Anywhere from p7–p8 to p6–10 are replaced, and exceptionally p2–p10.[6] An eccentric pattern of primary molt has been reported in first-cycle Green Kingfishers,[7] but if the molt patterns of this species mirror those of the Belted Kingfisher (as noted above) then the first-cycle molt may not be truly eccentric.

REFERENCES

1. Sibley and Alquist 1990; **2.** Cramp 1985; **3.** Woodall 2001; **4–6.** Cramp 1985; **7.** Pyle 1997b.

PICIDAE
Woodpeckers
(CBS; 22 species)

Woodpeckers are a distinctive and widespread family of landbirds (absent only from Australasia) that appear to be not especially closely related to other groups of birds. They occur in forested, wooded, and fairly open habitats, and all species nest in cavities. Some species are short- to medium-distance migrants, whereas others are among the most sedentary and hardy of birds, remaining in cold northern forests throughout the winter. The sexes and ages of woodpeckers vary somewhat in most species, but there are no seasonal changes in appearance. Adult plumage aspect is attained by the preformative molt. The altricial young hatch naked and molt directly into their juvenile plumage, a feature shared with several other cavity-nesting birds such as trogons and kingfishers. Woodpeckers fledge in 3 to 4 weeks, and most species first breed at 1 year of age.

As with most resident and short-distance migrant landbirds, the molting of woodpeckers follows the complex basic strategy. Among resident species and populations, the prebasic and preformative molts occur between the end of the breeding season and the onset of winter, but in migrants the molts may be suspended and complete on the wintering grounds. Primary molts follow the standard p1 out through p10 sequence typical of many birds. The molt of tail feathers, however, follows an uncommon sequence and usually ends with the central pair. The long, stiffened central tail feathers of woodpeckers are important as a prop for climbing, and when the time for molt comes, all of the tail feathers are at their most worn. This novel sequence provides new outer tail feathers for support when the central feathers are shed and recalls the strategy of hummingbird primary molt, where a new outermost primary (p10) is grown to support growth of the large and heavy p9. When your life depends on having optimal feather performance, seemingly small differences appear to be important.

PREFORMATIVE MOLTS. Preformative molts in woodpeckers are highly variable in extent and to a lesser degree in timing.[1] Most North American woodpeckers replace their primaries and tail feathers in the preformative molt, usually from summer into fall. Primary molt can even begin in the nest, before birds have fledged. Most species do not, however, renew their secondaries other than the relatively exposed tertials. Not surprisingly, the tertials are most often replaced by species living in relatively open and sunny habitats. While having a new, stronger tail may make sense for a bird that climbs trees, it is not known why the juvenile primaries should be replaced right away, in some cases before a bird has even flown.

Perhaps ancestral woodpeckers were vulnerable to nest predation and quickly grew functional but relatively weak juvenile primaries to get out of the nest. In particular, the inner one or two primaries of many juvenile woodpeckers are notably small and rudimentary. Such feathers would not last a year, and thus are replaced in the first cycle. In some woodpeckers, particularly those in temperate climates with short summer seasons, pressure to quickly complete the preformative molt might have pulled forward its timing so that it could even begin in the nest. In support of this conjecture, the preformative primary molt is more protracted in southern and tropical species and populations than in northern species and populations.[2]

Not all woodpeckers replace their juvenile primaries right away, however. The species with the most variable wing molt is the Red-headed Woodpecker; some individuals replace all flight feathers in their first winter (this species is one of the few that molts flight feathers in winter in temperate North America), whereas others retain the juvenile outer primaries and most or all of the secondaries. In Lewis's and Acorn woodpeckers, the juvenile flight feathers are not replaced until the second prebasic molt, when birds are about a year old. Given that these two species tend to occur in fairly open and sunny environments, not in cold shady forests, this strategy is even more surprising, and it has eluded satisfactory explanation. It might be an ancestral trait given that these two species, plus the Red-headed Woodpecker, are all in the genus *Melanerpes*. (The genus *Centurus*, which includes the Red-bellied and Gila woodpeckers, is merged into *Melanerpes* by some authors but appears sufficiently distinct in many ways, not just in molt, to be retained as separate.) Perhaps in part to allow stronger primaries to develop, the fledging period of Lewis's and Acorn woodpeckers is 10 to 15 percent longer than that of other North American woodpeckers, including larger species such as flickers and the Pileated Woodpecker.

In the case of migratory sapsuckers, there is a correlation between migration distance and the timing

The Gila Woodpecker (here a female, lacking the male's red cap) is typical of North American woodpeckers in molting its central tail feathers after the other tail feathers have grown. This sequence maintains the tail's function as a prop while climbing on trunks. *Nayarit, Mexico, 12 Jan. 2007. Steve N. G. Howell.*

The Red-headed Woodpecker has a variable preformative molt. Some birds replace all of their primaries and others retain their outer primaries; some birds acquire adultlike plumage colors by late fall and others retain juvenile head plumage into the winter. This first-cycle individual is growing p4 on each wing, with p5–p10 being juvenile feathers (p10 is a short feather on most woodpeckers); it has also replaced the outer two tail feathers on one side. *Frederick County, VA, 25 Nov. 2006. Steve N. G. Howell.*

of the preformative molt of head and body feathers. Thus, in the resident or short-distance migrant Red-breasted Sapsucker, juvenile head and body plumage is replaced quickly on or near the breeding grounds in late summer, as in most woodpeckers. In the short- to medium-distance migrant Red-naped Sapsucker, this molt occurs from summer into fall. And in the longer-distance migrant Yellow-bellied Sapsucker, the preformative molt of head and body feathers occurs from fall through the winter. Presumably Yellow-bellied Sapsuckers have the time to molt in late summer, but perhaps their resources at that time are better spent molting the primaries and getting into good shape for migration.

Again, the Red-headed Woodpecker stands out as different. The preformative head and body molt in this species occurs from late summer into winter. Thus, for example, in late November some first-cycle birds look adultlike while others alongside them are mostly in juvenile plumage. Are such differences correlated with fledging date, migration distance, interannual variation in food supply, or some combination of these and other factors?

What's So Special about Primary Coverts?

Woodpeckers are one of several groups in which "complete" preformative molts are often not truly complete. For some reason, even though the primaries are replaced in the preformative molt, most or all of the juvenile greater primary coverts are retained. Other birds that retain primary coverts in "complete" preformative molts include New World quail and flycatchers of the genus *Myiarchus,*[3,4] showing that this pattern is taxonomically widespread. Furthermore, most if not all species with eccentric preformative molts, such as the Yellow-breasted Chat and Lark Sparrow, replace primaries but retain most or all of their primary coverts, and in Franklin's Gull the first prealternate wing molt often includes more primaries than primary coverts.[5,6] In woodpeckers, most North American species even retain some to all of their juvenile primary coverts through their second prebasic molt, and primary coverts are sometimes retained in otherwise-complete adult prebasic molts.[7]

This phenomenon raises numerous questions

The preformative molt timing of sapsuckers varies among species and with migration distances. The short-distance migrant Red-breasted Sapsucker, shown here, quickly acquires adultlike plumage by early fall; the medium-distance Red-naped Sapsucker usually acquires adultlike plumage by late fall; and the longest-distance migrant Yellow-bellied Sapsucker may not acquire adultlike plumage until late winter. *Marin County, CA, 24 Nov. 2007. Steve N. G. Howell.*

but has received no satisfactory explanation. What traits might unite the diverse assemblage of birds that retain primary coverts? Are primary coverts disproportionately costly to produce? Are juvenile primary coverts relatively stronger than other feathers replaced in preformative molts? Are they sufficiently protected from the elements that they don't need to be replaced every year? For now, this puzzle remains just one of many in the world of molt.

REFERENCES

1. Pyle and Howell 1995; **2.** Winkler and Christie 2002; **3.** Pyle 1997b; **4.** Pyle 2008; **5.** Pyle 1997b; **6.** Pyle 2008; **7.** Pyle and Howell 1995.

SONGBIRDS

Songbirds, also known as passerines (that is, birds of the order Passeriformes), account for more than half of the living bird species on Earth and more than 40 percent of the species treated in this guide. They are united by several anatomical features and have relatively large brains, which may be related to their learning abilities and their evolution of complex vocalizations. The two main divisions of songbirds are the suboscines (with relatively simple, unmusical vocalizations) and the oscines, or "true" songbirds (with more complex and musical vocalizations). In North America, the only suboscines that occur north of Mexico are the tyrant-flycatchers; all other songbirds "north of the border" are oscines.

North American songbirds range from hulking ravens to tiny kinglets, from long-distance migrants such as the Northern Wheatear to the sedentary Wrentit, and from dazzling buntings to somber-hued wrens. Despite their great diversity in sizes,

OPPOSITE: Unless you're colorblind, the colors of a male Northern Cardinal are hard to miss in the verdant summer woodlands of the East. Like most North American songbirds, this species has the complex basic molt strategy and thus lacks an alternate plumage. *Uvalde County, TX, 16 Apr. 2001. Christopher L. Wood.*

life-history traits, and appearances, all passerines have either the complex basic or complex alternate molt strategy. Adult plumage aspect in most species is attained by the preformative molt, but in some species not until the second prebasic molt (see Delayed Plumage Maturation in the family account for wood-warblers). Young songbirds are altricial, and most hatch with sparse wispy down, fledge in 1 to 4 weeks, and can breed at 1 year of age. The large crows and jays fledge in 4 to 7 weeks and first breed at 2 to 4 years of age. Most of the fastest-fledging songbirds are found among a group known as the nine-primaried oscines, including the wood-warblers, tanagers, sparrows, and buntings. Young of these species can leave the nest within a week or so of hatching but typically wear a weak juvenile plumage that is soon replaced. Other songbirds, including some tyrant-flycatchers and swallows, have relatively strong juvenile plumages that can see them through migration.

A review of molt strategies among North American songbirds reveals some interesting patterns, which are touched on here and in some cases expanded upon in the family accounts that follow. Of the 31 families of North American songbirds, 16 exhibit only the complex basic molt strategy and 6 only the complex alternate strategy; 9 families exhibit both of these strategies (Table 1, pages 26–28). The preponderance of the complex basic molt strategy suggests that the passerine ancestor may have had this strategy and that prealternate molts have evolved independently in different families.

PREBASIC MOLTS. The prebasic molts of all residents and short-distance migrants occur in late summer and fall, between the end of the breeding season and the onset of winter. Most short-distance migrants undergo their prebasic molt on or near the breeding grounds, but a few species interrupt prebasic molt for migration.

Among medium-distance migrants, most species breeding in eastern North America molt before migrating in fall, whereas a fair number of species breeding in western North America molt after migration.[1] This broad-scale difference presumably reflects continental patterns of late-summer food—which is sufficient, even plentiful, in the humid East but relatively scarce in the dry West. Some of these western species migrate directly to their wintering grounds to molt or stage to molt in the Southwest, where late-summer monsoon rains provide a flush of food.[2] Other western species may remain in, or move upslope into, the relatively lush high mountains, which remain productive into fall.

Among the long-distance migrants (those wintering mainly in South America), tyrant-flycatchers and swallows migrate to South America before molting (some replace a few inner primaries before migration), whereas most vireos, thrushes, and wood-warblers molt before migrating. This may reflect feeding habits: tyrant-flycatchers and swallows are highly aerial foragers, and to maintain effective flight their wing molt is necessarily slow, which means it cannot be compressed into the short period available before migration.

PREFORMATIVE MOLTS. All species of North American songbirds undergo preformative molts, which vary in extent from partial to complete. Complete preformative molts are disproportionately common among families with an Old World origin, such as the Horned Lark, Verdin, Bushtit, Wrentit, European Starling, and House Sparrow. In families with a New World origin, complete preformative molts occur in some tyrant-flycatchers, the Phainopepla, Northern Mockingbird, Northern Cardinal, and some blackbirds. Such differences may reflect broad-scale patterns of environmental conditions, such as a preponderance of more open and abrasive habitats in the Old World, or they may simply be coincidences.

Preformative molts tend to be more extensive in birds of southern (versus northern) regions, and in birds of sunnier and more exposed habitats. This pattern crops up repeatedly within families, and even within species that have wide latitudinal ranges. It probably reflects, at least in part, the sunnier southern latitudes (which contribute to feather wear) combined with an earlier breeding season (and thus more time available for molt).

Preformative molts involving some or all flight feathers occur among a wide range of species (in at least 18 of the 31 families), although not in any residents of temperate forest habitats, where there is less need to replace flight feathers. These incomplete wing molts often follow a so-called eccentric sequence, whereby the protected juvenile inner primaries and outer secondaries are not replaced (see page 35). Such molts are not really eccentric but perfectly normal and make good sense. It's simply that people were at one time constrained by thinking that preformative molts were prebasic molts and would follow the standard prebasic sequence, starting with p1. Eccentric molts start at an inner or middle primary (not at p1) and continue out to p10 (or p9, in species with only nine primaries). They likely reflect time constraints on molting, but how a species could "know" at which primary it can start

What does the Yellow-billed Magpie, a localized species endemic to California, have in common with the Northern Cardinal, the quintessential bird of snowy Christmas card scenes in the East? Both exhibit the complex basic molt strategy, as do most North American songbirds. *Solano County, CA, 17 Dec. 2006. Steve N. G. Howell.*

and still have time to finish molt remains a mystery.

Among songbirds that replace some but not routinely all of their primaries in the preformative molt, 46 show eccentric patterns and only 8 show "typical" patterns.[3] Thus, "typical" patterns are atypical, and species reportedly showing them might bear reinvestigation. Do incomplete wing molts in typical sequence simply reflect bad judgment? Do they reflect instances in which a complete molt was compromised by food shortages, disease, or unfavorable weather?

Among resident and migrant songbirds, the preformative molts mostly mirror the prebasic molts in their timing and location. The few exceptions are found mainly in western species in which adults migrate to staging grounds for molting and the young remain on or near the breeding grounds to molt.

PREALTERNATE MOLTS. Fifteen of North America's 31 songbird families include some species that undergo prealternate molts. In many species, the prealternate molts bring about no appreciable change in appearance, as with tyrant-flycatchers, vireos, wrens, nuthatches, and some sparrows and finches. In others, a change in appearance, especially in males, can be brought about by a combination of wear and molt. A striking change in appearance caused by prealternate molt occurs in relatively few species, almost all of which are migratory and most of which breed in eastern and northern North America or have an Old World origin: gnatcatchers (males only), Old World thrushes (males only), some wagtails and pipits, some wood-warblers, some tanagers, a few sparrows and buntings, one icterid (the Bobolink), and one finch (the American Goldfinch).

Prealternate molts in some species are more extensive in first-cycle males than in adults, perhaps simply because adults migrate earlier to the breeding grounds and have less time available for molt; this timing can also allow young males to upgrade their plumage and look more like adults. Prealternate molts also tend to be more prolonged, and thus more extensive, in northern-breeding populations and species. These birds have more time for molt before the onset of their later breeding season relative to southern birds, as occurs in populations of the Orange-crowned Warbler[4] and the Chipping Sparrow.[5]

Resident species of temperate habitats generally lack prealternate molts, perhaps partly because their environment does not cause undue feather wear, and partly because late-winter and early-spring resources may be better invested in simply staying warm and in preparing for the breeding season, which tends to start earlier than it does for migrants.

REFERENCES

1–2. Rohwer et al. 2005; **3.** Pyle 1997a; **4.** Foster 1967; **5.** Willoughby 1991.

TYRANNIDAE
Tyrant-Flycatchers
(CBS, CAS; 35 species)

Although many birdwatchers associate tyrant-flycatchers with identification challenges, the family includes many spectacular and unmistakable species, such as the Scissor-tailed Flycatcher and the Great Kiskadee. Tyrant-flycatchers (often simply called flycatchers) are an exclusively New World family of birds that is highly diverse in the tropics. They are considered "primitive" songbirds (known as suboscines) on the basis of their relatively simple syrinx, which means their songs tend to be less musical to our ears than those of other songbirds. Other New World suboscines (such as antbirds) are also of tropical origin and none occurs north of Mexico, hence all other North American songbirds are considered "advanced" (or oscines).

Flycatchers live in a variety of habitats, and most of the 35 North American breeding species are migratory, withdrawing south for the winter. The brilliant male Vermilion Flycatcher and the handsome male Rose-throated Becard represent the only North American species that exhibit strong sexual dimorphism in appearance as well as delayed plumage maturation (a subject discussed in the family account for wood-warblers).

Despite their "special status" as suboscines, flycatchers have molts that follow much the same patterns as those of other North American songbirds and thus exhibit the complex basic and complex alternate molt strategies. Like all songbirds, flycatchers are vulnerable to nest predation, and their juvenile plumage tends to be a quick fix to get them out of the nest, although in several cases the juvenile plumage is strong enough to endure through fall migration. Prebasic and preformative molts can take place on the breeding or nonbreeding grounds and also can be interrupted between the two areas.

The inconspicuous Northern Beardless Tyrannulet epitomizes many people's idea of a flycatcher—drab with a bit of a crest and some pale wing edgings. The first-cycle molts of this species are not well known, and it has even been suggested that the preformative molt is complete. The outer tail feathers on this individual are growing, either as part of its molt or as an incidental replacement. *Nayarit, Mexico, 12 Jan. 2007. Steve N. G. Howell.*

In cases in which the preformative molt is interrupted, flight feathers are usually molted on the nonbreeding grounds, implying that food can be found more reliably in southern latitudes. Generally, the location of the preformative molt in a species agrees with that of the adult prebasic molt (see below). Prealternate molts occur mainly or entirely on the nonbreeding grounds and produce no distinct changes in appearance.

PREBASIC MOLTS. Flycatchers that undergo their prebasic molt on or near the breeding grounds are mainly residents and short-distance migrants, most of which breed in the Southwest, namely the Greater Pewee, Buff-breasted Flycatcher, phoebes (including Vermilion Flycatcher), Great Kiskadee, and most kingbirds. Phoebes are unusually hardy flycatchers that winter relatively far north, and they need to complete their prebasic molts before temperatures drop, much as do other songbirds that winter in temperate North America. Two other flycatchers, the Acadian and Great Crested, breed mainly in southeastern North America, where late-summer food is sufficient to fuel a complete late-summer molt.

Of the flycatchers breeding in northern and western North America, and for all of the long-distance migrants, the adult prebasic molt usually occurs on the nonbreeding grounds; this is the case for most pewees, most *Empidonax* and *Myiarchus* flycatchers, the Sulphur-bellied Flycatcher, and some kingbirds (e.g., the Eastern Kingbird, which winters in South America). In some cases the prebasic molt starts on the breeding grounds but then interrupts for migration, to be completed when more time and presumably more reliable food are available on the nonbreeding grounds.

An exception to this rule is the Hammond's Flycatcher, a western species that undergoes its prebasic (and preformative) molts in late summer and fall, before migrating south to Mexico. Conversely, Dusky Flycatchers, which often breed (and winter) in the same mountains as Hammond's, conform to the pattern of molting after migration.[1] Why this striking difference? Could it reflect slight variations in food supply and habitat on the breeding grounds? The western mountain forests where Hammond's breed are rich in food in late summer, and birds such as the Orange-crowned Warbler and the Western Tanager even move up from adjacent lowlands into the mountains to take advantage of the food resources and molt. Dusky Flycatchers, however, favor drier and scrubbier habitats, which may offer less food in late summer and fall than do the Mexican nonbreeding grounds.

PREFORMATIVE MOLTS. Preformative molts vary in extent from partial to complete. In all long-distance migrants (wintering in South America) the preformative molt is complete, or almost so, as in the Western Pewee, Eastern Kingbird, Alder Flycatcher, and Sulphur-bellied Flycatcher. Among residents and short-distance migrants this molt varies from partial, as in the Greater Pewee, Thick-billed Kingbird, Hammond's Flycatcher, and Great Kiskadee, to complete, or almost so, as in *Myiarchus* flycatchers and several kingbirds. These variations presumably reflect different balances between ancestry and life-history traits.

The preformative molt of several species involves outer primaries,[2] which in adults of some species (such as kingbirds) are structurally modified for sound production, especially in males. Are preformative molts that include the outer primaries more frequent in male than in female kingbirds? And how does the breeding success of males that acquire modified outer primaries compare with that of males that retain juvenile outer primaries?

PREALTERNATE MOLTS. Prealternate molts occur in some North American flycatchers (especially those of more open habitats and sunnier latitudes) but produce no appreciable change in a bird's appearance. The extent of prealternate molts in flycatchers is not well understood, and accounts of molt for many species in the literature are confusing. Often it seems that little consideration was given to the defining tenet of a molt: how many times per cycle is a feather replaced? Thus we find statements such as "The adult prealternate molt includes . . . those flight feathers not replaced during the adult prebasic molt" (for the Yellow-bellied Flycatcher[3]) or "The adult prebasic molt sometimes is complete, but 1-4 middle secondaries . . . often . . . are retained until the adult prealternate molt" (for the Western Kingbird[4]). In both cases the data can be more simply interpreted as protracted prebasic (or preformative) molts that overlap in timing with prealternate molts, a phenomenon that is frequent among many nonpasserines such as shorebirds and terns.

Prealternate molts in flycatchers are usually partial, involving only head and body feathers, but sometimes they include one or more tertials and rarely may include central tail feathers. Prealternate molts appear to be lacking, or their existence and extent require confirmation, in pewees, phoebes (including Vermilion Flycatchers), Great Kiskadees, and Sulphur-bellied Flycatchers. Among kingbirds, the concealed colored crown patches of Tropical, Cassin's, and Western kingbirds are acquired by the first prealternate molt.[5] Is this true of other kingbirds?

REFERENCES

1. Johnson 1963; **2.** Pyle 1998; **3.** Pyle 1997b:220; **4.** Ibid.:261; **5.** Dickey and Van Rossem 1938.

LANIIDAE
Shrikes
(CAS; 2 species)

Shrikes may be among the canaries in the coal mine we call Earth. They appear to be in decline worldwide for reasons that remain unclear but which may be linked to pollution. Being small predators, shrikes may be more susceptible to the effects of pesticide buildup than are larger predators. These predatory songbirds feed mainly on large insects, lizards, and small mammals and birds, and their hooked bills bring to mind miniature hawks. They are primarily an Old World family, widespread in Eurasia and Africa, with two representatives breeding in the New World, south to Mexico. Shrikes favor open and often fairly scrubby habitats and are most diverse in warm regions. Both North American species are short-distance migrants in which the sexes look alike.

Shrikes exhibit the complex alternate molt strategy, although prealternate molts may not occur in all individuals. Prebasic and preformative molts occur between the end of the breeding season and the onset of winter.[1] Migrants complete their molts before heading south or, at least in some populations of Loggerhead Shrikes, the prebasic molt may be suspended over migration and completed on the wintering grounds.[2] Prealternate molts apparently occur on the wintering grounds but produce no appreciable change in appearance.

PREFORMATIVE MOLTS. Preformative molts in the Loggerhead Shrike, an inhabitant of mid-latitudes, are variable in extent. Especially in more-southerly populations, some tertials, tail feathers, and even primaries can be replaced[3]—all feathers that tend to be relatively exposed and prone to bleaching and wear. It would not be too surprising if further study (such as in the deserts of Baja California) reveals that some birds undergo a complete or near-complete preformative molt. The Northern Shrike, which rarely experiences strong sunlight, molts head and body feathers, wing coverts, and sometimes tertials in the preformative molt, but no flight feathers.

Among Northern Shrikes there is an interesting dichotomy in the appearance of first-cycle birds in their first winter. Some are brownish and distinctly barred below, not too different in appearance from a juvenile, whereas others are grayer overall and less distinctly barred, more adultlike in appearance. Might such a difference, as in male Phainopeplas, be related to the timing of the preformative molt relative to hormone levels that determine color? Or does it reflect the age of the birds, as has been suggested in male Red-winged Blackbirds? Thus, do the browner birds molt earlier and the grayer birds later? Or are the grayer birds simply a little bit older?

PREALTERNATE MOLTS. A limited, facultative prealternate molt of feathers on the head and sometimes the underparts occurs in some Loggerhead Shrikes.[4] As with preformative molts, the prealternate molt might prove to be more extensive in southern populations, and it would not be surprising to find that central tail feathers or tertials are occasionally replaced. Among Northern Shrikes, the prealternate molt may be more extensive in first-cycle birds than in adults, but few data are available.[5]

REFERENCES

1. S. N. G. Howell, pers. obs., *contra* Pyle 1997b for the Northern Shrike; **2.** Pérez and Hobson 2006; **3–4.** Miller 1928; **5.** Miller 1931.

The plumage aspect of first-winter Northern Shrikes varies for reasons that may relate to age and to molt timing. Some birds, like this individual, are brownish-toned and resemble a juvenile, whereas others are gray-toned overall, more like an adult. *Solano County, CA, 17 Dec. 2006. Steve N. G. Howell.*

VIREONIDAE
Vireos
(CBS, CAS; 14 species)

Vireos are an exclusively New World family of small, understated birds that are perhaps best known for their monotonous singing. They inhabit a variety of forested and scrubby habitats and are arboreal foragers that eat mainly insects and fruit (the latter especially in winter). Of the 14 North American breeding species, only 1 (Hutton's) is ostensibly a permanent resident; 10 are medium-distance migrants that winter mainly in Mexico and Central America, and 3 (the Red-eyed Vireo group) are long-distance migrants that winter in South America. The sexes differ appreciably in only one species, the Black-capped Vireo.

The molt strategy of vireos varies from complex basic to complex alternate, although the facultative prealternate molts generally produce no appreciable change in appearance. Prebasic and preformative molts of most species usually occur on or near the breeding grounds in late summer and fall, before migration or winter. Prealternate molts occur mainly on the nonbreeding grounds.

PREBASIC MOLTS. An interesting look at the molts of Warbling Vireos[1] showed that adults of western populations migrate to northwestern Mexico to molt their flight feathers, whereas eastern populations undergo a complete molt before migrating. This dichotomy is similar to that of Bullock's and Baltimore orioles and reflects underlying differences in the distribution of food resources in North America in late summer and fall: the monsoon rains of late summer in the Southwest and adjacent northwestern Mexico result in more abundant food there than can be found in the relatively dry western U.S. and Canada at that time—and western Warbling Vireos, among other birds, take advantage of this. Together with distinct differences in song and habitat, the molt differences between adult western and eastern Warbling Vireos constitute another piece of evidence that they represent two species, and not subspecies as usually classified.

PREFORMATIVE MOLTS. Preformative molts in all species of vireos include head and body feathers, usually wing coverts, and sometimes tertials. They are most extensive in species (and populations) with southerly breeding distributions and in species that live in scrubby rather than arboreal habitats, such as White-eyed, Bell's, and Gray vireos; the first two of these species often replace outer primaries in the preformative molt. Both western and eastern Warbling Vireos undergo their preformative molt prior to migration, perhaps because the relatively weak and lax juvenile plumage of western juveniles would not be optimal for the sustained migratory flight that adults make to their molting grounds.

PREALTERNATE MOLTS. Prealternate molts have been reported in several vireos and usually involve head and body feathers, sometimes tertials, and occasionally central tail feathers. Interestingly, prealternate molts have not been reported in the thicket-loving White-eyed and Bell's vireos, suggesting that exposure to sunlight (more common in arboreal and open habitats) may drive these molts rather than exposure to vegetation.

The Red-eyed Vireo Problem

A study of molt in adult Red-eyed Vireos[2] illustrates some difficulties in determining molt homol-

Adults of the western Warbling Vireo, which realistically represents a species distinct from the eastern Warbling Vireo, migrate to northwestern Mexico to molt their flight feathers, whereas eastern birds undergo a complete molt before migrating. Another difference is that first-cycle eastern birds replace their juvenile greater coverts in the preformative molt, and western birds don't. The relatively worn and somewhat buff-tipped greater coverts on this individual are retained juvenile feathers, which identify it as a first-cycle western Warbling Vireo. *Marin County, CA, 18 Sept. 2008. Steve N. G. Howell.*

ogy and in naming molts of long-distance migrants. In fall, prior to migration, young Red-eyed Vireos have a typical preformative molt, but adults usually molt only head and body feathers, tertials, and sometimes central tail feathers. After arrival on their nonbreeding grounds in South America, adult Red-eyed Vireos undergo an extensive and perhaps complete molt, which may thus replace feathers molted in fall—but data remain equivocal on this point. This strategy may be driven by a protracted breeding season combined with a long-distance migration that requires the tracking of fruit resources.

If the winter and summer molts are exactly complementary and no feathers are replaced more than once, then this represents a prebasic molt that is interrupted for migration. If, however, the winter molt is complete or at least replaces some feathers again, then two molts are involved. The problem then comes down to how one defines a molt, a simple question that has never received a satisfactory answer. Some options for naming the two putative molts of Red-eyed Vireos have been discussed,[3] but the following possibility was overlooked. Could it be that the complete or near-complete winter molt represents simultaneous prebasic (which was interrupted for migration) and prealternate molts? These molts would be effectively inseparable unless one knew the history of every feather follicle and how often it had been activated within a cycle. (Overlapping prebasic and prealternate molts are frequent in some groups of birds, such as gulls, but rarely have been acknowledged in songbirds; see the family accounts for tyrant-flycatchers and finches for other possible examples among songbirds.)

While it may be tempting to view this pattern as a partial prebasic molt and a complete prealternate molt, such an interpretation may not reflect the homology of the molts. Check the Old World Warbler family account for a discussion of molt in the Arctic Warbler, an unrelated long-distance migrant whose molting recalls that of the Red-eyed Vireo. Transequatorial migration, combined with a bird's need to track food resources, clearly throws an interesting twist into the puzzles of homology and naming molts. There must be an answer, but for now it may be unknowable. It would be interesting to learn how the molts of North American Red-eyed Vireos compare with those of ostensibly resident populations of Red-eyed Vireos in northwestern South America. For example, do South American populations have two molts a year or just one? A comparative study of molts in these other populations could shed light on the Red-eyed Vireo problem.

REFERENCES

1. Voelker and Rohwer 1998; **2–3.** Mulvihill and Rimmer 1997.

CORVIDAE
Crows and Jays
(CBS; 18 species)

Crows, jays, and their allies (including magpies, nutcrackers, and ravens) are among the hardiest of songbirds, with some ravens remaining above the Arctic Circle in winter. They are found worldwide and are collectively termed corvids, an abbreviation of their family name. Corvids are often noisy and social and occur mainly in wooded to fairly open habitats throughout the Americas; crows reach their southern limit in Mexico, whereas jays occur all the way south to Argentina. Almost all corvids are residents, and the sexes look alike in New World species.

Despite their varied appearances, from the flashy tropical Green Jay and the striking pied magpies to the hulking Northern (or Common) Raven, which is the largest of all songbirds, the corvids are relatively "boring" and predictable in their molting, and all exhibit the complex basic strategy. Prebasic and preformative molts occur between the end of the breeding season and the onset of winter, and migrants typically complete their molts before heading south.

Clark's Nutcracker has an unusually protracted prebasic molt that overlaps with its breeding season. Why might this be? *Boulder County, CO, 20 Jan. 2008. Bill Schmoker.*

The American Crow is a common species in many areas, but how much do you really look at it? The next time you see a flock of crows in summer, check to see how many are in wing molt. Note how the juveniles have uniformly fresh wings that are not in molt, whereas older ages show different wing-molt timings that may reflect their age or breeding status. *Marin County, CA, 21 Oct. 2008. Steve N. G. Howell.*

Being relatively large and long-lived birds, some corvids do not breed in their first summer and can molt earlier than do breeding adults. Thus, if you watch a flock of crows or ravens in summer and through the fall you may see two schedules of wing molt, with presumed breeding adults molting up to a month later than 1-year-olds (and other nonbreeders or perhaps failed breeders); like most songbirds, juvenile corvids do not molt their wings and tail in the first fall.

PREBASIC MOLTS. Because corvids are relatively large and require more time to grow all their wing feathers than do smaller songbirds, some species (especially those living in colder climates) overlap their breeding and molting. An interesting example of this occurs in the Clark's Nutcracker, which has an unusually protracted prebasic molt (starting in March–April and ending in September–October) combined with early nesting (most eggs are laid in March–April, with young fledging in April–May[1]). Why should its molt take so long? Do cold nights and the need for good insulation constrain the number of feathers that can molt at any time? Does the overlap reflect the nutritional value and seasonality of pine seeds, the nutcracker's dietary staple? And do breeding birds suspend molt during the most energy-demanding phases of breeding?

PREFORMATIVE MOLTS. Preformative molts follow the usual patterns among resident songbirds of temperate regions and are mostly partial in extent, although individuals of some species replace tail feathers (mainly the central pair). The preformative molt is least extensive in northern and cold-climate species, such as the Gray Jay and Clark's Nutcracker, and most extensive in lowland and tropical species, such as the Green Jay (which may even replace some flight feathers[2]). Even within species, such as the Blue Jay, preformative molts may be less extensive in northern than in southern populations, and also less extensive in late-fledged southern birds, which have less time available to molt before winter.[3]

REFERENCES

1. Tomback 1998; **2.** Pyle 1997a; **3.** Tarvin and Woolfenden 1999.

ALAUDIDAE
Larks
(CBS; 1 species)

Larks are an ostensibly Old World family of ground-loving birds, with only one species—the Horned Lark (which is known in Europe as the Shore Lark)—having made its home in North America. At least 25 subspecies of Horned Lark occur in open habitats from the Canadian tundra to Mexican deserts, making this one of the most widespread and geographically variable of all North American birds. Northern populations tend to be migratory, while southern populations are resident. Sexual dimorphism in the family as a whole ranges from absent to striking, and the Horned Lark falls between these extremes.

Despite its vast geographic range, the Horned Lark appears to be consistent in its molt strategy, which is complex basic. Prebasic and preformative molts occur on the breeding grounds, before the onset of winter and before migration.

PREFORMATIVE MOLTS. Preformative molts in larks are complete, unlike the partial preformative molts typical of most songbirds. While this might not be surprising for populations living in the harsh conditions of sunny Mexican deserts, it is also true of tundra-breeding populations. Other songbirds breeding in the far north, such as the Lapland Longspur, have partial preformative molts, so why a complete molt in larks? Nobody knows for sure, but it may be simply an ancestral trait. As a family, larks are most diverse in Africa and the warmer regions of Asia, where complete preformative molts make sense in sunny and scrubby habitats. Thus, while a complete molt may not be "necessary" in high-latitude populations, there has not been enough pressure to change. And if the northernmost populations of Horned Larks can find enough food for a complete preformative molt, why don't longspurs and Snow Buntings do the same? Are there clear-cut ecological differences among these species, or are the differences in molt strategy simply ancestral?

PREALTERNATE MOLTS. A prealternate molt would not be surprising for birds living in such open and often sunny habitats, but only a few cases of prealternate molts in larks have been reported—in three African species and in one Asian population.[1,2]

REFERENCES

1. Willoughby 1971; **2.** De Juana et al. 2004.

While it is not surprising that Horned Larks inhabiting harsh desert environments would have a complete preformative molt, why would the same be true for populations breeding in cold northern habitats? *Boulder County, CO, 18 Feb. 2006. Bill Schmoker.*

HIRUNDINIDAE
Swallows
(CBS; 8 species)

Swallows are sleek songbirds that occur worldwide and spend most of their life gliding and darting through the air to snatch insects on the wing. A reliance on flying insects means that swallows cannot withstand persistent cold weather, when there are no aerial insects. Some species of swallows may be nomadic in the nonbreeding season, while others are perhaps the longest-distance migrant songbirds in the New World. Tree Swallows, however, have supplemented their diet with berries, which enables them to winter relatively far north. The sexes usually look alike but differ in Purple Martin, Violet-green Swallow, and in some immature Tree Swallows.

Like most songbirds, swallows have the complex basic molt strategy, but their need for full-time aerial feeding has greatly affected their molting. Swallows are thus among the few songbirds that regularly molt while migrating, a trait perhaps made easier by being daytime migrants that can forage as they go (most other songbirds migrate at night). Or are swallows daytime migrants *because* they need to feed while molting and migrating? When and where they molt mainly reflects migration distance (see below). The timing of preformative molts is similar to that of prebasic molts, although averaging slightly later.

Despite their small size, swallows take a long time to molt, and it's not difficult to understand why: they need to catch small flying insects and cannot afford to impair their flight any more than is absolutely necessary. The molt duration for a Barn Swallow wintering in warm tropical latitudes has been estimated at 5 to 6 months.[1] For a Tree Swallow wintering farther north, molt may be compressed into 4 months or less,[2] before the onset of winter.

Within North American swallows, the shorter-distance migrants (Tree, Violet-green, Northern Rough-winged, and Cave swallows) mostly complete molt before winter, whereas the longer-distance migrants (Bank, Barn, and Cliff swallows, plus Purple Martin) molt mainly after migration, although some start molt on the summer grounds and interrupt it for migration—it's all a matter of balancing time and food. Among Northern Rough-winged Swallows, eastern birds start molt about a month earlier than do western birds, a difference that may reflect the former's need to have full and fresh wings before they cross the Gulf of Mexico (western birds do not cross such appreciable water barriers).[3]

The Violet-green Swallow is among the minority of North American swallows that show obvious sexual dimorphism in appearance. The female is appreciably drabber than the stunning male, shown here. In juvenile plumage, however, the sexes look similar and the adult plumage aspect is attained via the complete preformative molt. *Mono County, CA, 17 June 2006. Larry Sansone.*

The molt of a Tree Swallow, which winters farther north than other swallows, may be compressed into 4 months or less, before the onset of winter. This first-cycle individual has just started to molt its inner primaries as part of its complete preformative molt. *Monterey County, CA, 24 July 2008. Steve N. G. Howell.*

PREFORMATIVE MOLTS. Swallows are unusual among songbirds in having complete preformative molts. Presumably these are needed because swallows live in exposed and often sunny habitats, in which juvenile flight feathers would become too worn to function after a year or so. Given the necessarily protracted nature of wing molt in swallows, if preformative molts did not include the primaries then it could be more than 18 months before the juvenile outer primaries finally got replaced in the second prebasic molt, late in a bird's second winter. This scenario parallels the complete preformative molts of other aerial-feeding, long-distance migrants such as jaegers and some terns. And as with larks (see the preceding family account), there may also be an ancestral component, particularly if swallows originated in the hotter regions of Africa. Thus it may be that northern populations of the Tree Swallow could survive with incomplete primary molts (retaining some of the inner, protected feathers), but there has not (yet) been sufficient pressure to change an ancestral pattern.

Among Barn Swallows, a shift in molt timing has been noted for some first-cycle birds, with their preformative molts being delayed until spring and summer.[4] This shift corresponds to an increase in wintering Barn Swallows in western North America, which in turn may be linked to warming climate. It is difficult to see what benefit a delayed molt may have, since the birds compromise their first prospective breeding season by undergoing molt instead of breeding. But if such birds migrated south early, could they be responsible for burgeoning breeding populations of Barn Swallows in Argentina?[5] The appearance of these reversed-season populations has mirrored in timing the wintering trend in North America. How birds switch hemispheres to establish new breeding populations at the "opposite" season remains an enigma—for they also have to reverse their molt timing. Do they add an extra molt, or do they wait longer than usual before the next molt?

PREALTERNATE MOLTS. The existence and extent of prealternate molts in swallows need study. Limited facultative prealternate molts may occur, as suggested by the finding that 34 percent of Tree Swallows examined from February to April were molting chin feathers.[6] However, such molting might also represent the last stages of protracted (or suspended) prebasic or preformative molts.

REFERENCES

1. Ginn and Melville 1983; **2.** Stutchbury and Rohwer 1990; **3.** Yuri and Rohwer 1997; **4.** Howell et al. 2006; **5.** Martínez 1983; **6.** Stutchbury and Rohwer 1990.

PARIDAE
Chickadees and Titmice
(CBS; 12 species)

Among the hardiest of small songbirds, chickadees and titmice remain through the winter in bleak and bare northern woods where their excited buzzy calls add a refreshing breath of energy to many a winter's day. Chickadees and titmice are part of a large family of birds found across temperate regions of the Northern Hemisphere, with a distinct offshoot in tropical Africa. The family originated in the Old World (where most species are known simply as tits) and spread into the Americas as far south as the mountains of Mexico. Most species are residents of forested and wooded habitats. The sexes look alike in North American species.

Like most resident songbirds in the temperate regions of North America, chickadees and titmice exhibit the complex basic molt strategy, and their prebasic and preformative molts occur between the end of the breeding season and the onset of winter. Although no prealternate molts have been described for chickadees and titmice, given that these birds often poke into holes and crevices it would not be too surprising if some individuals had a limited molt of feathers on the head and underparts, as occurs in nuthatches.

PREFORMATIVE MOLTS. Spending their lives in relatively cool forested habitats, chickadees are not exposed to much strong sunlight and hence their feathers are not prone to wear. Thus their preformative molts are partial, typically involving head and body feathers and some wing coverts. Some chickadees, however, especially those in more southerly populations, replace tertials and central tail feathers (which act as coverts for the closed wings and tail, respectively). In the titmice, which have a more southerly distribution than chickadees and often live in more open and sunnier habitats, the preformative molt more often includes tertials and some to all tail feathers. Some Black-crested Titmice may replace inner primaries in their preformative molts[1] or may even have complete preformative molts,[2] but data are few and these molt patterns require confirmation.

REFERENCES

1. Pyle 1997a; **2.** Grubb and Pravosudov 1994.

Like most residents of temperate wooded habitats, chickadees such as this Chestnut-backed Chickadee have the complex basic molt strategy. *Marin County, CA, 20 Dec. 2007. Steve N. G. Howell.*

A hardy fluffball inhabitant of southwestern deserts, the Verdin often replaces some outer primaries in its preformative molt, which is presumably an adaptation to the abrasive scrubby vegetation it inhabits. *Pima County, AZ, 13 Dec. 2007. Steve N. G. Howell.*

REMIZIDAE
Verdin
(CBS; 1 species)

This tiny fluffball of southern deserts is a taxonomic oddity: it has often been placed with the chickadees, but now it is usually placed in a family of Old World species that includes the penduline tits. From a birder's point of view, Verdins are hardy residents of scrubby desert habitats in the Southwest and northern Mexico, where their presence is often given away by bulky stick nests built both for nesting and roosting. The sexes look alike and there is no seasonal variation in appearance.

Regardless of its vexed taxonomic status, the Verdin has the complex basic molt strategy typical of resident passerines in temperate regions. Given its harsh scrubby environment, it is not surprising that the Verdin's preformative molt is extensive;[1] only the protected inner primaries and outer secondaries are not molted, a pattern typical of many songbirds living in such habitats. Prebasic and preformative molts occur between the end of the breeding season and the onset of winter.

Interestingly, the preformative molt of Old World penduline tits varies from partial to complete, being partial in northern populations and complete in southern populations that are resident in hotter regions.[2] Thus, it might not be surprising to find that some Verdins (perhaps early-hatched young exposed longer to the summer sun) have complete preformative molts. See the Wrentit family account for another example of a species that has a variable preformative molt—and note the ecological similarities with Verdin and also how shared ancestry may be a factor.

REFERENCES

1. Taylor 1970; **2.** Cramp and Perrins 1993.

AEGITHALIDAE
Bushtit
(CBS; 1 species)

A flock of Bushtits moving fussily through the foliage is the epitome of nervous energy, with birds flitting about, hanging upside down like titmice, picking under leaves and among cobwebs, and then moving on quickly. Although traditionally grouped with chickadees and titmice, the tiny Bushtit is actually related to the long-tailed tits of the Old World, a small family whose members share with the Bushtit a sociable nature, large globular nests woven with moss and cobwebs, thick lax plumage—and molt characteristics. Bushtits are residents of chaparral and oak woodland in western North America, from southwestern British Columbia south to the highlands of Guatemala. Although the sexes look alike in California but different in Mexico, all populations have the same molting patterns.

As in most resident songbirds, the molt of Bushtits follows the complex basic strategy. But unlike most resident songbirds in temperate regions, Bushtits have a complete preformative molt, replacing all of their wing and tail feathers within a month or two of leaving the nest. This complete preformative molt could have been a clue that bushtits were not related to chickadees (see below). Prebasic and preformative molts occur between the end of the breeding season and the onset of winter.

PREFORMATIVE MOLT. Preformative molt in the Bushtit recalls that of the Wrentit, another New World relic of an Old World family. Interestingly, the Bushtit and Wrentit often experience similar environmental factors and even occur side by side in parts of California—so perhaps an abrasive scrubby habitat is part of the reason for their complete preformative molts. But Song Sparrows and Chestnut-backed Chickadees that live alongside Bushtits and Wrentits don't have complete preformative molts, so habitat is unlikely to be the whole reason. A bigger part of the answer may lie in the Bushtit's ancestry. The Old World long-tailed tits also have complete preformative molts, suggesting that this is a shared ancestral trait. The Old World species of long-tailed tits are mainly tropical species (the handsome Long-tailed Tit [*Aegithalos caudatus*] familiar in Europe is another outlier), and be-

The Bushtit is unusual among resident North American songbirds in having a complete preformative molt, which it shares with other species in its family—all of which inhabit the Old World. The brownish crown of this individual indicates it is one of the West Coast (rather than interior) subspecies, and the pale eye marks it as a female. *Marin County, CA, 22 Dec. 2007. Steve N. G. Howell.*

Even though "black-eared" populations of the Bushtit in the mountains of Mexico look quite different from West Coast populations, they share the same molt strategy, which is complex basic. This dark-eyed and fully black-cheeked individual is a male. *Chiapas, Mexico, 2 Mar. 2007. Steve N. G. Howell.*

cause complete preformative molts tend to be more frequent in tropical species, it has been suggested that the family had a tropical origin.[1]

As has been found in Wrentits, might occasional Bushtits have incomplete preformative molts, with the inner primaries and outer secondaries not always being replaced? And if incomplete molts occur, might they be more frequent in the cold highlands of Mexico and Guatemala, where food stress related to climate might be more common than in the milder Mediterranean climate of coastal California?

REFERENCES

1. Cramp and Perrins 1993.

Unlike most songbirds residing in temperate forest habitats, nuthatches, such as this Pygmy Nuthatch, have a limited prealternate molt. Why might this be? *Marin County, CA, 22 Dec. 2007. Steve N. G. Howell.*

SITTIDAE
Nuthatches
(CAS; 4 species)

Nuthatches share many ecological similarities with chickadees, and the two groups often roam together in winter flocks. Also like chickadees, nuthatches as a group originated in the Old World and spread into the Americas, where their southern range limit lies in the mountains of Mexico. New World nuthatches are tied to trees, where they forage on the trunks and branches and roost and nest in cavities. They are residents or short-distance migrants, and two of the North American species exhibit slight sexual dimorphism in crown color.

Given the similarities between chickadees and nuthatches, it would not be surprising if they shared the same molt strategy. Nuthatches, however, have prealternate molts and thus exhibit the complex alternate molt strategy. Prebasic and preformative molts occur between the end of the breeding season and the onset of winter, and migrants mostly complete their molt before heading south. Prealternate molts occur in spring, on the wintering or breeding grounds.

PREFORMATIVE MOLTS. Preformative molts in nuthatches are partial and vary slightly in extent along the lines of a common pattern. Thus, the Brown-headed Nuthatch, a species of sunnier southern habitats, has a more extensive preformative molt than do other New World nuthatches, which inhabit cooler northern and montane regions. In the Old World, some tropical nuthatches even replace flight feathers in the preformative molt.[1]

PREALTERNATE MOLTS. Prealternate molts of variable extent have been found in all four species of North American nuthatches, being most frequent and most extensive in the migratory and pine-loving Red-breasted Nuthatch (where pine sap may damage plumage) and least extensive in the mostly sedentary White-breasted Nuthatch, which favors deciduous trees.[2] These facultative molts are partial in extent and mostly replace feathers on the throat and underparts, which may accumulate extra wear given the tree-hugging lifestyles of nuthatches.

REFERENCES

1. Cramp and Perrins 1993; **2.** Banks 1978.

CERTHIIDAE
Brown Creeper
(CBS; 1 species)

The unobtrusive Brown Creeper blends well with the flaky tree bark it favors for foraging and nesting. More lightly built and less hardy than nuthatches, creepers withdraw from the coldest parts of their range in winter, but they are still ostensibly birds of temperate forests. As with chickadees and nuthatches, this family originated in the Old World (where they are called treecreepers) and spread into the Americas south to the limit of native pines, which today means the mountains of Nicaragua. The sexes look alike, and there is no seasonal variation in appearance.

As a small, largely resident songbird of northern temperate forests, the Brown Creeper has the complex basic molt strategy. Prebasic and preformative molts occur in late summer and fall, before the onset of winter, and migrants complete their molt before heading south. An interesting aspect of creeper molt is that the longest central tail feathers are replaced after the other tail feathers,[1] a sequence opposite that of most songbirds but the same as that of woodpeckers. Why might this be? An explanation is suggested in the family account for woodpeckers.

Preformative molts include head and body feathers but no greater coverts or tertials. The tail feathers, which have stiffened shafts and act as a prop when birds hitch up trunks, have also been reported as being replaced in this molt[2]; this has not been found in the two European species of treecreeper[3] and requires confirmation for the Brown Creeper. An eccentric molt of outer primaries has, however, been reported for some Short-toed Treecreepers (*Certhia brachydactyla*) in southern Europe.[4]

REFERENCES

1. Cramp and Perrins 1993; **2.** Dwight 1900b; **3.** Cramp and Perrins 1993; **4.** Jenni and Winkler 1994.

The Brown Creeper is an unobtrusive inhabitant of temperate forests and woodlands. Like woodpeckers, it replaces its tail feathers in an outward-to-inward sequence, but whether birds replace their tail feathers in the preformative molt remains unclear. *Marin County, CA, 2 Jan. 2008. Steve N. G. Howell.*

TROGLODYTIDAE
Wrens
(CBS, CAS; 9 species)

From tumbling silvery cadences in western redwood forests and bright ringing repetitions shouted from eastern woodlands to garrulous chatters in southwestern deserts, wrens are well known for their varied voices, which juxtapose with their generally somber plumage tones. These accomplished songsters form a widespread and almost exclusively New World family whose range extends from western Alaska to Tierra del Fuego. Most wrens are resident except in North America, where five species withdraw from the colder and more northerly parts of their ranges in winter. The sexes look alike, and there is no seasonal variation in appearance.

Like the majority of resident or mostly resident songbirds in the temperate regions of North America, most wrens exhibit the complex basic molt strategy, but at least two species have the complex alternate strategy. Prebasic and preformative molts occur between the end of the breeding season and the onset of winter or migration. The location of prealternate molts is not well known; they may occur on both the wintering and breeding grounds.

PREFORMATIVE MOLTS. Given that wrens live in habitats ranging from brush-pile tangles to dense reedbeds and sunny deserts, you might predict that their feathers could become more worn than those of species such as chickadees and nuthatches, which live in protected forest environments. And the extent of the preformative molt bears out this prediction. Thus, the preformative molt in North American wrens often includes some to all tail feathers, tertials, and outer primaries. Conforming to a common pattern in songbirds, the preformative molt is usually more extensive among southern and western populations, as in the Carolina, Bewick's, and Marsh wrens.[1] Might some individuals even have complete preformative molts? This would be difficult to show, in practice, but it seems a possibility with at least the Rock and Cactus wrens, two species that live in open and harsh sunny environments.

The Marsh Wren in North America realistically comprises two species, eastern and western. The two apparently differ in the extent of their prealternate molts—western birds, such as this individual, have little if any prealternate molt. In the breeding season, eastern birds often look brighter and more richly colored than do western birds. Is this a difference in fundamental coloration, or might it be a function of fresh alternate feathers in eastern birds versus faded basic feathers in western birds? *Ventura County, CA, 7 Apr. 2007. Larry Sansone.*

No prealternate molt has been reported for the boldly marked Cactus Wren, but such a molt seems possible given the sunny and abrasive nature of its desert habitat. Like many wrens, this conspicuous species builds nests year-round for roosting. *Pima County, AZ, 10 Dec. 2007. Steve N. G. Howell.*

PREALTERNATE MOLTS. Prealternate molts have been reported only in the Marsh and Sedge wrens, and the new alternate feathers look much like the old basic ones in color and pattern. These two species are migratory and live in abrasive habitats associated with water, all factors that can wear plumage and drive the need for an extra molt. A complete prealternate molt has been reported for Marsh Wrens in a Georgia salt marsh,[2] whereas Marsh Wrens in California usually have at most a partial prealternate molt.[3] Some authors report an eccentric, and sometimes complete, preformative or first prealternate molt for the Sedge Wren.[4] Other authors, however, report that both Marsh and Sedge wrens have incomplete prealternate molts, usually including tertials and some to all tail feathers, and that preformative primary molts of Sedge Wrens follow a typical rather than eccentric sequence.[5,6] Clearly, more study is needed to clarify the molts of these two species. And do populations inhabiting salt marshes average more-extensive prealternate molts than those in freshwater marshes? Might there also be differences between eastern and western Marsh Wrens, which realistically constitute separate species? And given that all wrens poke into holes and crevices, might some individuals of other species also have a facultative prealternate molt of head and throat feathers, as has been reported in nuthatches?

REFERENCES

1. Pyle 1997b; **2.** Kale 1966; **3.** Unitt et al. 1996; **4.** Dwight 1900b; **5.** Pyle 1997a; **6.** Pyle 1997b.

CINCLIDAE
American Dipper
(CBS; 1 species)

The American Dipper is the most aquatic of North American songbirds, living along rushing streams and rivers where it often swims and dives for food. The species' scientific name, *mexicanus*, is a clue that several species we think of as native to western North America were actually first discovered and named from Mexico, much of which was explored many years before pioneers reached the western United States. (Other examples include the Canyon Wren and House Finch.) Five species of dippers occur around the world in the temperate regions of Eurasia and the Americas. Dippers are largely resident, the sexes look alike, and there are no seasonal changes in appearance.

To suit their remarkable lifestyle, dippers have dense plumage and a thick layer of underlying down—but their molt strategy appears to be complex basic, as is typical of songbirds resident in temperate regions. Also "as usual" in such cases, the prebasic and preformative molts occur between the end of the breeding season and the onset of winter. A partial prealternate molt has been alluded to,[1] but there is no published evidence for this, and prealternate molts have not been found in the well-studied European Dipper (*Cinclus cinclus*). Preformative molts are partial and involve the head and body feathers and sometimes one or two tertials.

PREBASIC MOLT. One molt modification that sets dippers apart from most songbirds is that the inner five or six primaries tend to be shed synchronously, rendering the birds flightless for a brief period;[2] the longest outer primaries then grow sequentially, as in most birds. Because dippers are heavy-bodied but run well and are semiaquatic, a synchronous molt of the inner primaries may be more efficient than having flight impaired over the course of a prolonged sequential molt. This compromise is similar to that seen in other heavy-bodied aquatic birds such as ducks, which also risk a short period of flightlessness rather than a long period of labored flight.

REFERENCES

1. Pyle 1997b; **2.** Sullivan 1965.

The American Dipper is the most aquatic of North American songbirds, swimming and diving readily for food in rushing water. The broad, blunt-tipped primary coverts of this individual indicate an adult rather than a first-cycle bird. *Wasatch County, UT, 25 June 2008. Steve N. G. Howell.*

In basic plumage the sexes of Blue-gray Gnatcatcher look alike, so this could be either a male or a female. Males acquire a distinct black eyebrow in their prealternate molt, putting them in the minority of North American songbirds that appreciably change their appearance seasonally by molt. *Marin County, CA, 23 Dec. 2007. Steve N. G. Howell.*

POLIOPTILIDAE
Gnatcatchers
(CAS; 4 species)

These active, slender, and long-tailed little birds are common in many parts of North America but are easily overlooked, being fairly plain in coloration and generally soft in voice. Gnatcatchers are also a bit of a taxonomic oddity and have led what can only be described as a tortured taxonomic history.[1] They have often been considered as a New World offshoot of the Old World warblers but are now usually placed in their own family, most closely allied to the wrens. About 15 species of gnatcatchers occur in wooded and scrubby habitats of warm regions from North America south to Argentina. The northernmost species, the Blue-gray Gnatcatcher, is the only one with strong migratory tendencies. The sexes of North American species look different, especially in spring and summer, when males undergo seasonal changes in appearance.

Gnatcatchers, at least those in North America, have the complex alternate molt strategy, although it is not immediately apparent from their life-history traits why this should be the case. Prebasic and preformative molts occur between the end of the breeding season and the onset of winter, and migrants typically complete their molts before heading south. Prealternate molts of migrants occur mostly on the nonbreeding grounds.

PREFORMATIVE MOLTS. Given their sunny and often scrubby environments, it is not surprising that gnatcatchers have preformative molts that can include more than just the head and body feathers—often the tertials and sometimes tail feathers are molted.[2] Individuals of some species living in sunny and scrubby habitats also replace outer primaries, as has been found in the Cuban Gnatcatcher (*Polioptila lembeyei*)[3] and recently in the Black-capped Gnatcatcher.[4]

PREALTERNATE MOLTS. Gnatcatchers are unusual

among North American songbirds in that males have distinct-looking alternate plumages. Unusual? Well, yes, if we look at all songbird families with breeding representatives in North America, in only three others (the colorful New World wood-warblers, the colorful New World grosbeaks, and the mainly Old World wagtails and pipits) are there appreciable proportions of species with distinct-looking alternate plumages; three other families include one or two such species. Although the main change in appearance in gnatcatchers is limited to black feathering on the head, the prealternate molts (which occur in both sexes) also can include tertials and central tail feathers. Interestingly, in tropical species of gnatcatchers the males look the same year-round, even those with black caps. As a rule, prealternate molts appear to be more common in temperate-zone and migratory species than in tropical residents, which raises the question: do tropical gnatcatchers have prealternate molts but not change their appearance? The Cuban Gnatcatcher, at least, has a prealternate molt of head and body feathers[5] and does not change its appearance, but what of other tropical gnatcatchers? Might some have the complex basic molt strategy?

REFERENCES

1. Atwood and Lerman 2006; **2.** Pyle and Unitt 1998; **3.** Pyle et al. 2004; **4.** W. Leitner, unpub. photos/P. Pyle, pers. comm.; **5.** Pyle et al. 2004.

The preformative molt in North American populations of the Blue-gray Gnatcatcher has not been found to include primaries, although the tertials and tail are often replaced. The dull brownish-washed primary coverts and alula of this individual indicate a first-cycle bird (the alula of adults in winter would be darker and fresher, as well as neatly edged with white). The tail and tertials look fresh (the shortest tertial is missing or misarranged) and presumably were replaced in the preformative molt. However, the contrast between the frayed brown tip of p4 (projecting just past the tertial tips) and the fresher and darker p5 suggest that the longest juvenile outer primaries have also been molted in an eccentric pattern (see page 35). This individual is from a resident tropical population, on Cozumel Island in Mexico, in which preformative primary molt may be of regular occurrence. *Quintana Roo, Mexico, 9 Dec. 2007. Steve N. G. Howell.*

Tiny, hyperactive inhabitants of temperate woodlands, kinglets such as this Golden-crowned Kinglet have the complex basic molt strategy typical of other species with similar life-history traits. *Marin County, CA, 30 Dec. 2006. Steve N. G. Howell.*

REGULIDAE
Kinglets
(CBS, CAS?; 2 species)

Among the smallest of North American birds, kinglets breed mainly in temperate coniferous forests and are best known to most of us in fall and winter. At these seasons, the Ruby-crowned in particular can be a common, hyperactive member of mixed-species foraging flocks in all types of woodland. Kinglets (known colloquially as "crests" in the Old World) inhabit temperate regions across Eurasia and North America and are resident or short-distance migrants. They have led a checkered taxonomic history (often having been grouped with Old World warblers) but are now usually treated as their own family, a small one of uncertain affinities. The sexes look alike except for the concealed flame-colored crown patches of males, and there are no seasonal differences in appearance.

Given that kinglets are small, short-distance migrants of mostly enclosed temperate habitats, their molt strategy would be expected to be complex basic, which indeed it is. Prebasic and preformative molts occur between the end of the breeding season and the onset of winter, and migrants typically complete their molts before heading south. Preformative molts include head and body feathers but usually no tertials or tail feathers.

A limited prealternate molt has been reported in the Ruby-crowned Kinglet, but whether this is of regular occurrence or simply an occasional adventitious replacement of a few feathers requires confirmation.[1]

REFERENCES

1. Ingold and Wallace 1994.

SYLVIIDAE
Old World Warblers
(CAS; 1 species)

Only one species of this large and taxonomically vexed Old World family, the Arctic Warbler, has a foothold in North America as a breeding species. And it is not here long, with most of the breeding population arriving in Alaska in June and departing in August. The sexes of Old World warblers look alike or different, but there are no seasonal changes in appearance. The Arctic Warbler apparently has a complex alternate molt strategy, but it poses a conundrum for molt terminology (see below). As in most songbirds, preformative molts occur on or near the breeding grounds and involve head and body feathers and sometimes central tail feathers.

Arctic Warblers are present on their breeding grounds only for the short northern summer and have relatively little time to nest there, let alone molt. Hence, adults undergo almost all of their molting on the nonbreeding grounds, where they spend up to 7 months a year. The same is not true, however, for two other Old World species that nest alongside Arctic Warblers and migrate as far, or even farther, to their nonbreeding grounds: the Bluethroat and Northern Wheatear. These two small thrushes undergo complete prebasic molts in Alaska in late summer, so what traits in their life history, or ancestry, are responsible for this difference? Observations of essentially flightless molting Bluethroats and wheatears[1,2] suggest that terrestrial feeding allows them to have an intense, temporally compressed molt, whereas the arboreal, insect-gleaning Arctic Warbler needs to maintain flight capabilities and does not have this option.

PREBASIC AND PREALTERNATE MOLTS. The molt of adult Arctic Warblers on the breeding grounds in fall involves only head and body feathers, and sometimes tertials and some tail feathers—a pattern typical of prealternate molts. Both adult and first-cycle birds then appear to have a complete molt on the nonbreeding grounds, from January to April.[3]

So, assuming two molts occur, has the prebasic molt really become reduced in extent and the prealternate molt become more extensive? Or do the second stages of an interrupted prebasic molt overlap in timing with a partial prealternate molt to merge as what seems to be a single overwinter molt? That the unrelated Red-eyed Vireo, another long-distance migrant, exhibits a similar molting pattern suggests that life-history traits such as long-distance migration (rather than ancestry) are shaping these patterns, and that the second option (which some might argue is only semantically different from the first) is what we are seeing.

REFERENCES

1. Haukioja 1971; **2.** Williamson 1975; **3.** Cramp 1992.

The Arctic Warbler breeds in taiga woodlands across Eurasia and into western Alaska and winters mainly in tropical Southeast Asia. Its molts remain something of a conundrum. *Denali, AK, 20 June 2004. Christopher L. Wood.*

In fresh fall plumage, the brick red underparts on an American Robin can be veiled with whitish and gray feather tips that wear off by spring. *Cook County, MN, 24 Oct. 2004. Christopher L. Wood.*

TURDIDAE
Thrushes
(CBS, CAS; 15 species)

This large and almost worldwide family (mostly absent from Australia) contains some renowned singers, including the solitaires and *Catharus* thrushes of North America. As in many songbird groups, the taxonomic relationships of thrushes are unresolved. Two main subdivisions often recognized are the "typical" thrushes, which are mainly forest-based, and the Old World chats (not to be confused with the Yellow-breasted Chat), which often occur in more open habitats and which may in fact be more closely related to the Old World flycatchers.[1] Among the chats only two species (the Bluethroat and Northern Wheatear) breed peripherally in the New World, in western Alaska. A third, much smaller, subdivision of thrushes includes the solitaires and bluebirds. Most North American thrushes are migratory, with a few species wintering in South America. The sexes look alike in some species, different in others, and there are no appreciable seasonal changes in appearance except in males of the two chats.

Given their wide geographic range and diverse habitats, thrushes might be expected to exhibit considerable variation in their molting. North American thrushes, however, follow the complex basic strategy typical of many forest-based songbirds, whereas the two chats apparently exhibit the complex alternate strategy. Prebasic and preformative molts occur between the end of the breeding season and the onset of winter, and migrants (even those species wintering in South America and Africa) typically complete their molts before heading south.

PREFORMATIVE MOLTS. Preformative molts usually involve only head and body feathers, and sometimes tertials, but in the bluebirds and solitaires (which spend more time in open country) some tail feathers can be replaced. Southern populations of American

Few North American songbirds are as eye-catching as the sky-blue male Mountain Bluebird. Like other North American thrushes, bluebirds complete their prebasic and preformative molts before winter sets in, when they will need a full coat of feathers to stay warm. *Routt County, CO, 7 Apr. 2008. Christopher L. Wood.*

Two thrushes of Old World origin, the Northern Wheatear, shown here, and the Bluethroat, breed in Alaska and undergo long-distance migrations to tropical Africa and Asia for the winter. Unlike other North American thrushes, these two species reportedly have a partial prealternate molt, although much of their seasonal change in appearance is brought about by wear. *Baja California, Mexico, 24 Oct. 2008. Steve N. G. Howell.*

Robins (early-hatched birds, at least) might be expected to have more-extensive preformative molts than do birds from northern populations, although this seems not to have been investigated.

PREALTERNATE MOLTS. The two Old World chats reportedly show evidence of limited prealternate molts, perhaps resulting from their more open and exposed habitats in combination with long-distance migration, two factors that could contribute to the evolution of a prealternate molt. Many other chats change their appearance seasonally through the wearing away of paler feather tips to reveal bolder underlying patterns, and this cost-effective strategy also occurs to a degree in wheatears and the Bluethroat. Are prealternate molts, which appear to be more extensive in males, enhancing the patterns brought about by wear? And are the birds with more-extensive prealternate molts more successful as breeders, or are these simply the birds with the longest migrations and those that live in the most exposed habitats?

REFERENCES

1. Collar 2005.

TIMALIIDAE
Wrentit
(CBS; 1 species)

The elusive Wrentit, heard far more often than it is seen, has been termed the bouncing-ball ghost of the chaparral, a reference to the male's song, which is given year-round. The Wrentit is also a ghost of continents past: it is the only New World representative of the babbler family, a diverse assemblage widespread in tropical Asia, with some species also found in Africa. Wrentits are resident in scrub habitats from southern Oregon to northern Baja California, Mexico. The sexes look alike and there is no seasonal change in appearance.

The Wrentit is among the most sedentary of North American birds, and despite its unique taxonomic status, its molt follows the complex basic strategy typical of many resident songbirds. Prebasic and preformative molts occur between the end of the breeding season and the onset of winter. Perhaps because they have little need to fly and they know well the ins and outs of their territory, adults get by with a single annual molt, despite heavy abrasion of the wing and tail feathers.

PREFORMATIVE MOLTS. Like other babblers, Wrentits have relatively soft plumage, live in scrubby abrasive habitats that are hard on feathers, and have preformative molts that traditionally have been viewed as complete.[1,2] Young birds have weaker feathers than adults and tend to wander more in their first year, two reasons why they might have a complete preformative molt rather than the partial preformative molt typical of most songbirds. Wrentits also may have a complete preformative molt because that is what their babbler ancestors had and there has been no pressure to change.

A 2000 study has shown, however, that some Wrentits have incomplete preformative molts and can retain a number of their juvenile inner primaries and outer secondaries through the first year.[3] Given that the preformative molt involves whatever a bird needs to molt to cross the bridge from juvenile plumage into the adult cycle, it's not too surprising that these relatively protected wing feathers are sometimes not molted. But why do some and

Heard far more often than it is seen, the Wrentit is a sedentary denizen of poison-oak thickets where pairs remain together year-round. *Marin County, CA, 18 Mar. 2008. Steve N. G. Howell.*

The Wrentit is the only New World representative of the diverse Old World babbler family, members of which typically have a complete preformative molt. Some Wrentits, however, retain some of their juvenile flight feathers through the first year. Why might this be? *Marin County, CA, 30 Dec. 2006. Steve N. G. Howell.*

not all individuals do this? Are these late-hatched young that have relatively fresher feathers than young that fledged months earlier and also have less time for a complete molt before winter? Are they the fittest young that grew the strongest feathers, so they don't need to invest energy in molting? Is the Wrentit diverging from the Old World babblers and evolving toward an incomplete preformative molt? Or might closer study reveal that some Old World babblers also do not always have complete preformative molts?

REFERENCES

1. Cramp and Perrins 1993; **2.** Pyle 1997b; **3.** Flannery and Gardali 2000.

MIMIDAE
Mockingbirds and Thrashers
(CBS, CAS?; 10 species)

The mockingbirds, thrashers, and allies (including the Gray Catbird) are exclusively New World in distribution and are collectively called mimids, an abbreviation of their family name. Many species are well known as accomplished vocal mimics, and the family has its center of diversity in Mexico, the Southwest, and the Caribbean. Most mimids are permanent residents, although some of the more northerly breeding species are short- to medium-distance migrants that withdraw from colder regions in winter. Their habitats range from open deserts to broadleaf woodlands, and their plumages are mostly somberly hued and do not vary in appearance by sex or season.

There is nothing exceptional about the lifestyle of mimids to suggest they would have any unusual molt strategies, and this is indeed the case. Like most resident songbirds, mimids exhibit the complex basic molt strategy. Prebasic and preformative molts usually occur between the end of the breeding season and the onset of winter, although migrants may interrupt their molts for migration and complete them on the nonbreeding grounds. A prealternate molt of some body feathers has been

The Gray Catbird is the longest-distance migrant among the mimids breeding in North America, and it is the only species for which a prealternate molt has been suggested. *Cape May County, NJ, 6 May 2008. Christopher L. Wood.*

The fresh fall plumage of a Sage Thrasher looks much brighter and "sexier" to us than the worn and faded plumage of a breeding summer bird. But for many birds, having feathers in good condition may be more important for migration and for surviving the winter than for showing off in spring—maybe that's why thrashers have such great songs? *Los Angeles County, CA, 1 Nov. 2003. Brian L. Sullivan.*

reported in the Gray Catbird,[1] which is the longest-distance migrant among the mimids, but data concerning this putative molt appear not to have been published.

PREFORMATIVE MOLTS. Preformative molts vary in their extent, and this variation often appears to correlate with habitat: not surprisingly, birds in sunnier and more exposed habitats tend to have more-extensive preformative molts. Thus, species of wetter and more enclosed wooded habitats, such as the Gray Catbird and Brown Thrasher, only occasionally replace tertials and central tail feathers, whereas desert species such as Bendire's, Curve-billed, and Crissal thrashers usually replace tertials, some to all tail feathers, and sometimes outer primaries. How do these and other species know at which primary they can start an eccentric wing molt (see page 35) and still have time to finish it? Is it by subtle differences in daylight length or light intensity? This intriguing puzzle has to date eluded satisfactory explanation.

The preformative molt in the Northern Mockingbird varies from partial to complete,[2] but the primaries are replaced in typical sequence and some individuals apparently replace only one or two inner primaries and one outer secondary. But why replace these protected feathers in preference to more exposed outer primaries? Are such birds "programmed" to attempt a complete molt but then have to arrest it because of inadequate time or sickness? If daylight length is the cue to eccentric molt, is the ability to judge daylight faulty in some birds? Or do these molts occur only in urban populations, where artificial lights might disrupt molt cues?

REFERENCES

1. Cimprich and Moore 1995; **2.** Michener 1953 *contra* Pyle 1997b.

Despite living in relatively open and sunny desert environments, thrashers such as this Curve-billed Thrasher do not have prealternate molts. Yet sparrows, which tend to live in grassy habitats, often have prealternate molts. In these cases it thus seems that abrasion from vegetation may be more damaging to feathers than mid-latitude sunlight. *Pima County, AZ, 26 Jan. 2003. Larry Sansone.*

STURNIDAE
Starlings and Allies
(CBS; 1 species)

Although reviled by many birders as an introduced pest in North America, the European Starling is, to an independent observer, a strikingly beautiful and successful bird with a remarkable vocal repertoire. It is a northern outlier of a large Old World family that includes mynas, another handsome group of accomplished vocalists that have been introduced to many parts of the world. Starlings and mynas have their greatest diversity in tropical Africa and Australasia—which may be a clue to the starling's molt patterns. Starlings inhabit mostly open country, northern populations withdraw south in winter, and the sexes are alike in plumage.

The molt strategy of starlings is complex basic, typical of temperate-zone songbirds that are residents and short-distance migrants. Prebasic and preformative molts occur mainly on or near the breeding grounds, between the end of the breeding season and the onset of winter.

An interesting aspect of a starling's appearance is that the glossy purple and green "breeding plumage" is brought about simply by feather wear. In other words, the breeding plumage is simply a rather worn basic or formative plumage. The pale spotting of fresh plumage wears off over the winter, although these pale tips may have a genetically determined "shelf life" so that most disappear at just the right time. This is a more cost-effective way of changing appearance than undergoing a molt, and it occurs in other species such as the Snow Bunting and House Sparrow.

Heavily into its complete preformative molt, this first-cycle European Starling shows an obvious contrast between the faded plain juvenile feathers (such as on the head) and the incoming formative feathers (such as on the flanks). *Monterey County, CA, 8 Aug. 2007. Brian L. Sullivan*

PREFORMATIVE MOLTS. Preformative molts are typically complete in the European Starling, after which first-year birds look very similar to adults. Given that starlings live a generally unexceptional life, why would they have a complete preformative molt? It doesn't seem that their plumage would be exposed to excessive wear, and they don't live in intensely sunny latitudes. Part of the answer may lie in their ancestry: starlings and mynas may have had a tropical origin, and many species in the family have complete preformative molts (perhaps a reflection of their open habitats and intense tropical sun). Starlings may simply molt the way they do because there has not been enough pressure to change an ancestral trait. If it's not broken, don't fix it.

Avian Lab Rats?

Starlings are seemingly hardy birds, which, with the low esteem they are often afforded, may have led to them being chosen for several laboratory studies that have examined how diet and breeding cycles relate to molt. In one experiment, food deprivation caused a delayed onset of molt after breeding, and the later a bird started to molt, the more quickly it molted.[1] Some of the latest-starting adults even skipped molting one or two inner primaries, which mirrors the eccentric molt patterns of first-cycle songbirds and suggests that eccentric patterns could have developed as a response to time constraints.

In free-living birds, those molting later tend to molt more quickly, which has been assumed to relate to waning daylight hours that signal the coming of winter and a period unsuitable for molt. However, the same study of food deprivation kept starlings under a constant regime of daylight, and yet late-molting birds still molted faster than early-molting ones.[2] This suggests that starlings use some cue other that daylight length to recognize that they are molting "late" and hence molt faster in order to finish molt in a timely manner.

Another study of captive starlings found that food deprivation caused the growth rate of primary feathers to decrease, although the rates at which primaries were shed was similar to that of birds on an unrestricted diet.[3] In contrast, a study of White-crowned Sparrows fed a nutrient-poor diet found that birds shed primaries more slowly but that feathers grew at the normal rate.[4] This suggests that quantity and quality of food may have different effects on the timing and speed of molt.

Yet another lab study on starlings accelerated their molt by reducing daylight hours at an unnatu-

The upperparts of this male European Starling (sexed by the blue bill base, which is pinkish in females) are distinctly spotted. These pale spots will soon wear away to reveal an ever glossier breeding aspect, but no molt is involved in the starling's change of appearance from winter to summer. *Sonoma County, CA, 26 Feb. 2008. Steve N. G. Howell.*

rally fast rate.[5] These birds completed their molt in 73 days (versus 103 days in normal birds), and this quicker molt produced significantly poorer quality feathers. The growth rate of feathers was not measured in this study, so it is unknown to what extent molt was shortened by faster growth rates as opposed to more feathers being grown at any one time. The difference between experimental and normal birds was extreme, and unlikely to be found in free-living birds. Nonetheless, the authors suggested that late-breeding birds with limited time for molt may suffer consequences (such as decreased flight performance and poorer insulation) through having poorer-quality feathers. They considered this the first demonstration of how the "cost of breeding" may be deferred beyond the current breeding season, and thus epitomize the traditional (myopic?) view of many ornithologists that breeding is more important than molt.

REFERENCES

1–2. Meijer 1991; **3.** Swaddle and Witter 1997; **4.** Murphy and King 1984; **5.** Dawson et al. 2000.

MOTACILLIDAE
Pipits and Wagtails
(CAS; 5 species)

Pipits and wagtails compose an Old World family of birds that have "invaded" the Americas on two or more occasions. An earlier wave of pipit colonists reached South America, and the Sprague's Pipit of North America may have been part of that invasion. Another wave involved the American Pipit, which also has breeding populations in Asia. Three other species, including two wagtails, breed in North America only in western Alaska and return in winter to Asia. Pipits and wagtails are birds of open habitats, and all North American species are migratory, moving south in winter away from colder regions. The sexes look alike in pipits but different in wagtails, and several species change their appearance seasonally.

Preformative and prebasic molts occur mainly on the breeding grounds, but sometimes they are interrupted to be completed at staging sites or on the nonbreeding grounds. Preformative head and body molt may even continue during migration, as

Recently, an eccentric pattern of preformative primary molt was documented in Sprague's Pipit, shown here. Might this occur in other New World pipits of open grasslands, a habitat that may cause heavy wear to the outer primaries? *Kern County, CA, 9 Oct. 2008. Larry Sansone.*

First-cycle pipits often show a molt contrast in the upperwing coverts. The uniform-looking coverts of this American Pipit suggest it may be an adult bird, even though it landed on a boat well offshore—something more expected of misoriented young birds in fall migration. *Marin County, CA, 13 Oct. 2006. Steve N. G. Howell.*

in the Sprague's Pipit.[1] Prealternate molts occur on the nonbreeding grounds.

In many ways, pipits recall larks in their open-country lifestyle, and the two groups often associate together, especially in the nonbreeding season. Thus we might predict that these two families would have similar molt strategies. However, pipits and wagtails exhibit the complex alternate strategy, whereas larks have the complex basic strategy.

So why the difference in molt strategies between larks and pipits, given that both families seemingly experience similar ecological conditions? Well, nobody really knows. It may be that these underlying patterns are atavistic traits, ghosts of conditions that prevailed upon the ancestors of larks, which likely evolved in Africa, and of pipits and wagtails, which are believed to have evolved in eastern Asia.[2] Both strategies evidently work for the families concerned, and apparently there has not been enough pressure to modify them under present-day conditions.

PREFORMATIVE MOLTS. Unlike larks, which have complete preformative molts, pipits and wagtails have partial preformative molts, as do most songbirds. Not surprisingly, the preformative molts of these open-country birds can include tertials and sometimes central tail feathers. It has recently been found that Sprague's Pipits can also replace some outer primaries in an eccentric pattern (see page 35).[3]

PREALTERNATE MOLTS. Prealternate molts of both first-cycle birds and adults include head and body feathers, often one or more tertials, and sometimes central tail feathers. Thus, using molt contrasts (see pages 59–64) to determine the age of a pipit or wagtail in spring and summer may not be possible. In some Old World species, however, a prealternate molt may be absent.[4] Larks generally lack prealternate molts.

REFERENCES

1. Robbins and Dale 1999; **2.** Tyler 2004; **3.** Pyle et al. 2008; **4.** Alström and Mild 2003.

BOMBYCILLIDAE
Waxwings
(CBS; 2 species)

Waxwings are a small, Northern Hemisphere family of temperate forest birds. They are known for their sleek and elegantly patterned plumage and their habit of devouring berries in winter parks and gardens, where they may descend like locusts and quickly strip bushes bare of fruit. Waxwings move south in winter but are only short-distance and somewhat nomadic migrants, whose southern winter-range limits reflect the availability of their favored berries. The predominantly fruit-based diet of waxwings fuels them well to grow a soft dense plumage that keeps them warm throughout the year, and also provides carotenoid pigments that color the waxy red appendages to the secondaries, which give the family its name. The sexes look alike, and there is no seasonal change in appearance.

Given their northern temperate origin and forest habitat, it is not surprising that waxwings exhibit the complex basic molt strategy. Preformative and prebasic molts occur mainly on the breeding grounds in the Bohemian Waxwing, which winters in cold northern climes. Cedar Waxwings, however, winter farther south and in warmer climates and often undergo much or all of their molting on the nonbreeding grounds. The preformative molts are partial and include few if any wing coverts and no tertials, a pattern typical of other temperate forest birds such as chickadees.

The tail feathers of waxwings are typically tipped yellow, but starting in the 1960s, increasing numbers of birds with orange-tipped feathers have been noted in parts of the East—almost all of them first-cycle birds. This trend has been linked to increased plantings of ornamental honeysuckles, whose berries contain large amounts of the red pigment found in the orange tail tips. These berries are available mainly in June–July, before adults molt but at a time when young are growing their juvenile plumage in the nest, which helps explain the age bias in orange-tipped birds.[1]

REFERENCES

1. Mulvihill et al. 1992.

Unlike most landbirds of temperate regions, the Cedar Waxwing, shown here, apparently undergoes its prebasic molt mainly after migrating south to the nonbreeding grounds. The more northerly-wintering Bohemian Waxwing molts before migrating south for the winter. *Oaxaca, Mexico, 20 Mar. 2007. Steve N. G. Howell.*

PTILOGONATIDAE
Phainopepla
(CBS; 1 species)

Phainopeplas inhabit the deserts of the Southwest and Mexico, where they perch like satin sentinels marking the locations of fruiting mistletoe clumps. They are the northernmost representative of the silkies (or silky-flycatchers), a small family of birds that has its center of diversity in Central America. The closest relatives of the silkies appear to be the waxwings, and some authors merge these two families into one.[1] Indeed, in many ways the silkies are the southern counterparts of waxwings, being nomads or short-distance migrants that wander in search of berries. Silkies show no seasonal change in appearance, but, unlike in waxwings, the sexes look appreciably different.

Given the similarities between waxwings and silkies, we might predict that the molt strategy of the latter family is also complex basic—and it is. Preformative and prebasic molts occur on the breeding grounds or can be suspended until birds reach the nonbreeding grounds.

PREFORMATIVE MOLTS. Preformative molts in the Phainopepla are much more variable in extent than in waxwings, and sometimes even complete, which presumably reflects a southerly breeding distribu-

The patchy gray-and-black patterns often shown by male Phainopeplas in their first year do not result from retained juvenile feathers but from variation in the appearance of the formative plumage. That said, some of the soft-looking gray plumage on this bird may still be juvenile feathering that has yet to be molted. *Baja California, Mexico, 24 Oct. 2008. Steve N. G. Howell.*

Like satin-black sentinels, male Phainopeplas stake out mistletoe clumps in the desert. The uniformly glossy plumage, bright red eye, and all-dark bill of this individual point to it being an adult, although some first-cycle birds can look similar if they undergo a complete preformative molt. *Pima County, AZ, 10 Dec. 2007. Steve N. G. Howell*

tion and desert habitat. It has also been reported that Phainopeplas have an extra preformative molt (which has been called a presupplemental molt), immediately upon fledging in May–July, whereby the soft juvenile body plumage is replaced, before being replaced again a short time later by a second preformative molt.[2] This phenomenon is discussed further in the family account for cardinals and allies.

An interesting study by Alden Miller[3] on the preformative molt of the Phainopepla raised questions that remain unanswered and provided evidence for the rapid hormonal shift that relates to pigment deposition on incoming feathers (see below). Some Phainopeplas appear to be resident in the Sonoran Desert, where they breed early (with eggs laid mainly in March), whereas others are summer residents that breed later in oak and sycamore canyons north to central California (with eggs laid mainly in late May to mid-June). Miller found that 65 percent of first-cycle birds in the northwestern populations had partial molts (involving no primaries), but for desert populations the proportion was only 25 percent, and some of these might have been migrants from the northwest. He concluded that a longer period available for molting (because of earlier nesting and no migration) was likely the reason for this difference. Nobody seems to know where the northwestern birds winter, but might the proportion of birds with partial molts in desert areas in winter be an index of the number of migrants occurring there? And if some desert breeders do have partial or incomplete preformative molts, are these birds from late nestings?

Two desert birds in Miller's study had incomplete preformative molts, with three to eight inner primaries replaced and the juvenile outer primaries retained. Were these birds that had interrupted molt for migration, or had they truly arrested it?

The patchy gray-and-black patterns often shown by first-cycle male Phainopeplas do not result from retained juvenile feathers but from variation in the appearance of their formative plumage, which never quite achieves the satin black of an adult male. Miller found that early-molted feathers are gray and later-molted feathers black, and that within feathers the basal areas may be black and the tips gray; thus the hormones that determine color can change during the growth of a single feather.

REFERENCES

1. Sibley and Monroe 1990; **2.** Thompson and Leu 1994; **3.** Miller 1933.

PEUCEDRAMIDAE
Olive Warbler
(CBS, CAS?; 1 species)

The taxonomically enigmatic Olive Warbler is an inhabitant of highland pine forests in northern Central America and Mexico, and its range extends into the U.S. only in the mountains of southeast Arizona and adjacent New Mexico. This small insectivorous songbird is largely resident in temperate habitats, although some individuals apparently withdraw south in winter from the northern edge of the range. Olive Warblers exhibit no seasonal change in appearance, but the sexes look different.

Regardless of its unique taxonomic status, the Olive Warbler has molts similar to those of other small nonmigratory songbirds, and it appears to exhibit the complex basic strategy. Study is needed, though, on whether a limited prealternate molt occurs (see below). Preformative and prebasic molts occur in late summer, before winter and the onset of any migration. The preformative molt is partial, at least in North American populations, and includes few if any median and greater wing coverts, no tertials, and no tail feathers.

PREALTERNATE MOLT. It has been suggested that some males, perhaps especially first-cycle birds, may have a limited prealternate molt that includes a few throat feathers,[1] but no data have been provided to support this. If a molt occurs only in first-cycle birds and lacks a counterpart in the adult cycle, then it could represent a second preformative molt. An interesting feature of the formative plumage of male Olive Warblers is that its coloration varies somewhat by latitude, with northern birds resembling adult females in appearance and birds farther south usually resembling adult males.[2] This might lead one to ask: does the putative "first prealternate" molt occur only in northern populations, so that they may more closely resemble adult males? As is so often the case in molt studies, baseline data are lacking to answer simple but intriguing questions.

REFERENCES

1. Pyle 1997b; **2.** Webster 1958.

The enigmatic Olive Warbler floats in taxonomic limbo. Long treated as a slightly odd member of the wood-warbler family, it is now classified in its own family, and its closest relatives remain unclear. The relatively tapered and abraded tail feathers of this individual suggest it is a first-cycle bird. *Jalisco, Mexico, 12 Feb. 2008. Steve N. G. Howell.*

PARULIDAE
Wood-Warblers
(CBS, CAS; 52 species)

"Black-throated Blue down low, Magnolia above it, and a male Blackburnian! Wow!" Trees dripping with bright spring warblers are a sought-after sight from the Texas coast to the shores of the Great Lakes. Usually known simply as "warblers" in North America, this large family of insectivorous birds is popular among many birders, and justifiably so. Wood-warblers have their origins in the New World tropics and are usually associated with five other families (tanagers, sparrows and allies, cardinals and allies, blackbirds and orioles, and finches) that share the characteristic of nine functional primaries (most other songbirds, indeed, most other birds, have ten functional primaries). Relationships among these so-called "nine-primaried oscines" are unresolved, and species are frequently shifted back and forth among families as new studies elucidate relationships.[1,2] North American wood-warblers inhabit forest, woodland, and scrub, and almost all are migratory to some degree. The sexes look alike in some species, and different in others, and several species exhibit seasonal changes in appearance.

Almost all tropical-breeding wood-warblers are resident, and almost all show little or no difference in appearance relating to sex or to season. And even though many birders associate warblers with bright spring males, in only 6 of the 52 North American breeding species are there striking seasonal changes in appearance brought about by molt (in 10 others there are lesser changes that result from limited molts combined with wear). As is often the case, birders have keyed in on the exception rather than the rule.

North American breeding warblers have both complex basic and complex alternate molt strategies. Preformative and prebasic molts of most species occur on the breeding grounds, although in some cases they may be completed during migration or on the nonbreeding grounds. At least some Lucy's Warblers migrate to the wintering grounds to molt,[3] and in West Coast populations of Orange-crowned Warblers there is a late summer exodus of birds from the dry coastal slopes up into the wetter, food-rich mountains to molt.[4] Prealternate molts occur mainly on the nonbreeding grounds but may be continued through spring migration.

PREFORMATIVE MOLTS. The juvenile head and body plumage of most warblers is soft and fluffy. It is replaced quickly in preformative molts that typically also include the wing coverts, and occasionally one or more tertials. More-extensive preformative molts (often including tertials and sometimes tail feathers) occur in some early-breeding western species and populations that inhabit fairly open, sunny, and scrubby habitats, such as Orange-crowned, Nashville, and Virginia's warblers. Lucy's Warblers have eccentric preformative molts that may be complete on some individuals,[5] and in Common Yellowthroats the preformative molt also may include outer primaries.[6] Earlier reports of such molts in yellowthroats, however, were questioned by researchers looking at northern-breeding populations.[7] Are eccentric molts more common in southern populations of Common Yellowthroat, and might they prove more common in populations exposed to the corrosive potential of salt water?

The widespread Yellow-rumped Warbler (here of the western population known as Audubon's Warbler) is one of the few species of wood-warblers that undergo striking seasonal changes in their appearance through molt. *Yolo County, CA, 14 Oct. 2007. Lyann A. Comrack.*

An eccentric preformative molt has been well documented in one "warbler"—the Yellow-breasted Chat[8]—but this enigmatic bird is almost certainly not a wood-warbler, although its affinities remain unresolved.[9] Chats of both eastern and western populations typically replace outer primaries and inner secondaries but apparently at most only the central tail feathers. What aspects of their life history (or ancestry?) might account for this extensive molt?

PREALTERNATE MOLTS. Extensive prealternate molts (involving head and body feathers, wing coverts, and sometimes tertials, but not flight feathers) are limited to several migratory species in the genus *Dendroica*, all of which exhibit sexual dimorphism, such as Magnolia, Blackpoll, and Yellow-rumped warblers. A trend among wood-warblers (also seen in icterids) is that males and females of temperate-breeding species, regardless of whether they have a prealternate molt, are often sexually dimorphic in appearance, whereas the sexes of tropical-breeding species look alike. In birds that are present only a short period on their breeding grounds, having bright plumages to signal sex and perhaps fitness as potential mates may save time and speed up the process of pairing. But how did the bright plumage aspects develop? It may be that some ancestral prealternate molts occurred when prebreeding hormone levels were high. And if these hormones affected feather pigmentation, the new feathers might have grown in with different colors and patterns. Sexual selection took it from there. At least, this has been the conventional way to view the phenomenon, and it may well be what happened. An alternative is that both sexes originally were brightly colored and that dull-colored plumages subsequently developed in females and nonbreeding males (see Which Came First—Dull or Bright? in the family account for blackbirds and orioles, page 237).

The prebasic molt can be intensive in many songbirds, and sometimes the tail feathers are shed synchronously, as in this male Hooded Warbler. *Santa Barbara County, CA, 22 July 2007. Steve N. G. Howell.*

The juvenile head and body plumage of wood-warblers is generally weak and loosely textured, as on this Common Yellowthroat. This "quick fix" feathering allows young to leave the vulnerability of the nest, after which they can acquire a stronger plumage via their preformative molt. *Marin County, CA, 8 June 2008. Steve N. G. Howell.*

Several other warblers have prealternate molts limited to some head and throat feathers, such as the nectar-feeding warblers of the genus *Vermivora*, which may need to replace these feathers from getting them sticky and worn over the winter. A 1967 study found that among Orange-crowned Warblers, northern-breeding populations have more-extensive prealternate molts than do southern-breeding populations (even within the same subspecies), which may reflect less time available to the latter for molt before the start of their earlier nesting season.[10] The same study found that within populations, first-cycle males undergo more-extensive prealternate molts than do adult males, perhaps for the same reason: adult males arrive first on the breeding grounds and thus have less time available for a prealternate molt. This extra time for prealternate molt also enables first-cycle males to acquire full orange crown patches, which are not always acquired in the preformative molt; adult males acquire full crown patches in their prebasic molt.

It might be tempting to look at species such as the Black-throated Blue Warbler and the Hooded Warbler and wonder if they developed year-round sexual dimorphism through prealternate molts that have since been lost. However, this interpretation overlooks the fact that color and molt are controlled independently. Instead, it seems more likely that hormones controlling color are produced in the fall and coincide with the prebasic and preformative molts, as has been suggested for ducks. But why would these warblers want bright male plumages in winter? Well, at least in these two species, males defend winter feeding territories, and since they spend

The American Redstart is one of a small number of North American songbirds in which first-cycle males, like this individual, do not achieve the bright plumage aspect of an adult male. Why this might be has spawned a number of ideas. *Jalisco, Mexico, 24 Jan. 2007. Steve N. G. Howell.*

more of their life on the wintering grounds than on the breeding grounds, having a strong plumage signal may be at least as important for winter as for summer.

Delayed Plumage Maturation (DPM)

The formative plumages of sexually dimorphic warblers typically resemble the adult basic plumages of the respective sex—with one notable exception. For some reason, first-cycle male American Redstarts look femalelike, although they acquire small spots of black (malelike) feathers by the spring.[11] Although these males can breed in their first summer, they do not attain the plumage aspect of adult males until their second prebasic molt at about 1 year of age. This trait is known in academic circles as "delayed plumage maturation," or DPM, although in strict H-P terms it might be more accurately described as delayed *aspect* maturation. DPM occurs in several North American songbirds, from Purple Martins to Red-winged Blackbirds. Several hypotheses and much debate have been spawned about how DPM evolved to serve functions that confer costs and benefits that relate to food and breeding.[12,13,14,15] And are these plumages adaptations to the winter season or to the first breeding season?

While such hypotheses may be intellectually stimulating, almost all overlook one fundamental possibility: DPM may not have evolved to serve social functions in either winter or summer, but instead may simply reflect the fact that some young birds cannot physiologically produce adult colors at the age when molt occurs. Thus the birds may not be "choosing" their plumage coloration with a view to, say, reducing competition on the wintering grounds; they may simply be constrained by their physiology. Subsequent aspects of behavior would thus be secondary consequences, but not the overriding reason for the existence of subadult plumage aspects. One feature common to many songbirds with DPM is that they do not immediately develop the extensive black, bright red, or bright blue coloration of adult males. Might these pigments (or feather structures, in the case of blue) be simply too costly to produce at a young age in these species? This idea, known as the molt constraints hypothesis, has been mooted for breeding-season plumages[16] but not fully explored for the first-winter period. Yet to an observer living outside academia it might seem like the most obvious explanation.

This is not to say that aspects of various DPM theories are invalid, such as young birds honestly signaling their subordinate status—only that there

The preformative molt of wood-warblers doesn't always include all of the wing coverts. Within the greater coverts of this Bay-breasted Warbler, note the contrast between the six inner coverts (which are formative feathers; longer with bolder whitish tips) and the three outer feathers (which are retained juvenile feathers; shorter and more frayed, with narrower whitish tips). This molt contrast points to a first-cycle bird, in which case the pink-washed flanks indicate a male. *Santa Cruz County, CA, 28 Sept. 2008. Steve N. G. Howell.*

might be some confusion between cause and effect. Hypotheses concerning DPM are also difficult to test because directly comparing individuals with and without DPM is usually not possible, since a species either exhibits DPM or it doesn't. A unique opportunity to test the potential costs and benefits of DPM has been provided by the Mute Swan, although in this case the situation might better be called advanced plumage maturation. Most Mute Swans have a brown juvenile plumage and molt into a dusky-gray formative plumage, but a minority have a white juvenile plumage and molt into a white formative plumage, thus appearing older than they "should." This difference is produced by a recessive genetic trait that occurs to differing degrees in different populations.

In one study[17] it was found that dusky-gray young Mute Swans received more parental care and 87 percent survived from hatching to fledging, whereas white young were chased off earlier by their parents and only 73 percent survived to fledging. However, only about 10 percent of the dusky birds nested in their first breeding season at age 2, whereas about 30 percent of white birds nested. So the cost of "advanced plumage maturation" is less chance of surviving until fledging, while the benefit is having a greater chance of nesting at a younger age. The authors concluded that "the status-signaling hypothesis . . . appears to be the only hypothesis that can explain the evolution of subadult plumage in Mute Swans."[18] But did the "subadult plumage" really evolve simply to signal status, or did status signaling result from having a subadult plumage?

Studies such as those debating DPM also tend to assume that the molt patterns we see today are "it" and thus represent a perfect state which must be explainable. Yet if evolution occurs, as many people believe, then presumably molt strategies and plumage patterns today are simply points on a continuum. Might the ancestral plumage aspect for both sexes of American Redstarts have been "female-like," with a bright male plumage aspect developing later? And is DPM in American Redstarts simply a way-point to what we see in Black-throated Blue Warblers, or in Blackpoll Warblers? Thousands of years from now, might male American Redstarts attain adultlike plumage aspect by their preformative molt, or perhaps by an extensive first prealternate molt?

REFERENCES

1. Klicka et al. 2000; **2.** Lovette and Bermingham 2002; **3.** Rohwer et al. 2007; **4.** Greenberg et al. 1974; **5.** Rohwer et al. 2007; **6.** Pyle 1997a; **7.** Ewert and Lanyon 1970; **8.** Phillips 1974; **9.** Lovette and Bermingham 2002; **10.** Foster 1967; **11.** Rohwer et al. 1983; **12.** Lyon and Montgomerie 1986; **13.** Rohwer and Butcher 1988; **14.** Rohwer et al. 1980; **15.** Thompson 1991; **16.** Rohwer and Butcher 1988; **17.** Conover et al. 2000; **18.** Ibid.:198.

THRAUPIDAE
Tanagers
(CAS, CBS?; 4 species)

"What is a tanager?" is not a new question, but there is still no agreed-upon answer. Fast-moving flocks of glittering and twittering birds in the canopy of a South American forest are the epitome of tanagers, which compose a diverse family characteristic of the New World tropics. But the four traditional North American "tanagers" (in the genus *Piranga*) almost certainly aren't true tanagers and are actually more closely related to cardinals and their allies.[1] (After this book was written, the AOU also decided that *Piranga* "tanagers" aren't tanagers and moved them into the Cardinalidae; the tanager family account has been retained here because, for now, most birders still think of *Piranga* species as "tanagers"—and their molts haven't changed!)[2] Besides genetic clues, the songs, streaked juvenile plumages—and molts—of *Piranga* are all more like those of species such as the Rose-breasted and Black-headed grosbeaks than they are of tanagers. *Piranga* comprises birds of forest and woodland that live mostly in the canopy. *Piranga* species are migratory to varying degrees, with eastern populations and species being longer-distance migrants than western ones, a pattern repeated among many groups of North American breeding birds. The sexes look different, and some species have seasonal changes in appearance.

The species of *Piranga* exhibit the complex alternate molt strategy, although the extent of their prealternate molts is variable (see below), and some species or populations might have the complex basic strategy. Preformative and prebasic molts occur on the breeding grounds, prior to migration, except for some adult Western Tanagers that molt at staging sites on their way south.[3] Juvenile Western Tanagers may also move slightly to molt, migrating upslope to wetter mountain forests with richer food, but their soft and fluffy juvenile plumage may not be able to sustain a prolonged flight to the Southwest.[4] Prealternate molts occur mostly on the nonbreeding grounds but may continue through spring migration.

PREFORMATIVE MOLTS. Preformative molts in *Piranga* include head and body feathers, plus wing coverts and sometimes tertials, but no tail feathers. Following these molts, male Scarlet and Western tanagers appear intermediate between adult males and females, whereas male Summer and Hepatic tanagers resemble adult females.

PREALTERNATE MOLTS. The extent of prealternate molts in the genus *Piranga* generally correlates with migration distance, but critical study is still needed in some cases. The Scarlet Tanager, the longest-dis-

The stunning breeding aspect of a male Western Tanager is brought about by a combination of molt and wear. *Fremont County, CA, 17 May 2008. Bill Schmoker.*

As well as having delayed plumage maturation, the Hepatic Tanger (shown here) and the migratory Summer Tanager lack the distinct seasonal changes in appearance shown by the migratory Scarlet and Western tanagers. What aspects of life history, or ancestry, might account for these differences? This first-cycle Hepatic can be aged by the molt contrast among its greater coverts: the two outermost coverts are retained juvenile feathers, being shorter than the adjacent formative feathers and having abraded whitish edgings, much like those on the juvenile tertials. *Nayarit, Mexico, 14 Jan. 2009. Steve N. G. Howell.*

tance migrant, clearly has an extensive prealternate molt, which changes the male's appearance strikingly. Particularly in first-cycle birds, this molt often includes tertials and tail feathers. At the other end of the migration cline, the resident or short-distance migrant Hepatic Tanager may lack a prealternate molt, or it may be reduced to little more than a protracted facultative molt of a few head and body feathers through the winter.

Among intermediate-distance migrants, eastern-breeding populations of the Summer Tanager have a fairly extensive prealternate molt in late winter,[5] paralleling that of the Scarlet Tanager (although adult male Summers undergo no change in appearance). In the Western Tanager, the first prealternate molt may be fairly extensive, but adults also achieve some color change through the wearing off of paler feather tips, and their prealternate molt is less extensive than in first-cycle birds. Do western-breeding populations of the Summer Tanager, which are shorter-distance migrants (akin to the Western Tanager), have less-extensive prealternate molts than their eastern counterparts? Does the extent of prealternate molt in migrants reflect food resources in late winter, perhaps in combination with the intensity of solar exposure? Or might it reflect the time available between molt and breeding, which may be less for western birds that breed earlier?

Why do male Scarlet Tanagers have a striking seasonal change in appearance when male Summer and Hepatic tanagers look the same year-round? Part of the answer may lie with the food resources available in late winter, and whether they are sufficient to fuel an extensive molt. Another possibility may be the nature of winter territoriality, as with several wood-warblers that have bright male plumages year-round, such as the Hooded Warbler and American Redstart. Thus, are Summer Tanagers territorial in winter, whereas Scarlet Tanagers are not?

REFERENCES

1. Klicka et al. 2000; **2.** Kratter et al. 2009; **3–4.** Butler et al. 2002; **5.** Parkes 1967.

EMBERIZIDAE
Sparrows and Allies
(CBS, CAS; 49 species)

The sparrows and allies (including seedeaters, towhees, the Lark Bunting, and juncos) compose a large and diverse family of primarily seed-eating birds. They are found throughout the Americas (where the family is most diverse) and also in Eurasia (where most species belong to the large genus *Emberiza*, from which the family gets its name). The longspurs and Snow Buntings are also usually placed in this family, although genetic evidence suggests they are not closely related to sparrows and that their similarities are due to convergence;[1] their patterns of molt and plumage-aspect change suggest they might be finches that have adapted to a terrestrial lifestyle. North American sparrows live in a variety of open, semiopen, and wooded habitats, and all of the northern-breeding species and populations are migratory, with many species wintering in the southern U.S. and northern Mexico.

Of the 49 North American species, 2 are sexually dimorphic (Spotted and Eastern towhees) and 4 are seasonally dimorphic (Black-chinned, Chipping, Golden-crowned, and Harris's sparrows). Eight others are both sexually and (in males) seasonally dimorphic, with the White-collared Seedeater and Lark Bunting changing their appearance by molt, whereas the longspurs and Snow Bunting change mostly or wholly by wear. At least 28 other species may undergo prealternate molts but do not appreciably change their appearance.

The molt strategies of sparrows are complex basic and complex alternate, and the molts of some species have been studied in considerable detail. Preformative and prebasic molts occur mostly on the breeding grounds, before the onset of winter or migration. In some cases, though, preformative molts occur mostly at migration staging sites and on the wintering grounds, where food is more plentiful. This strategy is most frequent in western species and populations (food on the dry, late-summer western breeding grounds is often reduced compared with that in wetter eastern regions) and occurs in Lark, Chipping, Clay-colored, Brewer's, Grasshopper, Le Conte's, and Baird's sparrows and in the Lark Bunting. Prealternate molts occur mostly on the wintering grounds but may continue through spring migration.

PREBASIC MOLTS. The duration of prebasic molt in Pacific Coast populations of the White-crowned Sparrow has been linked to latitude, which translates to the time available for molt between the end of breeding and the onset of winter.[2] The prebasic molt of resident birds in southern California (35° N) requires about 11 weeks, whereas the preba-

The extent or even the existence of prealternate molts among towhees, such as this skulking male Spotted Towhee, require elucidation. *Marin County, CA, 18 Mar. 2008. Steve N. G. Howell.*

The Grasshopper Sparrow is one of several species of grassland sparrows known to migrate in juvenile plumage, which this bird is wearing. Might this be an adaptation to escaping from habitats that could dry up or burn, thus offering little fuel to a molting bird? *Marin County, CA, 13 Aug. 2007. Steve N. G. Howell*

The handsome Black-throated Sparrow, here in fresh basic or formative plumage, is among several species of North American sparrows that lack a prealternate molt. *Baja California Sur, Mexico, 26 Oct. 2008. Steve N. G. Howell.*

sic molt of potentially migratory birds in southern British Columbia (49° N) requires only 6 to 7 weeks; on average, the period shortens northward by 2.6 days per degree of latitude. Such patterns are likely frequent among other species with wide latitudinal breeding ranges. The shorter period may reflect the minimum time required; the longer period may reflect the absence of pressure to complete molt quickly before migration, and so the energy demands for molt are spread over a longer period.

PREFORMATIVE MOLTS. Preformative molts in sparrows range from partial to complete. One oddity among sparrows is that an extra first-cycle molt has been reported in some species,[3,4] during which most or all of the soft juvenile body plumage is replaced just after fledging—only to be replaced again a short time later by the main preformative molt. Similar "extra" molts have been reported in other species, and their true nature remains to be elucidated (see page 21 and the family account for cardinals and allies for further discussion of this phenomenon).

Among northern-breeding species, the main preformative molts are partial, and in mid-latitude species and populations they sometimes include tertials and central tail feathers. More-extensive and sometimes eccentric preformative molts occur in several species in the genera *Aimophila*, *Spizella*, and *Ammodramus*, and in Black-throated and Lark sparrows, which all live in open, sunny, and abrasive grassland, desert, and saltmarsh habitats that may contribute to plumage wear. The preformative molt can be almost complete in some Field and Lark sparrows and complete in most Bachman's, Cassin's, and Grasshopper sparrows.

PREALTERNATE MOLTS. Prealternate molts, like preformative molts, are variable in extent, which again usually correlates to habitat and latitude. Prealternate molts apparently are absent or all but absent in several species, including the Olive Sparrow, towhees, and Vesper, Black-throated, Sage, and Song sparrows. Why would they be absent in some desert species such as Black-throated and Sage sparrows but present in species that breed alongside them, such as Brewer's and Lark sparrows? Could it reflect the resources available in late winter to fuel prealternate molts? Or that the former species are residents and shorter-distance migrants that start nesting earlier and have no time for a prealternate molt? This latter scenario may apply to resident first-cycle White-crowned Sparrows in coastal California (the subspecies *nuttalli*), which start breeding appreciably earlier than other populations. First-summer *nuttalli* are characterized by having only partially black-and-white crowns, rather than the full, adultlike crown patterns acquired by first-cycle birds of other White-crowned populations, all of which breed later and thus may have time for a more-extensive molt (see pages 51–53).

In many sparrows of temperate and more protected habitats (such as juncos and the Fox Sparrow), prealternate molts appear to be facultative, limited to some head feathers, and perhaps even lacking in some individuals. More-extensive prealternate molts occur in many species living in less protected and often sunnier habitats, such as sparrows of the genus *Spizella*, in which the relatively exposed tertials and central tail feathers are usually replaced. In some species the prealternate molts can be extensive but bring about no change in appearance, as in several *Aimophila* sparrows, which inhabit sunny shrub-grasslands and are likely to experience appreciable wear of their plumage. Conversely, the striking change between nonbreeding and breeding aspects of the Snow Bunting is brought about not by molt but by the wearing away of pale feather tips to reveal a bold, underlying black-and-white pattern. The same thing happens with longspurs, although they may also undergo limited prealternate molt of some head feathers.

Prealternate molt changes the appearance of Black-chinned, Chipping, Golden-crowned, and Harris's sparrows, so that the head and chest patterns of both sexes become bolder. Such changes prompt the unanswered question: why do prealternate molts in related species (such as Clay-colored and adult White-crowned sparrows) produce no change in appearance? This might also be phrased as: why do the prebasic molts of Chipping and Golden-crowned sparrows result in duller patterns when the molts of Clay-colored and White-crowned sparrows do not?

The White-collared Seedeater is the northernmost member of a widespread genus of Neotropical grassland birds whose breeding is often tied to seasonal seeding, which in turn is linked to seasonal rains—the timing of which can vary considerably from year to year. Males have a colorful "breeding plumage" and a femalelike "nonbreeding plumage" that usually are presumed to be alternate and basic, respectively. However, because males in breeding plumage can be found at any time of year in Mexico (with a marked summer peak corresponding to the rainy season[5]), their breeding aspect might not correlate strictly to an alternate plumage. Thus, if local rains come at the time of the prebasic molt, might breeding hormones be stimulated and cause the incoming basic plumage to look like the typical alternate plumage? Strange as it may sound, this phenomenon has been documented in an African grassland species, the Black-chested Prinia (*Prinia flavicans*), which also breeds during seasonal rains,[6] and it may be more common than is recognized.

This male Snow Bunting in bold "breeding" plumage is actually in worn basic plumage—plumage wear, but no molt, is involved in the striking seasonal changes in appearance that occur in this species. Recent genetic studies indicate that longspurs and the Snow Bunting are not closely related to emberizid sparrows. Their molts are much like those of many finches. Are these similarities simply convergence, or might longspurs actually be terrestrial finches? *Spitsbergen, Norway, 21 June 2008. Lyann A. Comrack.*

SPIZELLA SPARROWS

A detailed study of molt in sparrows of the genus *Spizella* by Ernest Willoughby[7] highlights how little we really know about the fine-level details of molt in most species of songbirds. The patterns of molting by which we categorize birds are simply the best fits, or the patterns that apply to most individuals—if we look closely there will be exceptions, or patterns that do not fit with time-honored hypotheses. Here are two questions that arise from Willoughby's study. First, why would Field Sparrows (which winter in relatively humid and temperate habitats) but not Clay-colored and Brewer's sparrows (which winter farther south and in sunnier, scrubbier habitats) replace primaries in the preformative molt? Is it because they can, and they have sufficient resources in fall that the western species are lacking? Or might larger samples of the western species show that they too sometimes replace primaries?

A second question is: why would some Brewer's Sparrows apparently replace protected inner primaries and outer secondaries in their (presumed) prealternate molt? Might this instead represent the first stages of an interrupted prebasic molt, even though conventional wisdom tells us that prebasic molts occur after breeding?

Willoughby's study shows that, while we would like to partition the molts of songbirds into discrete prealternate molts in spring and prebasic molts in fall, the reality isn't necessarily that clear-cut. In *Spizella* sparrows, molting appears to be an almost continuous, year-round activity with peaks in spring and fall, and with suspensions during other energy-demanding phases of the life cycle, such as migration or perhaps egg production and laying. Prealternate molts can be more extensive in northern-breeding populations, which might be due to those individuals having more time to molt than do lower-latitude breeders, which start nesting earlier. Southern populations of Black-chinned Sparrows have the most protracted breeding season and also have a low-level prealternate body molt that continues through the summer, a strategy that has been found in Cassin's and Bachman's sparrows.[8] Overlap between breeding and molting has been found rarely in temperate songbirds, but it is more common in tropical songbirds with potentially protracted breeding seasons.[9] Northern populations of the Black-chinned Sparrow, and other species of *Spizella* (which breed farther north), molt little during their shorter breeding seasons.

Low-level molt through the breeding season in these other species was limited mainly to face and

The molting of sparrows of the genus *Spizella*, such as this American Tree Sparrow, has been studied in considerable detail. An interesting finding is that these sparrows molt their chin and throat feathers as many as six times per cycle, perhaps as an adaptation to reduce infestations by lice. *Tompkins County, NY, 21 Dec. 2008. Christopher L. Wood.*

throat feathers, which in Field and Chipping sparrows (the two species studied in most detail) were molted four to six times between February and September. But why molt these feathers rather than the forehead feathers, which are not replaced but which tend to suffer the greatest degree of wear while foraging? One possibility is that these molts are adaptations to avoid infestations by lice.[10,11,12] These parasites attach their egg cases mainly to the chin and throat feathers (which ordinarily would be relatively protected from wear and from having the egg cases lost), and their breeding may be timed to increase the chances of infesting nestling birds. But through continuous breeding-season molt of chin and throat feathers, which are small and probably not costly to grow, the adult sparrows may reduce the chances of parasitism.

Willoughby had difficulty applying the H-P system to his observations of extra face molts in *Spizella* sparrows, although in theory it should not be too difficult. The initial face molt in March, corresponding to the rest of the prealternate molt, and to the homologous prealternate molts in related species, would be part of the prealternate molt. Subsequent face molts through the spring and summer (which replace breeding-aspect feathers with breeding-aspect feathers) represent extra molts (that is, additional to the prebasic and prealternate molts) and thus are presupplemental molts, up until the face molt that corresponds to the prebasic molt of other head (and body) feathers, which would thus be part of the prebasic molt. But knowing which generation of face feathers you are seeing on a breeding bird is not possible, without knowing the history of each feather follicle. This is another case of how two different systems can be applied. The H-P system informs us how many molts occur, and how they relate to the molts of other species. The life-year system tells us of how the plumage aspect relates to the life history—in this case, the presupplemental face molts are simply extra "nuptial molts" as long as they produce feathers with the same color and pattern as the breeding plumage. The H-P system is informative in terms of evolution but often impractical to use in the field, whereas the life-year system is relevant to what we actually see, as birders and field ornithologists.

REFERENCES

1. Klicka et al. 2000; **2.** Mewaldt and King 1978; **3.** Sutton 1935; **4.** Willoughby 1986; **5.** S. N. G. Howell, pers. obs.; **6.** Herremans 1999; **7.** Willoughby 1991; **8.** Willoughby 1986; **9.** Foster 1975; **10.** Foster 1969a; **11.** Foster 1969b; **12.** Willoughby 1991.

CARDINALIDAE
Cardinals and Allies
(CAS, CBS; 10 species)

The quintessence of primary colors in North American birds, cardinals and their allies have long attracted attention from birders and ornithologists. Even their molts have been examined in relative detail, although some studies have stirred rather than cleared the waters. Cardinals and allies are mainly arboreal, and most species inhabit woodland and scrubby habitats, where they eat seeds (especially in winter) and insects (mainly in summer). They vary from sedentary and hardy residents such as the Northern Cardinal to nomadic long-distance migrants such as the Dickcissel; most North American species are short- to medium-distance migrants that withdraw south of the U.S. in winter. Cardinals and allies are another exclusively New World family in the enigmatic assemblage of nine-primaried oscines (see the wood-warbler family account for more on this subject). The sexes look different in all species, and some species also have seasonal changes in appearance.

The molt strategies of cardinals and allies vary from complex basic to complex alternate. Preformative and prebasic molts in most species occur on the breeding grounds or are interrupted for migration to complete on the wintering grounds. But in the Blue Grosbeak, Lazuli Bunting, western populations of Painted Bunting, and perhaps Black-headed Grosbeak, these molts often occur at staging sites or perhaps on the wintering grounds, where food is more plentiful than on the dry, late-summer western breeding grounds. Prealternate molts occur mainly on the wintering grounds, but in some species they can continue through spring migration.

PREFORMATIVE MOLTS. One intriguing aspect of this family is that several species (at least those in the genera *Cardinalis*, *Passerina*, and the tropical genus *Cyanocompsa*) apparently have an "extra" preformative molt (which has been called a presupplemental molt). In this putative molt, some or all of the soft juvenile body plumage is replaced just after fledging—only to be replaced again a short time later by the main preformative molt, which usually includes flight feathers. In some species this can be followed shortly by a prealternate molt, such that a

The Rose-breasted Grosbeak may be a colorful and well-known summer bird in eastern North America, but does it have a prealternate molt? Could this adult male in basic plumage attain the bold breeding plumage aspect simply by the wearing away of buff feather tips, or is some molt involved? The jury still seems to be out on this question. *Yucatán, Mexico, 30 Nov. 2007. Steve N. G. Howell.*

In its preformative molt, the Pyrrhuloxia (here a first-cycle male, with a small and relatively dull bill) replaces flight feathers in an eccentric pattern, whereas the closely related Northern Cardinal replaces flight feathers in a "typical" sequence. Why might these differences occur? *Pima County, AZ, 12 Dec. 2007. Steve N. G. Howell.*

bird may wear four plumages in its first 10 months of life! This was noted first in a careful study by George Sutton[1] and has since been reported in other species.[2,3,4,5] "Extra" preformative molts also occur among some sparrows and have been reported for the Phainopepla, yet their ephemeral nature and seeming redundancy have yet to be satisfactorily explained; functionally they act as a second phase of juvenile plumage. In fact, might these "extra" molts simply represent different phases of juvenile plumage, as have been found in many European songbirds?[6,7] Alternatively, given that two successive preformative molts occur in domestic fowl[8] and in turkeys,[9] might such a strategy be an ancestral state that has been suppressed or overlooked in other groups of birds?

The main preformative molt of North American cardinals and allies often involves some flight feathers, and it appears to be comparable to the preformative molts of other songbirds, such as blackbirds and orioles. Among Northern Cardinals, early-hatched birds often have complete molts, but late-hatched birds replace few or no flight feathers, with primary molts becoming progressively less extensive after July.[10] This trend indicates that time can be an important constraint on the extent of a molt. Among cardinals breeding in eastern North America, the preformative primary molt usually starts with the innermost primary and arrests with some juvenile outer primaries retained, whereas in Pyrrhuloxias the preformative primary molt is usually eccentric in sequence. Nobody has suggested a reason for this difference. Do Pyrrhuloxias use some daylight-related clue to "calculate" at which inner or middle primary they can start and still have time to molt the outermost primary? And is the sequence of preformative primary molt in southwestern populations of cardinals, which occur alongside Pyrrhuloxias, the same as that of eastern populations?

In the Black-headed and Rose-breasted grosbeaks the preformative molts have not been well studied, but sometimes they include tertials and tail feathers (mostly molted on the wintering grounds) but not other flight feathers. The molts of the Dickcissel are also not well known, but the preformative molt is partial, sometimes including tertials and central tail feathers. Buntings in the genus *Passerina* have complex molts and are considered below.

PREALTERNATE MOLTS. Prealternate molts occur only in migratory species, suggesting that more prolonged exposure to stronger sunlight, combined with the vagaries of migration (such as weather and potentially unfamiliar or altered habitats), may be forces that drive the evolution of prealternate molts. However, in several species the breeding aspect of bright males is brought about by a combination of wear and molt, although the relative importance of each factor is not well known. And in first-cycle Rose-breasted and Black-headed grosbeaks the main preformative molt and first prealternate molt may overlap, making it difficult to evaluate the extent of either. Critical studies on the wintering grounds could help resolve the molt strategies of these "well-known" birds.

PASSERINA BUNTINGS

The molts of buntings in the genus *Passerina* have received considerable attention, particularly in a seminal study of the Indigo Bunting by Sievert Rohwer[11] and in subsequent studies of Painted and Lazuli buntings.[12,13] The genus is most diverse in Mexico, where six of the seven species breed.

All species reportedly have an "extra" preformative molt,[14] which occurs in late summer and early fall on the breeding grounds, usually prior to migration. The appearance of all birds following this molt is "femalelike" and not greatly different from that of the juvenile plumage.

Among the three well-studied species, the extra preformative molt is least extensive in the Lazuli Bunting, some juveniles of which may molt directly into the main formative plumage.[15] This raises an interesting question: given the considerable overlap in timing that often occurs between the prejuvenile (first prebasic) molt and the extra preformative molt, between the two preformative molts, and between the main preformative molt and the first prealternate molt, do all individuals really have three molts added into their first cycle? Or might it be that lines are being drawn on a continuum based on whether birds hatched early or late, or on features such as the color of a plumage? Color and molt are controlled differently, and the former is not an infallible criterion for the latter (see pages 17–18).[16]

The main preformative molt occurs in fall and early winter and includes outer primaries, inner secondaries, and all tail feathers, as well as most or all body feathers. In Indigo Buntings this molt occurs mainly on the wintering grounds, in eastern Painted Buntings mainly on the breeding grounds, and in Lazuli and western Painted Buntings mainly at migration staging sites in the Southwest and northwestern Mexico. These differences in location likely reflect the relative availability of food resources, such as those produced by late-summer rains in the Southwest. This molt is most extensive in the Lazuli Bunting, in contrast to its limited extra preformative molt. In males, the plumage aspect produced by this molt resembles the basic plumage of adult males in Lazuli and Indigo buntings but continues to resemble the female in Painted and Varied buntings.

The first prealternate molts occur mainly on the wintering grounds and appear broadly similar to adult prealternate molts. In male Lazuli Buntings the prealternate molt is limited to a few head feathers, whereas male Indigo Buntings have a fairly extensive prealternate molt that, especially in first-cycle birds, may continue through spring migration. Adult male Painted Buntings have a partial prealternate molt even though there is little appreciable change in their appearance, whereas the less migratory Varied Bunting apparently lacks a prealternate molt (although it has not been studied in detail). First-cycle male Painted and Varied buntings continue to look femalelike through their first summer, a phenomenon known as delayed plumage maturation (see the wood-warbler family account for discussion of this subject).

This adult male Blue Grosbeak apparently attains its breeding aspect not by molt but by the wearing away of cinnamon-brown feather tips that veil the bright waxy-blue undercoat through the winter. *Jalisco, Mexico, 23 Jan. 2009. Steve N. G. Howell.*

Numerous questions are raised by the studies of molt in *Passerina* buntings. For example, why do the extra preformative molts exist? Are they even distinct molts or might they represent phases in the growth of juvenile plumage? Why do males of some species look adultlike in their first summer but others femalelike? How much relates to ancestry? How much to physiological responses to environmental factors? It is also tempting to make comparisons between these four *Passerina* buntings and the four North American breeding species of *Piranga* tanagers. Scarlet and Western tanagers acquire adultlike plumage in their first summer, as do Indigo and Lazuli buntings, whereas Summer and Hepatic tanagers have delayed aspect maturation, as do Painted and Varied buntings. There are also similarities (and a few notable differences) in these "species pairs" in their breeding ranges, migration distances, and the locations where different molts occur. Would a critical comparison of these two groups help resolve some of the puzzles? A better understanding of the common and "boring" patterns among related songbirds might also make the molt strategies of *Passerina* buntings easier to interpret, but for now they remain a pebble in the shoe of students of molt.

REFERENCES

1. Sutton 1935; **2.** Rohwer 1986; **3.** Thompson 1991; **4.** Thompson and Leu 1994; **5.** Young 1991; **6.** Dorsch 1993; **7.** Jenni and Winkler 1994; **8.** Lucas and Stettenheim 1972; **9.** Leopold 1943; **10.** Wiseman 1977; **11.** Rohwer 1986; **12.** Thompson 1991; **13.** Young 1991; **14.** Thompson and Leu 1994; **15.** Young 1991; **16.** Howell et al. 2004.

Buntings of the genus *Passerina* have been the subjects of several molt studies, but we still don't know exactly what is going on and how many molts there really are in a bird's first year. This first-cycle male Indigo Bunting will likely acquire more extensive blue head and body plumage as the spring progresses. Note the contrast between blue-edged formative outer primaries and dull juvenile inner primaries: this eccentric first-cycle primary molt is typical of *Passerina* buntings. *Petén, Guatemala, 5 Mar. 2008. Steve N. G. Howell.*

The Yellow-headed Blackbird (here an adult male) is unusual among North American blackbirds in that its preformative molt is incomplete, with the juvenile flight feathers being retained until the second prebasic molt. Is this an ancestral trait, or does it reflect aspects of the species' present-day life history? *Ward County, ND, 16 June 2008. Steve N. G. Howell.*

ICTERIDAE
Blackbirds and Orioles
(CBS, CAS; 20 species)

The blackbirds, orioles, and allies (including grackles, cowbirds, meadowlarks, and the Bobolink) are often collectively called icterids, an abbreviation of their family name. Icterids are another exclusively New World family of nine-primaried oscines (see the wood-warbler family account for more on this subject) with considerable diversity in the tropics. They can be divided for convenience into the mostly terrestrial blackbirds and allies of open and often marshy country (whose diet is mainly seeds and insects) and the arboreal orioles and allies of mostly wooded habitats (whose diet includes fruit and nectar). Many northern-breeding species are short-distance migrants, and one species, the Bobolink, is a long-distance migrant that winters in South America. In most temperate-breeding (including North American) species the sexes look different, often strikingly so, but in most tropical species the sexes look alike. Seasonal change in appearance occurs in a few species.

North American icterids mainly exhibit the complex basic molt strategy, although a limited prealternate molt may occur in several species and the Bobolink has a complete prealternate molt (see below). Preformative and prebasic molts of blackbirds occur mostly on the breeding grounds, before the onset of winter or migration. Orioles, however, molt mostly on the nonbreeding grounds—the exception being the Baltimore Oriole, which may need good-quality feathers for overwater migratory flights that western species do not make, and which also may have access to better food resources in its late-summer range. And many first-cycle Brown-headed Cowbirds, at least in the West, disperse widely while still in juvenile plumage or while molting. Prealternate molts occur mostly on the nonbreeding grounds but can continue through spring migration.

An interesting aspect of both the prebasic and preformative molts in the grackles is that the tail feathers are shed synchronously, typically when the outer primaries are being molted.[1,2] Flocks of "tailless" grackles flying around are thus a common sight in late summer and fall, and the longest central tail feathers are often the last feathers to grow back fully. Why might this strategy have developed? One possibility is that the distinctly structured, "keeled" tails of grackles are important for displays and

As in several species of blackbirds and orioles, the fresh plumage of this male Tricolored Blackbird has brownish tips that will wear away over the winter to reveal a glossy black undercoat. *Marin County, CA, 13 Oct. 2008. Steve N. G. Howell.*

may benefit from all being of one generation, and as fresh as possible, as occurs with the late-molted gorgets of hummingbirds, or the late-molted elongated central tail feathers of jaegers.

PREFORMATIVE MOLTS. Preformative molts in blackbirds and allies are ostensibly complete in most North American species, although the underwing coverts of several species can be retained;[3] certainly there seems little reason to replace these protected feathers, but this molt trait appears unique to blackbirds. In some cases a few juvenile middle secondaries and primary coverts are also retained, perhaps mainly in late-hatched birds that lack time to complete their molt. The Yellow-headed Blackbird and Bobolink are exceptions to this pattern, and their preformative molts are partial, involving only head and body feathers, wing coverts, and occasionally tertials. The Bobolink is a long-distance migrant that can molt again on its food-rich nonbreeding grounds, but why would the Yellow-headed Blackbird be an exception? Is its strategy an ancestral trait, or does it reflect aspects of the species' present-day life history?

In several species of blackbirds the preformative molt produces an adultlike appearance, but in some species the first-cycle males can vary greatly in appearance. In Red-winged Blackbirds, some first-cycle males resemble adult males whereas others resemble females, and others appear intermediate. It has been suggested that this variation depends, at least in part, on the age of the individual at the time of molt,[4] and it shows once again that the hormones controlling color and pattern are separate from those controlling molt.

Among orioles, partial to incomplete preformative molts are the rule. These molts include head and body feathers, wing coverts, and some to all tertials and tail feathers. At least in some southern and western species of open and sunnier habitats (Orchard, Hooded, and Scott's orioles), the preformative molts are often eccentric and in some cases may even be complete. Tertials and central tail feathers are replaced in the preformative molts of all orioles except the Baltimore, which replaces these feathers in what has been termed the first prealternate molt.[5] It could also be argued, however, that the molt of these feathers on the nonbreeding grounds is part of an interrupted preformative molt that overlaps in timing with the first prealternate molt.

An interesting and well-documented case of ancestry overriding present-day environmental conditions is found in Bullock's and Baltimore orioles. The ranges of these two species have spread from the dry West (the home of Bullock's) and the humid East (the home of Baltimore) to overlap on the Great Plains, where the two species breed alongside one another and even hybridize. Yet Bullock's Orioles in the overlap zone migrate away to molt, whereas Baltimore Orioles remain to molt.[6] Clearly there is enough food in such areas for an oriole to undergo its prebasic or preformative molt, but Bullock's Orioles still migrate away to their ancestral molting grounds. And if hybrid Bullock's x Baltimore orioles attempt to molt once on the breeding grounds and then again on the nonbreeding grounds, the stress of two molts could be a factor selecting against the survival of hybrids.[7]

PREALTERNATE MOLTS. Prealternate molts certainly occur in the Bobolink, which is unusual in having a complete prealternate molt (the only other North American breeding species with two complete molts a year is Franklin's Gull). After a complete prebasic molt in late summer in North America, Bobolinks migrate to the grasslands of South America, where there is enough food to undergo a complete prealternate molt prior to spring migration. Why do Bobolinks have two complete molts a year? Might it be because they inhabit open and sunny and grassy habitats year-round? Most long-distance migrants wintering in South America live for at least part of the year in wooded or forested habitats, where they are somewhat protected from the sun. Swallows are an exception to this, but they need to maintain highly efficient flight so are forced to undergo protracted molts and thus don't have time for two molts a year. To counter this, the slower growth of a swallow's flight feathers may allow more melanin pigments to be deposited for strengthening. Which raises the question: how "good" are the prebasic and prealternate flight feathers of a Bobolink? If each set has to last only half a year, are they as strong as the flight feathers of other icterids that last for an entire year? Or are they of sufficient quality to last for only 6 to 8 months?

Our knowledge of prealternate molts in other icterids is partial at best, and it may be that in some cases the changes brought about by protracted preformative molts or by wear have been attributed to prealternate molts. An interesting dichotomy in prealternate molts has been reported for Baltimore and Bullock's orioles.[8] Bullock's Orioles essentially lack a prealternate molt (and in their first summer, young males look much like females), whereas Baltimore Orioles have a prealternate molt that is most extensive in first-cycle males (which in their first summer look variably intermediate between adult females and adult males). Why don't Bullock's Orioles have prealternate molts? Are the late winter resources in western Central America and Mexico insufficient to fuel them, as may also be the case with *Piranga* tanagers and *Passerina* buntings? Or is there insufficient time for a late-winter or spring molt before migration?

The pattern of prealternate molts being most extensive in first-cycle males has been suggested for other icterids,[9,10,11] but is this the case? Might limited prealternate molts have been overlooked in adults? Might prealternate molts have once occurred in adults but have now been suppressed, yet their vestiges remain in first-cycle birds? And how might geographic variation play into this phenomenon? For example, in the Red-winged Blackbird, females and first-cycle males of populations breeding in northeastern North America usually have prealternate molts but adult males usually do not.[12] Yet among western populations of Red-winged Blackbirds, no evidence of prealternate molt has been found in any age or sex class.[13] Northeastern adult males return earlier to the breeding grounds than do females and immatures, and it has been suggested that their energy is better spent on territorial defense than on molt.[14] In western populations, might an earlier start to the nesting season mean that all ages and sexes have no time available for a prealternate molt?

Which Came First—Dull or Bright?

Among birds in general, if one sex is brighter and more colorful it is usually the male. Examples of North American birds in which males are often obviously more colorful or boldly patterned include ducks, grouse, quail, some shorebirds, hummingbirds, and numerous songbirds, including gnatcatchers, bluebirds, wagtails, some wood-warblers, tanagers, grosbeaks, buntings, finches, blackbirds, and several orioles. In some species the sexes look different year-round, and in others the differences exist only in the breeding season. There are some groups, such as phalaropes, in which things go the other way, but these are exceptions.

The conventional explanation for this phenomenon is that males need to be showy to impress females, and that males have thus evolved to become brighter. The implicit assumption has usually been that the ancestral bird was dull-colored and that sexual selection has driven males to become bright. And this may be what happened in many cases. However, it has long been noted that colorful plumages are frequent in both sexes of tropical wood-warblers, and that bold colors and patterns characterize both sexes of most New World orioles. Thus

The formative plumage of the male Red-winged Blackbird varies greatly in appearance. Some birds resemble females, and others, like this individual of the bicolored subspecies *mailliardorum*, are black overall and have virtually no red on the shoulders. *Marin County, CA, 19 Mar. 2008. Steve N. G. Howell.*

The conventional view is that bright colors have evolved in male songbirds for display. In orioles, however, it appears that the ancestral state was colorful, and instead it is the females, like this Orchard Oriole, that have changed by evolving a dull plumage aspect. *Nayarit, Mexico, 12 Jan. 2009. Steve N. G. Howell.*

it is reasonable to conclude that the ancestral state for such groups may have been brightly colored and boldly patterned in both sexes.[15,16] In North America, think of the Altamira Oriole in south Texas as a typical oriole. What this means, then, is that the females of the migratory northern orioles—Baltimore, Bullock's, Scott's, Orchard, and Hooded—have evolved duller plumages.

Whereas increased brightness in males is associated with sexual selection, why would females become duller? One possibility is to make them less conspicuous, and thus less prone to predation when nesting. But are predators any more of a concern in temperate environments than in the tropics? Another theory that has been proposed is that duller coloration could decrease male aggression and facilitate rapid pair formation in migratory species.[17] As with delayed plumage maturation (see the family account for wood-warblers), however, these theories almost imply a conscious adaptation on the part of the bird to become dull, or bright. But what might initially have triggered the changes in coloration so they could be selected for? And why are they prevalent in migratory orioles but not in resident species?

Tropical songbirds such as wood-warblers and orioles are more likely to remain paired year-round and share in territory defense, in which case there may be an advantage to both sexes being colorful. But in migratory species that pair only for a short breeding season, would the male care if the female was colorful or not? In orioles, then, might competition for food resources during the prebasic molt have pushed females to consume foods less rich in the carotenoid pigments that promote bright orange and yellow plumage coloration?

Whatever the cause for a shift from bright to dull plumages, this overlooked idea is a refreshing challenge to conventional dogma—and further work may shed more light on the subject. Recall, for example, that the bright male plumage of a Mallard is the "default" state for either sex, and that the ancestral state in waterfowl also may have been for both sexes to be brightly colored.[18] In what other groups of birds might a bright-to-dull transition have occurred?

REFERENCES

1. McIlhenny 1937; **2.** Selander 1958; **3–4.** Selander and Giller 1960; **5.** Rohwer and Manning 1990; **6.** Rohwer and Johnson 1992; **7–8.** Rohwer and Manning 1990; **9.** Greenwood et al. 1983; **10.** Selander 1958; **11.** Pyle 1997b; **12.** Greenwood et al. 1983; **13.** Unitt 2004; **14.** Greenwood et al. 1983; **15.** Hamilton 1961; **16.** Hofmann et al. 2008; **17.** Hamilton 1961; **18.** Kimball and Ligon 1999.

FRINGILLIDAE
Finches
(CBS, CAS; 16 species)

Finches compose a widespread family of birds typical of temperate regions, with their greatest diversity being in the Old World. In North America they range from the American Goldfinch, a familiar visitor at many backyard feeders, to the nomadic and taxonomically vexing crossbills. Finches occur mainly in wooded and scrubby habitats, feeding either in trees or on the ground, where their main food comprises a variety of seeds. Most North American species exhibit some migratory tendencies, and several are well known for their irruptive movements triggered by shortages of their favored foods. The sexes look different in most species. One species has pronounced seasonal changes in its appearance because of molt, whereas other species change their appearance seasonally by plumage wear.

Most finches follow the complex basic molt strategy, as is typical of resident or short-distance migrant songbirds in temperate habitats, but a few species exhibit the complex alternate strategy. Some species show interesting patterns with regard to the presence and extent of prealternate molts and to the extent of their preformative molts. Prebasic and preformative molts generally occur on the breeding grounds, prior to the onset of winter or migration, but in more-southerly species (mainly the goldfinches and Pine Siskin) the molts may be interrupted or protracted into winter. Prealternate molts occur on the wintering or breeding grounds.

The molts and plumages of the Red Crossbill have attracted considerable attention, and this species provides a good example of how the processes controlling pigmentation and molt are independent.[1,2,3,4] Basically, if male crossbills (both adults and first-cycle birds) molt in fall, say between July and October, then they acquire the classic red head and body plumage. If they molt head and body feathers at other times of year, the new feathers are yellowish. Because crossbills breed throughout much of the year, young hatched in late fall may undergo their preformative molt in late winter or spring, and thus attain a yellowish appearance. But young hatched in spring and summer molt in fall and acquire a "typical" red plumage. And since this species varies its breeding schedules in response to the timing of conifer seeding, then even adult males could conceivably be red in some years and yellow in others! In some cases the change from yellow to red coincides with a change from feeding on old cones to new cones, which has suggested to some that pigments derived from food may be involved.[5] Given that "Red Crossbills" in North America may represent six or more species, might differences in breeding schedules, molt timing, and food supply among these types mean that some types frequently include yellow males and others rarely or never?

PREFORMATIVE MOLTS. Preformative molts in finches are highly variable in extent. Among northern or cold-climate species such as the rosy-finches and Hoary Redpoll, the preformative molt involves head and body feathers, often some median and greater wing coverts, but no flight feathers. In other species (such as the Pine Siskin and Common Redpoll) it can also include tertials and central tail feathers; in crossbills it may be eccentric (perhaps reflecting their nomadic and aseasonal breeding systems); and in the trio of "southwestern" species that breed early and inhabit sunny regions (the House Finch and Lesser and Lawrence's goldfinches) it is often eccentric, and in some House Finches the preformative molt can be complete.

An interesting overlay to the typical "cold north to hot south" increase in the extent of preformative molts has been documented in House Finches in southern California,[6] where fledging spans April to September. Among early-hatched birds a complete preformative molt is common, and young males typically molt into a bright red plumage; among late-hatched birds partial molts are the rule, and males often molt into a dull yellowish plumage; be-

The familiar American Goldfinch of many backyard feeders is an exception among North American finches in that it has distinct-looking basic and alternate plumages. This adult male is in fresh basic plumage. *Frederick County, VA, 26 Nov. 2008. Steve N. G. Howell.*

The rosy-finches are typical of most North American finches in lacking a prealternate molt; they just grow a dense and strong basic plumage to carry them through a plumage cycle in which they may well see snow year-round. This brightly marked Brown-capped Rosy-Finch is an adult based on its blunt-tipped primary coverts with distinct pink edging; its bright plumage aspect indicates a male. *Boulder County, CO, 5 Apr. 2008. Bill Schmoker.*

tween these extremes lie almost all intermediates, with successively later primary molts initiating with middle and then outer primaries. Thus, the yellow "variant" of male House Finches, while often linked to diet, may also reflect age-related hormones, as with the shift from gray to black formative feathers in male Phainopeplas. And among House Finch populations in the southern Great Basin the formative plumage of males can be femalelike, with no color;[7] are such birds late-hatched, or is there some other reason (such as diet) for this anomaly within the species?

In contrast to the above study, a study of House Finches in Massachusetts (where they were introduced from southern California) found that only 20 percent of birds replaced some primaries in their preformative molt, none had complete molts, and partial molts were the norm.[8] Thus, environmental factors, such as a colder environment with shorter breeding and molting seasons, appear to have fairly quickly reduced the extent of the House Finch's preformative molt.

PREALTERNATE MOLTS. Well-documented and extensive prealternate molts in North American finches occur only in American and Lesser goldfinches (see below). Presumably these species can fuel two molts, and their plumage, or their ancestor's plumage, experienced enough wear for prealternate molts to evolve. Limited prealternate molts have been reported in other species, such as House and Purple finches, Lawrence's Goldfinch, Pine Siskin, Red Crossbill, and Evening Grosbeak. However, the "prealternate molt" of Lawrence's Goldfinch has been expunged[9]—the seasonal color change comes about through plumage wear—and if a prealternate molt occurs in the other species it is likely only a facultative or limited molt of some head feathers.

The Lesser Goldfinch Problem

The American Goldfinch is a textbook example of alternate plumage: a partial prealternate molt occurs in late winter and spring, producing a brighter plumage in which the birds breed, and a complete prebasic molt in fall produces a duller, nonbreeding plumage. The molts of the Lesser Goldfinch, however, result in no appreciable seasonal change in appearance. The prealternate molt of this species varies in extent depending on sex and region, and in some cases it may even be complete. An interesting paper by Ernest Willoughby[10] examines this complex situation in detail, and a simplified summary of his work follows.

The Lesser Goldfinch in North America comprises two types: males of northwestern populations have green backs and breed mainly during April–August, following winter rains; males of interior and southern populations have black backs and breed mainly during the June–October rainy season. Females of both populations look the same. In western birds, complete prebasic molts and incomplete to complete preformative molts occur from summer to early winter after breeding, and the prealternate molt in spring is limited in extent or perhaps even absent in some birds. This molt pattern is broadly similar to that of Lawrence's Goldfinch, which often breeds in the same areas.

In southern birds, prebasic molts and incomplete to complete preformative molts occur from fall to midwinter after breeding, in accord with a later breeding season. Prebasic and preformative molts may be interrupted in late winter and resume in April–June, when they overlap with an extensive prealternate molt that also can include flight feathers. Young males acquire their black backs during the prealternate molt, usually having had green or dark-mottled backs in formative plumage. Thus, interior birds enter the breeding season in fresh alternate plumage, while western birds enter it in worn formative or basic plumage—a situation that recalls the dichotomy between Baltimore and Bullock's orioles or between Indigo and Lazuli buntings.

An interesting twist to this story is the extent of flight-feather molt documented in the two molting periods of southern birds. Only 4 percent of birds replace 90 percent or more of their flight feathers in the post-breeding (October–February) molt, whereas 28 percent of birds do so in the prebreeding (April–June) molt, and the latter molt averages more extensive than the former. On average, birds replace each primary from 1.1 to 1.4 times per cycle. As in other cases of interrupted molts, the natural tendency might be to attribute each bout of molting to a discrete molt—which is functionally the case but which may not reflect homologies. It all comes back to the simple (but unresolved) question: what is *a* molt (see page 3). Another interpretation is that southern birds have insufficient time and food available for all flight feathers to be replaced in their prebasic molt before midwinter, so the molt is interrupted until conditions again become favorable. Thus the spring molt may represent a continuation of the prebasic molt overlapping with a prealternate molt. This interpretation is supported by the fact that the inner primaries are most often replaced in October–February and the outer primaries in April–June, with one molt seemingly continuing from where the preceding molt stopped; p3 is the feather replaced least fre-

This male Red Crossbill provides a great example of how plumage coloration can vary depending on the timing of a molt. Many birders who see a male crossbill with some yellow plumage patches might call it an immature male, but it is just as likely to be an adult that molted partly outside the "window" in which red pigment is produced. Conversely, many immatures molting at the "right time" acquire a solidly bright red plumage via their preformative molt (see Figure 22, page 17). *Boulder County, CO, 16 Dec. 2007. Bill Schmoker.*

The molts of the Lesser Goldfinch vary geographically in their timing and extent and have been the subject of a careful study. Northwestern populations have a molt schedule similar to that of Lawrence's Goldfinch (a female of which is on the left side of this feeder), whereas southern populations molt later in the fall and also have an extensive prealternate molt that northwestern populations lack. The middle bird, in fresh plumage, is presumably in its first cycle and contrasts markedly in appearance with the worn and molting male on the right. *Santa Barbara County, CA, 7 July 2007. Steve N. G. Howell.*

quently during the annual cycle. Other examples of these overlapping molts in songbirds may occur in some tyrant-flycatchers (see that family account for details).

In attempting to reconcile the molts of the Lesser Goldfinch with the H-P system, Willoughby identified homologies of color (plumage aspect) rather than homologies of molt.[11] He also did not recognize that the spring molt of eastern birds might represent a continuum of the prebasic molt overlapping a prealternate molt, owing to the more extensive replacement of flight feathers during the spring.[12] He thus suggested that western males had a complex basic molt strategy, western females a complex alternate strategy, and eastern males and females a simple alternate strategy. If we use the H-P system, however, then it is apparent that all Lesser Goldfinches have the complex alternate strategy, which evolved from the presumed ancestral complex basic strategy of other finches simply by the addition of a prealternate molt. The prealternate molt of the Lesser Goldfinch varies from limited (or even absent) to almost complete, and the colors produced during this and other molts are irrelevant to molt homologies.

REFERENCES

1. Tordoff 1952; **2.** Newton 1973; **3.** Phillips 1977; **4.** Jenni and Winkler 1994; **5.** Newton 1973; **6.** Michener and Michener 1940; **7.** Van Rossem 1936; **8.** Stangel 1985; **9.** Willoughby et al. 2002; **10.** Willoughby 2007; **11.** Howell et al. 2004; **12.** E. J. Willoughby, pers. comm.

PASSERIDAE
Old World Sparrows
(CBS; 2 species)

In North America, the familiar House Sparrow, or English Sparrow as it is sometimes known, often gets dismissed into the "trash-bird" bin along with a few other non-native species that have been introduced from Europe. Yet to an independently minded observer it is a resilient species, full of character, and the male is actually rather handsome. House Sparrows (and the Eurasian Tree Sparrow, introduced locally in North America) are not closely related to New World species such as the Song Sparrow and White-crowned Sparrow, but instead are members of the Old World sparrow family, which has its greatest diversity in Africa and the Middle East. These are generally chunky, thick-billed birds of open and semiopen dry habitats. Most species are sedentary, and the sexes look different.

Along with most other resident songbirds of temperate habitats, the House Sparrow has the "default" complex basic molt strategy. The seasonal change in its appearance, with males becoming brighter as the breeding season approaches, is brought about by the wearing away of pale feather tips that have veiled bold patterns through the winter. This cost-effective strategy is found in several other songbirds, including the European Starling and Snow Bunting. Preformative and prebasic molts occur in late summer and fall, before the onset of winter.

PREFORMATIVE MOLT. Unlike most North American songbirds, the House Sparrow has a complete preformative molt, after which first-cycle birds look very similar to adults. House Sparrows don't seem to have any exceptional life-history characters that might explain this complete molt, although their frequent coming and going from cavities in buildings and trees might wear down their plumage a bit. Still, chickadees and many other cavity-loving birds don't have complete preformative molts. As with the European Starling, Wrentit, Bushtit, and Horned Lark (all of which also have their origins with Old World families), part of the answer may be ancestral. A complete preformative molt for birds living in the dry, sunny, tropical regions of Africa makes sense, and distant as this may seem from the chirping birds in your garden, tropical Africa may have been the ancestral home of the House Sparrow. Apparently there has not been enough pressure to modify the ancestral strategy, so complete preformative molts still occur.

Because the House Sparrow attains its adult plumage aspect via the preformative molt, we can't say whether this handsome male is in formative plumage and less than a year old, or in basic plumage and perhaps as old as 5 years. *Marin County, CA, 17 May 2008. Steve N. G. Howell.*

ACKNOWLEDGMENTS

It was in the summer of 1996, when I had some time on my hands, that I started observing the molting patterns of Western Gulls on the coast near my home. What I saw didn't fit with the conventional wisdom concerning molt in large gulls, and I sought to resolve the puzzle. Little did I know that my simple field observations were to precipitate an overhaul of the Humphrey-Parkes system for describing and naming molts and plumages. So I could thank the Western Gull for altering the path of my life.

Foremost in my studies of molt I thank Chris Corben for sharing his remarkable insights and for reining in my sometimes outlandish ideas. Subsequent work with Chris and with Peter Pyle and Danny Rogers helped formulate my ideas about patterns of molting, and the many long discussions were lots of fun. Philip S. Humphrey and Ernest J. Willoughby offered thoughtful and encouraging criticisms along the way. The folks at Point Reyes Bird Observatory, especially at the Palomarin Field Station, were supportive throughout, and I thank Geoff Geupel and Tom Gardali for helping to maintain this productive work environment, which stimulated many thoughtful discussions. This is PRBO contribution number 1692.

For reviewing some or all of the text I thank Jessie Barry, Renée Cormier, Ted Floyd, Keith Hansen, Diana Humple, Al Jaramillo, Nina Karnovsky, Kim Kreitinger, Jim Kushlan, Jerry Liguori, Peter Pyle, Danny Rogers, Brian L. Sullivan, Ernest J. Willoughby, and Chris Wood. Other helpful discussions involved Dave DeSante, Catherine Hickey, Scott Jennings, Kirsten Lindquist, Libby Porzig, Melissa Pitkin, Nathaniel Seavy, Diana Stralberg, and Nils and Sarah Warnock. Parvaneh Abbaspour, Mark Herzog, Stella Moss, Marc Parisien, and Nils Warnock helped obtain and translate references.

For their help with obtaining and selecting photos I thank Tom Blackman, Patricia Briceño P., Renée Cormier, Ryan DiGaudio, Tom Gardali, Keith Hansen, Burr Heneman, Catherine Hickey, Mark Holmgren, Diana Humple, Terry Hunefeld, Robert Kirk, Kim Kreitinger, Ron Martin, Guy McCaskie, Brian Patteson, Melissa Pitkin, Eric Preston, Will Russell and WINGS, Larry Selman, Debi Shearwater, Nils Warnock, and Wendy Willis. I am particularly grateful to the other photographers who have contributed to this work: Mike Bowles, Lyann A. Comrack, Loretta Erickson, Jerry Liguori, Larry Sansone, Bill Schmoker, Brian L. Sullivan, and Christopher L. Wood.

Bolinas, California, January 2009

GLOSSARY

accelerated stepwise molt. A stepwise wing molt in which the waves of molt are "kick-started" by the addition of an extra wing molt relative to the basic cycles. See **normal stepwise molt** and pages 36–45.

adult. A bird with adult plumage aspect, i.e., a plumage (basic, alternate, supplemental) whose appearance does not change appreciably with age. The Humphrey-Parkes (1959) system uses the term "definitive" for this plumage stage. Does not necessarily reflect sexual maturity.

alternate plumage. Any second plumage in a cycle in addition to—and which alternates with—basic plumage. Attained by a prealternate molt. See page 22.

altricial. Describes young that hatch helpless, naked or with sparse down, and with their eyes closed. See **semi-altricial** and **precocial.**

alula. The small group of feathers attached to the thumb, or first digit of the hand bone, which move independently of the rest of the wing.

arrested molt. A molt that stops before being completed. Usually refers to cases in which the molt could be complete (any partial or incomplete molt is technically an "arrested molt"), but often misused in the literature for **suspended molt.** Also see **interrupted molt.**

aspect (plumage aspect). The overall appearance of a bird, which can be a composite of basic and alternate plumages. For example, large white-headed gulls with clean white heads manifest a breeding aspect, even though they may not be breeding or may have started their prebasic molt. European Starlings in glossy, mostly unspotted plumage manifest a breeding aspect, even though this is simply their worn basic plumage. Misused by Pyle (2008) in the sense of "basic-plumage aspect" (which presumably is intended to mean nonbreeding aspect); the terms "basic" and "alternate" were proposed by Humphrey and Parkes (1959) to be free from associations with plumage aspect.

basic plumage. The plumage attained by the prebasic molt (which is complete, or nearly so) and presumed to be homologous in all birds. See page 20.

bristle. A specially modified, hairlike feather, such as the rictal bristles around the gape of birds such as nightjars, flycatchers, and the American Redstart, which catch flying insects.

complete molt. A molt that involves all of the feathers.

complex alternate (molt) strategy. A molt strategy in which, in addition to the basic plumage, one or more plumages (alternate and supplemental) occur in each cycle and a formative plumage occurs in the first cycle. See pages 29–32.

complex basic (molt) strategy. A molt strategy in which only basic plumages occur in each cycle, except for the addition of a formative plumage in the first cycle. See pages 29–30.

contour feathers. The vaned feathers that cover the head, body, and wings and form the outline (or contours) of the bird. They vary in size and function, from primaries to crown feathers.

coverts. Feathers that overlie, and thus cover and protect, other feathers.

cycle. A regularly repeated phenomenon, such as a plumage cycle. A basic plumage cycle extends from the start of one prebasic molt to the start of the next prebasic molt. The first plumage cycle extends from the acquisition of juvenile plumage to the start of the second prebasic molt.

definitive (molt) cycle. Refers to molt cycles in which the sequence of molts does not change (thus, for example, there are no formative plumages). For most species the second molt cycle is the definitive cycle, but for species with the simple basic molt strategy the first cycle is definitive.

definitive plumage. A plumage whose aspect does not change with time. See **predefinitive plumage** and page 24.

dimorphic. Occurring in two forms that differ in size, shape, or color.

eccentric wing molt. A pattern of incomplete wing molt in which the outer primaries and sometimes inner secondaries are replaced but one or more of the relatively protected inner primaries and outer secondaries are retained.

fading. Whitening, bleaching, or becoming colorless as a result of exposure to sunlight.

filoplume. A specially modified, hairlike feather with an expanded tip, used in sensory functions. Often inconspicuous but sometimes contrastingly colored, as in the white plumes of Anhingas and some petrels.

fledge. A term whose meaning varies somewhat depending on which birds are involved. For nest-bound young of songbirds, hummingbirds, and petrels, refers to leaving the nest, which is when a young bird first flies and graduates from being a nestling to a fledgling. For species such as shorebirds, waterbirds, and hawks, a less easily defined term that refers to the time at which the young become independent or can first fly with proficiency. Often, though, there can be periods of extended parental care, which can last for some months or even up to a year in the case of frigatebirds.

flight feathers. The main feathers of the wings and tail; that is, the primaries, secondaries, and rectrices. In some literature used only for the primaries and secondaries, and at other times also includes the primary coverts.

formative plumage. Any plumage present only in the first (that is, formative) year of life and lacking a counterpart in subsequent cycles. Attained by the preformative molt. Most conventional "first basic" plumages are formative plumages. See pages 21–22.

genus. A taxonomic grouping of related species. In some cases a genus may contain only one species and is known as a monotypic genus.

homologous. Sharing common ancestry.

immature. A general term for any nonadult plumage, including juvenile plumage.

incomplete molt. A molt that involves head and body feathers plus some flight feathers (usually central tail feathers, tertials, and sometimes outer primaries).

interrupted molt. A general term that encompasses **arrested molt** and **suspended molt.** Used, for example, when a molt contrast can be seen but it is not known whether the molt has been suspended or arrested.

juvenile. A bird in juvenile plumage.

juvenile plumage. The first plumage of "true" or vaned (nondowny) feathers; often the plumage in which a bird fledges. Considered synonymous with "first basic plumage" in recent molt studies and attained by the prejuvenile (first prebasic) molt. Also termed "juvenal" plumage, a pseudo-academic distinction that is unnecessary.

long-distance migrant. A species that migrates long distances, typically to spend its "winter" (see **nonbreeding grounds**) in the tropics or in the opposite hemisphere of its breeding grounds.

medium-distance migrant. A species that migrates moderate distances, typically to winter in Mexico, Central America, or the Caribbean, generally in tropical or subtropical regions.

molt. The normal and regular growth process of feathers (i.e., molting), by which a new generation of feathers (that is, a plumage) is attained. Feather loss is a passive byproduct of molting.

molt contrast (molt limit). The point of contrast between two generations of feathers in a nonmolting bird; for example, between the alternate and basic primaries on a tern.

nonbreeding grounds. Refers to areas that include regions not associated with the Northern Hemisphere winter, such as the tropics and South America. For species wintering in the Northern Hemisphere, the term is sometimes used to include both **staging grounds** and **wintering grounds.**

normal stepwise molt. A stepwise wing molt in which the waves of molt develop in the same number of cycles as the prebasic molts. See **accelerated stepwise molt** and pages 36–45.

obligate stepwise molt. A pattern of stepwise molt shown by all members of a species.

opportunistic stepwise molt. A pattern of stepwise molt shown by only some members of a species.

partial molt. A molt that involves some to all head and body feathers but no flight feathers.

plumage. There are two meanings for this word. In the Humphrey-Parkes system, refers simply to a generation of feathers attained by a molt, and the color of the feathers is not relevant. In everyday usage, refers to a bird's coat of feathers and may be termed "male plumage," "immature plumage," "breeding plumage," etc; thus, the color and pattern of the feathers are relevant.

prealternate molt. The molt by which alternate plumage is attained.

prebasic molt. The molt by which basic plumage is attained.

precocial. Describes young that hatch covered in down, with their eyes open, and that are soon able to leave the nest and often feed themselves. For example, waterfowl and shorebirds. See **semi-precocial** and **altricial.**

predefinitive plumage. A plumage whose aspect is not fixed with age, such as distinct-looking juvenile or formative plumages. See page 25.

preformative molt. The molt by which formative plumage is attained. In traditional Humphrey-Parkes terminology this was called the first prebasic molt.

presupplemental molt. The molt by which supplemental plumage is attained.

primary (p). The wing feathers attached to the hand bones. Most birds have 10 functional primaries, numbered from p1 (innermost) to p10 (outermost).

primary coverts. Feathers on the wing that protect the bases of the primaries.

rectrices (singular: rectrix). The main tail feathers, numbering 12 on most birds (6 pairs each side of the central point; r1 is the central rectrix, r6 the outer rectrix). The bases of the rectrices are protected by tail coverts.

scapulars. A group of feathers that originate from a point at the base of the humerus and fan out to protect the base of the wings at rest; they form a seamless joint between the wings and body in flight.

secondaries. The secondary wing feathers attached to the forearm bone (numbering from 6 in hummingbirds to more than 30 in albatrosses, and 9 in most songbirds), and including the tertials.

secondary coverts (often simply called wing coverts). Feathers on the wing that protect the bases of the secondaries.

semi-altricial. Describes young that hatch downy, with their eyes closed, and do not leave the nest. For example, petrels, hawks, and owls. See **semi-precocial** and **altricial.**

semi-precocial. Describes young that hatch downy and with their eyes open but that remain in or near the nest and are cared for by their parents. For example, gulls and nightjars. See **semi-altricial** and **precocial.**

sequential wing molt. The typical pattern of wing molt in which the primaries are molted in sequence, from p1 out to p10. See pages 34–35.

short-distance migrant. A species that migrates relatively short distances. In this book, typically a North American breeding species that winters in temperate habitats of North America and northern Mexico.

simple alternate (molt) strategy. A molt strategy in which alternate plumages occur in definitive cycles and only a single molt (preformative or prealternate) is added into the first cycle. See pages 29–32.

simple basic (molt) strategy. A molt strategy in which only basic plumages occur in each cycle. See pages 25 and 29–30.

songbird. A bird in the order Passeriformes, also known as passerines or perching birds.

Staffelmauser. An academic term for **stepwise wing molt.**

staging grounds. An area visited by a species for feeding or molting, usually lying between the breeding and **nonbreeding grounds.**

stepwise wing molt. A molt strategy of birds that cannot replace all of their primaries by means of a **sequential wing molt** in the time available between breeding seasons. Usually found in large, long-winged birds but also in some birds of tropical and arid environments with unpredictable seasons. Successive waves of primary molt are set up so that all or most primaries can be replaced efficiently courtesy of two or more waves of molt. See pages 36–45.

supplemental plumage. Any third plumage in a cycle (attained by a presupplemental molt), additional to basic and alternate plumages. See pages 22–24.

suspended molt. A molt that starts at one time or place and is interrupted, to be completed or continued at another time or place. See **arrested molt** and **interrupted molt.**

synchronous wing molt. A wing molt strategy whereby all of the primaries are shed at about the same time, rendering a bird flightless. See page 36.

tail coverts. Feathers that protect the bases of the tail feathers.

tertials. Used in this book for the inner few secondaries, which act as coverts on the closed wing.

totipalmate. Having feet with all four toes connected by webbing. Used to refer collectively to tropicbirds, boobies and gannets, pelicans, cormorants, Anhingas, and frigatebirds.

wear (or plumage wear). The physical abrasion of feathers by contact with water, blowing sand, vegetation, etc.; compounded by weakening that is due to fading.

wing-loading. A measure of body weight per unit of wing area. Birds with higher wing-loading have relatively heavy bodies but small wing areas (such as ducks and grouse). Birds with low wing-loading have relatively light bodies and big wing areas (such as frigatebirds and vultures).

wintering grounds. The nonbreeding grounds of species that are short- or medium-distance migrants and that spend their winter in a region that has a northern winter season. See **nonbreeding grounds.**

BIBLIOGRAPHY

Ainley, D. G., and R. J. Boekelheide, eds. 1990. *Seabirds of the Farallon Islands.* Stanford, CA: Stanford University Press.

Ainley, D. G., T. J. Lewis, and S. Morrell. 1976. Molt in Leach's and Ashy storm-petrels. *Wilson Bulletin* 88:76–95.

Ainley, D. G., S. Morrell, and T. J. Lewis. 1974. Patterns in the life histories of storm-petrels on the Farallon Islands. *Living Bird* 13:295–312.

Alström, P., and K. Mild. 2003. *Pipits and Wagtails.* Princeton, NJ: Princeton University Press.

American Ornithologists' Union. 1983. *Check-list of North American Birds.* 6th ed. Washington, D.C.: American Ornithologists' Union.

———. 1998. *Check-list of North American Birds.* 7th ed. Washington, D.C.: American Ornithologists' Union.

Ammann, G. A. 1937. Number of contour feathers of *Cygnus* and *Xanthocephalus. Auk* 54:201–202.

Andersson, M. 1999. Phylogeny, behavior, plumage evolution, and neotony in skuas, Stercorariidae. *Journal of Avian Biology* 30:205–215.

Andres, B. A., and G. A. Falxa. 1995. Black Oystercatcher (*Haematopus bachmani*). No. 155 in The Birds of North America, ed. A. Poole and F. Gill. Philadelphia, PA: Birds of North America Inc.

Ankney, C. D. 1979. Does the wing molt cause nutritional stress in Lesser Snow Geese? *Auk* 96:68–72.

Ashmole, N. P. 1963. The biology of the Wideawake or Sooty Tern *Sterna fuscata* on Ascension Island. *Ibis* 103b:297–364.

———. 1965. Adaptive variation in the breeding regime of a tropical sea bird. *Proceedings of the National Academy of Sciences* 53:311–318.

———. 1968. Breeding and molt of the White Tern (*Gygis alba*) on Christmas Island, Pacific Ocean. *Condor* 70:35–55.

Atwood, J. L., and S. B. Lerman. 2006. Family Polioptilidae (Gnatcatchers). Pp. 350–377 in *Handbook of the Birds of the World,* vol. 11, ed. J. del Hoyo, A. Elliott, and D. A. Christie. Barcelona: Lynx Ediciones.

Baltosser, W. H. 1995. Annual molt in the Ruby-throated and Black-chinned hummingbirds. *Condor* 97:484–491.

Banks, R. C. 1978. Prealternate molt in nuthatches. *Auk* 95:179–181.

Battley, P. F. 2007. Plumage and timing of migration in Bar-tailed Godwits: a comment on Drent et al. (2003). *Oikos* 116:349–352.

Battley, P. F., D. I. Rogers, and C. J. Hassell. 2006. Prebreeding moult, plumage, and evidence for a presupplemental moult in the Great Knot *Calidris tenuirostris. Ibis* 148:27–38.

Beebe, C. W. 1914. Notes on the ontogeny of the White Ibis, *Guara alba. Zoologica* 1:241–248.

Bernstein, N. P., and S. J. Maxson. 1981. Notes on moult and seasonably variable characters of the Antarctic Blue-eyed Shag *Phalacrocorax atriceps bransfieldensis. Notornis* 28:35–39.

Billard, R. S., and P. S. Humphrey. 1972. Molts and plumages of the Greater Scaup. *Journal of Wildlife Management* 36:765–774.

Bloom, P. H., and W. S. Clark. 2001. Molt and sequences of plumages of Golden Eagles and a technique for in-hand ageing. *North American Bird Bander* 26:97–116.

Boss, W. R. 1943. Hormonal determination of adult characters and sex behavior in Herring Gulls (*Larus argentatus*). *Journal of Experimental Zoology* 94:181–209.

Boyd, W. S., and J. R. Jehl, Jr. 1998. Estimating the abundance of Eared Grebes on Mono Lake, California, by aerial photography. *Colonial Waterbirds* 21:236–241.

Bradstreet, M. S. W. 1982. Pelagic feeding ecology of Dovekies, *Alle alle,* in Lancaster Sound and Western Baffin Bay. *Arctic* 35:126–140.

Bridge, E. S., A. W. Jones, and A. J. Baker. 2005. A phylogenetic framework for the terns (Sternini) inferred from mtDNA sequences: implications for taxonomy and plumage evolution. *Molecular Phylogenetics and Evolution* 35:459–469.

Bridge, E. S., and I. C. T. Nisbet. 2004. Wing molt and assortative mating in Common Terns: a test of the molt-signaling hypothesis. *Condor* 106:336–343.

Bridge, E. S., G. Voelker, C. W. Thompson, A. W. Jones, and A. J. Baker. 2007. Effects of size and migratory behavior on the evolution of wing molt in terns (Sternae): a phylogenetic comparative study. *Auk* 124:841–856.

Brown, R. E., and D. K. Saunders. 1998. Regulated changes in body mass and muscle mass in molting Blue-winged Teal for an early return to flight. *Canadian Journal of Zoology* 76:26–32.

Bull, E. L., and C. T. Collins. 1993. Vaux's Swift (*Chaetura vauxi*). No. 77 in The Birds of North America, ed. A. Poole and F. Gill. Philadelphia, PA: Birds of North America Inc.

Burtt, E. H., Jr. 1979. Tips on wings and other things. Pp. 75–110 in *The Behavioral Significance of Color,* ed. E. H. Burtt Jr. New York: Garland STPM Press.

———. 1986. *An Analysis of Physical, Physiological, and Optical Aspects of Avian Coloration with Emphasis on Wood-Warblers.* Ornithological Monographs no. 38. Washington, D.C.: American Ornithologists' Union.

Burtt, E. H., Jr., and J. M. Ichida. 1999. Occurrence of feather-degrading bacilli in the plumage of birds. *Auk* 116:364–372.

Butler, L. K., M. G. Donahue, and S. Rohwer. 2002. Molt-migration in Western Tanagers (*Piranga ludoviciana*): age effects, aerodynamics, and conservation implications. *Auk* 119:1010–1023.

Byrkjedal, I., and D. B. A. Thompson. 1998. *Tundra Plovers.* London: Poyser.

Cannell, P. F., J. D. Cherry, and K. C. Parkes. 1983. Variation and migration overlap in flight feather molt of the Rose-breasted Grosbeak. *Wilson Bulletin* 95:621–627.

Chandler, R. J., and J. H. Marchant. 2001. Waders with non-breeding plumage in the breeding season. *British Birds* 94:28–34.

Chapman, F. M. 1905. A contribution to the life history of the American Flamingo (*Phoenicopterus ruber*), with remarks upon specimens. *Bulletin of the American Museum of Natural History* 21:53–77.

Cimprich, D. A., and F. R. Moore. 1995. Gray Catbird (*Dumetella carolinensis*). No. 167 in The Birds of North America, ed. A. Poole and F. Gill. Philadelphia, PA: Birds of North America Inc.

Cink, C. L., and C. T. Collins. 2002. Chimney Swift (*Chaetura pelagica*). No. 646 in The Birds of North America, ed. A. Poole and F. Gill. Philadelphia, PA: Birds of North America Inc.

Collar, N. J. 2005. Family Turdidae (Thrushes). Pp. 514–807 in *Handbook of the Birds of the World,* vol. 10, ed. J. del Hoyo, A. Elliott, and D. A. Christie. Barcelona: Lynx Ediciones.

Combs, D. L., and L. H. Fredrickson. 1995. Molt chronology of male Mallards wintering in Missouri. *Wilson Bulletin* 107:359–365.

Conover, M. R., J. G. Reese, and A. D. Brown. 2000. Costs and benefits of subadult plumage in Mute Swans: testing hypotheses for the evolution of delayed plumage maturation. *American Naturalist* 156:193–200.

Cooper, J. 1985. Biology of the Bank Cormorant, part 2: morphometrics, plumage, bare parts, and moult. *Ostrich* 56:79–85.

Cramp, S., ed. 1985. *Handbook of the Birds of Europe, the Middle East, and North Africa.* Vol. 4. Oxford: Oxford University Press.

———. 1988. *Handbook of the Birds of Europe, the Middle East, and North Africa.* Vol. 5. Oxford: Oxford University Press.

———. 1992. *Handbook of the Birds of Europe, the Middle East, and North Africa.* Vol. 6. Oxford: Oxford University Press.

Cramp, S., and C. M. Perrins, eds. 1993. *Handbook of the Birds of Europe, the Middle East, and North Africa.* Vol. 7. Oxford: Oxford University Press.

———. 1994. *Handbook of the Birds of Europe, the Middle East, and North Africa.* Vol. 8. Oxford: Oxford University Press.

Cramp, S., and K. E. L. Simmons, eds. 1977. *Handbook of the Birds of Europe, the Middle East, and North Africa.* Vol. 1. Oxford: Oxford University Press.

———. 1980. *Handbook of the Birds of Europe, the Middle East, and North Africa.* Vol. 2. Oxford: Oxford University Press.

———. 1983. *Handbook of the Birds of Europe, the Middle East, and North Africa.* Vol. 3. Oxford: Oxford University Press.

Dawson, A. 2003. A detailed analysis of primary feather moult in the Common Starling *Sturnus vulgaris*—new feather mass increases at a constant rate. *Ibis* 145(2):E69–E76.

Dawson, A., S. A. Hinsley, P. N. Ferns, R. H. C. Bonser, and L. Eccleston. 2000. Rate of moult affects feather quality: a mechanism linking current reproductive effort to future survival. *Proceedings of the Royal Society, London, Series B.* 267:2093–2098.

de Boer, B. A. 1979. *Flamingos on Bonaire and in Venezuela.* Stinapa Documentation Series no. 3. Curaçao: Stichting National Park.

De Juana, E., F. Suárez, and P. G. Ryan. 2004. Family Alaudidae (Larks). Pp. 496–601 in *Handbook of the Birds of the World,* vol. 9, ed. J. del Hoyo, A. Elliott, and D. A. Christie. Barcelona: Lynx Ediciones.

DeKorte, J., and T. DeVries. 1978. Moult of primaries and rectrices in the Greater Frigatebird, *Fregata minor*, on Genovesa, Galápagos. *Bijdragen tot Dierkunde* 48:81–88.

Delacour, J. 1954. *The Waterfowl of the World.* Vol. 1. London: Country Life Ltd.

del Hoyo, J., A. Elliott, and J. Sargatal, eds. 1992–2007. *Handbook of the Birds of the World.* Vols. 1–12. Barcelona: Lynx Ediciones.

———. 1994. *Handbook of the Birds of the World.* Vol. 2. Barcelona: Lynx Ediciones.

Diamond, A. W. 1972. Sexual dimorphism in breeding cycles and unequal sex ratios in Magnificent Frigatebirds. *Ibis* 114:395–398.

———. 1975a. The biology of tropicbirds at Aldabra Atoll, Indian Ocean. *Auk* 92:16–39.

———. 1975b. Biology and behavior of frigatebirds *Fregata* spp. on Aldabra Atoll. *Ibis* 117:302–323.

Diamond, A. W., and E. A. Schreiber. 2002. Magnificent Frigatebird (*Fregata magnificens*). No. 601 in The Birds of North America, ed. A. Poole and F. Gill. Philadelphia, PA: Birds of North America Inc.

Dickey, D. R., and A. J. Van Rossem. 1938. The Birds of El Salvador. *Zoological Series, Field Museum of Natural History* 23:1–609.

Dietz, M. W., S. Daan, and D. Masman. 1992. Energy requirements for molt in the Kestrel *Falco tinnunculus*. *Physiological Zoology* 65:1217–1235.

Dittman, D. L., and S. W. Cardiff. 2009. The alternate plumage of the Ruby-throated Hummingbird. *Birding* 41(5):32–35.

Dittman, D. L., and D. K. Demcheck. 2006. Contributions to our knowledge of molt in the Calliope Hummingbird. *Birding* 38(6):32–40.

Dorsch, H. 1993. Zur Entwicklung der dritten Federgarnitur bei Jungvöglen einiger Passeres-Arten. *Vogelwarte* 37:19–25.

Dorward, D. F. 1962. Comparative biology of the White Booby and the Brown Booby *Sula* spp. at Ascension. *Ibis* 103b:174–220.

Dorward, D. F., and N. P. Ashmole. 1963. Notes on the biology of the Brown Noddy *Anous stolidus* on Ascension Island. *Ibis* 103b:447–457.

Dwight, J. 1900a. The moult of the North American Tetraonidae (quails, partridge, and grouse). *Auk* 17:34–51,143–166.

———. 1900b. The sequence of plumages and moults of the passerine birds of New York. *Annals of the New York Academy of Sciences* 13:73–360.

———. 1901. The sequence of moults and plumages of the Laridae (gulls and terns). *Auk* 18:49–63.

———. 1914. The moults and plumages of the scoters—genus *Oidemia*. *Auk* 31:293–308.

Dyck, J. 1979. Winter plumage of the Rock Ptarmigan: structure of the air-filled barbules and function of the white colour. *Dansk Ornitologisk Forenings Tidsskrift* 73:41–58.

Earnst, S. L. 1992. The timing of wing molt in Tundra Swans: energetic and non-energetic constraints. *Condor* 94:847–856.

Edwards, A. E., and S. Rohwer. 2005. Large-scale patterns of molt activation in the flight feathers of two albatross species. *Condor* 107:835–848.

Emslie, S. D., R. P. Henderson, and D. G. Ainley. 1990. Annual variation of primary molt with age and sex in Cassin's Auklet. *Auk* 107:689–695.

Erickson, R. C., and S. R. Derrickson. 1981. The Whooping Crane. Pp. 104–118 in *Crane Research Around the World,* ed. J. C. Lewis and H. Masatomi. Baraboo, WI: International Crane Foundation.

Espie, R. H. M., P. C. James, I. G. Warkentin, and L. W. Oliphant. 1996. Ecological correlates of molt in Merlins. *Auk* 113:363–369.

Ewert, D. N., and W. E. Lanyon. 1970. The first prebasic molt of the Common Yellowthroat (Parulidae). *Auk* 87:362–363.

Fernández, G., P. D. O'Hara, and D. B. Lank. 2004. Tropical and subtropical Western Sandpipers (*Calidris mauri*) differ in life history strategies. *Ornitologia Neotropical* 15 (supplement):385–394.

Filardi, C. E., and S. Rohwer. 2001. Life history implications of complete and incomplete primary molts in Pelagic Cormorants. *Condor* 103:555–569.

Flannery, M. E., and T. Gardali. 2000. Incomplete first prebasic molt in the Wrentit. *Western Birds* 31:249–251.

Folk, M. J., S. A. Nesbitt, J. M. Parker, M. G. Spalding, S. B. Baynes, and K. L. Candelora. 2008. Feather molt of nonmigratory Whooping Cranes (*Grus americana*) in Florida. *Proceedings of the North American Crane Workshop* 10:128–132.

Foster, M. S. 1967. Molt cycles of the Orange-crowned Warbler. *Condor* 69:169–200.

———. 1969a. The eggs of three species of mallophaga and their significance in ecological studies. *Journal of Parasitology* 55:453–456.

———. 1969b. Synchronized life cycles in the Orange-crowned Warbler and its mallophagan parasites. *Ecology* 50:315–323.

———. 1975. The overlap of molting and breeding in some tropical birds. *Condor* 77:304–314.

Friesen, V. L., A. J. Baker, and J. F. Piatt. 1996. Phylogenetic relationships within the Alcidae (Charadriiformes: Aves) inferred from total molecular evidence. *Molecular Biology and Evolution* 13:359–367.

Garrett, K. L., and J. B. Dunning, Jr. 2001. Parrots and Allies. Pp. 326–331 in *The Sibley Guide to Bird Life and Behavior,* ed. C. Elphick, J. B. Dunning, and D. A. Sibley. New York: Knopf.

Gates, R. J., D. F. Caithamer, T. C. Tacha, and C. R. Paine. 1993. The annual molt cycle of *Branta canadensis interior* in relation to nutrient reserve dynamics. *Condor* 95:680–693.

George, W. G. 1973. Molt of juvenile White-eyed Vireos. *Wilson Bulletin* 85:327–330.

Ginn, H. B., and D. S. Melville. 1983. *Moult in Birds.* British Trust for Ornithology guide no. 19. Tring, U.K.: British Trust for Ornithology.

Greenberg, R., T. Keeler-Wolf, and V. Keeler-Wolf. 1974. Wood warbler populations in the Yolla

Bolly Mountains of California. *Western Birds* 5:81–90.

Greenwood, H., P. J. Weatherhead, and R. D. Titman. 1983. A new age- and sex-specific molt schedule for the Red-winged Blackbird. *Condor* 85:104–105.

Greenwood, J. G. 1983. Post-nuptial moult in Dunlin. *Ibis* 125:223–228.

Greij, E. D. 1973. Effects of sex hormones on plumages of the Blue-winged Teal. *Auk* 90:533–551.

Grubb, T. C., Jr. 2006. *Ptilochronology: Feather Time and the Biology of Birds.* Oxford: Oxford University Press.

Grubb, T. C., Jr., and V. V. Pravosudov. 1994. Tufted Titmouse (*Parus bicolor*). No. 86 in The Birds of North America, ed. A. Poole and F. Gill. Philadelphia, PA: Birds of North America Inc.

Hackett, S. J., R. T. Kimball, R. Sushma, et al. 2008. A phylogenomic study of birds reveals their evolutionary history. *Science* 320:1763–1768.

Hamilton, T. H. 1961. On the functions and causes of sexual dimorphism in breeding plumage characters of North American species of warblers and orioles. *American Naturalist* 95:121–123.

Hardy, L. 2003. The peculiar puzzle of the pink Ring-billed Gulls. *Birding* 35:498–504.

Harrington, B. A., B. Winn, and S. C. Brown. 2007. Molt and body mass of Red Knots in the eastern United States. *Wilson Journal of Ornithology* 119:35–42.

Harris, M. P. 1973. The biology of the Waved Albatross *Diomedea irrorata* on Hood Island, Galapagos. *Ibis* 115:483–510.

Haukioja, E. 1971. Flightlessness in some moulting passerines in northern Europe. *Ornis Fennica* 48:101–117.

Hedenström, A., and S. Shunada. 1998. On the aerodynamics of moult gaps in birds. *Journal of Experimental Biology* 202:67–76.

Henny, C. J., R. A. Olson, and T. L. Fleming. 1985. Breeding chronology, molt, and measurements of *Accipiter* hawks in northeastern Oregon. *Journal of Field Ornithology* 56:97–112.

Herremans, M. 1999. Biannual complete moult in the Black-chested Prinia *Prinia flavescens. Ibis* 141:115–124.

Higgins, P. J., and S. J. J. F. Davies, eds. 1996. *Handbook of Australian, New Zealand, and Antarctic Birds.* Vol. 3. Melbourne: Oxford University Press.

Hill, G. E., and K. J. McGraw, eds. 2006. *Bird Coloration.* Vol. 1. Cambridge, MA: Harvard University Press.

Hockey, P. A. R., J. K. Turpie, and C. R. Velásquez. 1998. What selective pressures have driven the evolution of deferred northward migration by juvenile waders? *Journal of Avian Biology* 29: 325–330.

Hofmann, C. M., T. W. Cronin, and K. E. Omland. 2008. Evolution of sexual dichromatism, 1: convergent losses of elaborate female coloration in New World orioles (*Icterus* spp.). *Auk* 125:778–789.

Hohman, W. L. 1996. Prevalence of double wing molt in free-living Ruddy Ducks. *Southwestern Naturalist* 41:195–198.

Höhn, E. O., and C. E. Braun. 1980. Hormonal induction of feather pigmentation in ptarmigan. *Auk* 97:601–607.

Holder, K., and R. Montgomerie. 1993. Rock Ptarmigan (*Lagopus mutus*). No. 51 in The Birds of North America, ed. A. Poole and F. Gill. Philadelphia, PA: Birds of North America Inc.

Holmes, R. T. 1971. Latitudinal differences in breeding and molt schedules of Alaskan Red-backed Sandpipers (*Calidris alpina*). *Condor* 73:93–99.

Holthuijzen, A. M. A. 1990. Prey delivery, caching, and retrieval rates in nesting Prairie Falcons. *Condor* 92:475–484.

Höst, P. 1942. Effect of light on the moults and sequences of plumage in the Willow Ptarmigan. *Auk* 59:388–403.

Howell, S., and B. Patteson. 2007. Moult and Fea's Petrel identification. *Birding World* 20(5):201–202.

Howell, S. N. G. 1994. Magnificent and Great Frigatebirds in the Eastern Pacific: a new look at an old problem. *Birding* 26:400–415.

———. 1999. A basic understanding of moult: what, why, when, and how much? *Birders Journal* 8:296–300.

———. 2001a. A new look at moult in gulls. *Alula* 7:2–11.

———. 2001b. Feather bleaching in gulls. *Birders Journal* 10:198–208.

———. 2001c. Molt of the Ivory Gull. *Waterbirds* 24:438–442.

———. 2004. South Polar Skuas off California. *Birding World* 17(7):288–297.

———. 2005. Large skuas off California. *Birding World* 18(7):290–296.

———. 2006. Primary molt in the Black-footed Albatross. *Western Birds* 37:241–244.

———. 2007. A review of moult and ageing in jaegers (smaller skuas). *Alula* 13(3):98–113.

———. 2008. Moult of southern skuas. *Birding World* 21(1):28.

Howell, S. N. G., J. H. Barry, and J. E. Pike. 2006. Delayed preformative molt in the Barn Swallow. *Western Birds* 37:61–63.

Howell, S. N. G., and C. Corben. 2000a. Molt cycles and sequences in the Western Gull. *Western Birds* 31:38–49.

———. 2000b. A commentary on molt and plumage terminology: implications from the Western Gull. *Western Birds* 31:50–56.

———. 2000c. Retarded wing molt in Black-legged Kittiwakes. *Western Birds* 31:123–125.

Howell, S. N. G., C. Corben, P. Pyle, and D. I. Rogers. 2003. The first basic problem: a review of molt and plumage homologies. *Condor* 105:635–653.

———. 2004. The first basic problem revisited: reply to commentaries on Howell et al. (2003). *Condor* 106:206–210.

Howell, S. N. G., and J. L. Dunn. 2007. *A Reference Guide to Gulls of the Americas.* Peterson Reference Guide Series. Boston: Houghton Mifflin.

Howell, S. N. G., J. R. King, and C. Corben. 1999. First prebasic molt in Herring, Thayer's, and Glaucous-winged gulls. *Journal of Field Ornithology* 70:543–554.

Howell, S. N. G., and J. B. Patteson. 2008. Variation in the Black-capped Petrel—one species or more? *Alula* 14:70–83.

Howell, S. N. G., and P. Pyle. 2002. Ageing and molt in nonbreeding Black-bellied Plovers. *Western Birds* 33:268–270.

———. 2005. Molt, age determination, and identification of puffins. *Birding* 37(4):412–418.

Howell, S. N. G., and S. Webb. 1995. *A Guide to the Birds of Mexico and Northern Central America.* Oxford: Oxford University Press.

Howell, S. N. G., and C. Wood. 2004. First-cycle primary moult in Heermann's Gulls. *Birders Journal* 75:40–43.

Hudon, J., and A. H. Brush. 1990. Carotenoids produce flush in Elegant Tern plumage. *Condor* 92:798–801.

Humphrey, P. S., and Parkes, K. C. 1959. An approach to the study of molts and plumages. *Auk* 76:1–31.

———. 1963a. Plumages and systematics of the Whistling Heron (*Syrigma sibilatrix*). *Proceedings of the XIII International Ornithological Congress*: 84–90.

———. 1963b. Comments on the study of plumage sucession. *Auk* 80:496–503.

Imber, M. J. 1971. Filoplumes of petrels and shearwaters. *New Zealand Journal of Marine and Freshwater Research* 5:396–403.

Ingold, J. L., and G. E. Wallace. 1994. Ruby-crowned Kinglet (*Regulus calendula*). No. 119 in The Birds of North America, ed. A. Poole and F. Gill. Philadelphia, PA: Birds of North America Inc.

Jacobsen, E. E., Jr., C. M. White, and W. B. Emison. 1983. Molting adaptations of Rock Ptarmigan on Amchitka Island, Alaska. *Condor* 85:420–426.

James, D. J. 2004. Identification of Christmas Island, Great, and Lesser Frigatebirds. *Birding Asia* 1:22–38.

Jehl, J. R., Jr. 1987. Moult and moult migration in a transequatorially migrating shorebird: Wilson's Phalarope. *Ornis Scandinavica* 18:173–178.

Jenni, D. A. 1996. Family Jacanidae (Jacanas). Pp. 276–291 in *Handbook of the Birds of the World,* vol. 3., ed. J. del Hoyo, A. Elliott, and J. Sargatal. Barcelona: Lynx Ediciones.

Jenni, D. A., and T. R. Mace. 1999. Northern Jacana (*Jacana spinosa*). No. 467 in The Birds of North America, ed. A. Poole and F. Gill. Philadelphia, PA: Birds of North America Inc.

Jenni, L., and R. Winkler. 1994. *Moult and Ageing of European Passerines.* London: Academic Press.

Jiguet, F. 2007. Brown Skua in the North Atlantic. *Birding World* 20:327–333.

Johns, J. E. 1964. Testosterone-induced nuptial feathers in phalaropes. *Condor* 66:449–455.

Johnson, A., F. Cezilly, and V. Boy. 1993. Plumage development and maturation in the Greater Flamingo *Phoenicopterus ruber roseus. Ardea* 81:25–34.

Johnson, N. K. 1963. Comparative molt cycles in the Tyrannid genus *Empidonax. Proceedings of the XIII International Ornithological Congress:* 870–884.

Johnson, O. W. 1985. Timing of primary molt in first-year golden-plovers and some evolutionary implications. *Wilson Bulletin* 97:237–239.

Johnson, O. W., and P. G. Connors. 1996. American Golden-Plover (*Pluvialis dominica*) and Pacific Golden-Plover (*Pluvialis fulva*). Nos. 201–202 in The Birds of North America, ed. A. Poole and F. Gill. Philadelphia, PA: Birds of North America Inc.

Johnston, R. F. 1962. Precocious sexual competence in the Ground Dove. *Auk* 79:269–270.

Jongsomjit, D., S. L. Jones, T. Gardali, G. R. Geupel, and P. J. Gouse. 2007. *A Guide to Nestling Development and Aging in Altricial Passerines.* Biological technical publication, FWS/BTP-R6008-2007. Washington, D.C.: U.S. Department of the Interior, Fish and Wildlife Service.

Jukema, J., and T. Piersma. 1987. Special moult of breast and belly feathers during breeding in Golden Plovers *Pluvialis apricaria. Ornis Scandinavica* 18:157–162.

———. 2000. Contour feather moult of Ruffs *Philomachus pugnax* during northward migration, with notes on homology of nuptial plumages in scolopacid waders. *Ibis* 142:289–296.

———. 2006. Permanent female mimics in a lekking shorebird. *Biology Letters* 2:161–164.

Jukema, J., I. Tulp, and L. Bruinzeel. 2003. Differential moult patterns in relation to antipredator behavior during incubation in four tundra plovers. *Ibis* 145:270–276.

Kale, H. W., III. 1966. Plumages and molts in the Long-billed Marsh Wren. *Auk* 83:140–141.

Kear, J., and N. Duplaix-Hall, eds. 1975. *Flamingos.* Berkhamsted, England: Poyser.

Keitt, B. S., B. R. Tershey, and D. R. Croll. 2000. Dive depth and diet of the Black-vented Shearwater (*Puffinus opisthomelas*). *Auk* 117:507–510.

Keppie, D. M., and R. M. Whiting, Jr. 1994. American Woodcock (*Scolopax minor*). No. 100 in The Birds of North America, ed. A. Poole and F. Gill. Philadelphia, PA: Birds of North America Inc.

Kimball, R. T., and J. D. Ligon. 1999. Evolution of avian plumage dichromatism from a proximate perspective. *American Naturalist* 154:182–193.

Kinsky, F. C., and J. C. Yaldwyn. 1981. The bird fauna of Niue Island, southwest Pacific, with special notes on the White-tailed Tropicbird and Golden Plover. *National Museum of New Zealand Miscellaneous Series* 2:1–49.

Klicka, J., K. P. Johnson, and S. M. Lanyon. 2000. New World nine-primaried oscine relationships: constructing a mitochondrial DNA framework. *Auk* 117:321–326.

Knappen, P. 1932. Number of feathers on a duck. *Auk* 49:461.

Knopf, F. L. 1975. Schedule of presupplemental molt of White Pelicans with notes on the bill horn. *Condor* 77:356–359.

Komar, O. 1997. Communal roosting behavior of the Cave Swallow in El Salvador. *Wilson Bulletin* 109:332–337.

Kratter, A. W., et al. 2009. Fiftieth supplement to the American Ornithologists' Union *Check-list of North American Birds. Auk* 126:706–714.

Kushlan, J. A., and K. L. Bildstein. 1992. White Ibis (*Eudocimus albus*). No. 9 in The Birds of North America, ed. A. Poole, P. Stettenheim, and F. Gill. Philadelphia, PA: Birds of North America Inc.

Lande, R. 1980. Sexual dimorphism, sexual selection, and adaptation in polygenic characters. *Evolution* 34:292–305.

Langston, N. E., and N. Hilgarth. 1995. Molt varies with parasites in Laysan Albatrosses. *Proceedings of the Royal Society of London, Series B* 261:239–243.

Langston, N. E., and S. Rohwer. 1995. Unusual patterns of incomplete primary molt in Laysan and Black-footed albatrosses. *Condor* 97:1–19.

———. 1996. Molt-breeding tradeoffs in albatrosses: life history implications for big birds. *Oikos* 76:498–510.

Lenton, G. M. 1984. Molt of Malaysian Barn Owls *Tyto alba. Ibis* 126:188–197.

Leopold, A. S. 1939. Age determination in quail. *Journal of Wildlife Management* 3:261–265.

———. 1943. The molts of young wild and domestic turkeys. *Condor* 45:133–145.

LeValley, R., and P. Pyle. 2007. Notes on plumage maturation in the Red-tailed Tropicbird. *Western Birds* 38:306–310.

Ligon, J. D. 1968. The biology of the Elf Owl *Micrathene whitneyi. Miscellaneous publications of the Museum of Zoology, University of Michigan* 136:1–70.

Littlefield, C. D. 1970. Flightlessness in Sandhill Cranes. *Auk* 87:157.

Livezey, B. C. 1991. A phylogenetic analysis and classification of recent dabbling ducks (tribe Anatini) based on comparative morphology. *Auk* 108:471–507.

Lovette, I. J., and E. Bermingham. 2002. What is a wood-warbler? Molecular characterization of a monophyletic Parulidae. *Auk* 119:695–714.

Lovvorn, J. R., and J. A. Barzen. 1988. Molt in the annual cycle of Canvasbacks. *Auk* 105:543–552.

Lucas, A. M., and P. R. Stettenheim. 1972. Avian Anatomy—Integument. P. 1. Agriculture handbook no. 362. Washington, D.C.: U.S. Department of Agriculture.

Lyon, B. E., and R. D. Montgomerie. 1986. Delayed plumage maturation in passerine birds: reliable signaling by subordinate males? *Evolution* 40:605–615.

Lyon, D. L. 1962. Comparative growth and plumage development in *Coturnix* and Bobwhite. *Wilson Bulletin* 74:5–27.

Marchant, S., and P. J. Higgins, eds. 1990. *Handbook of Australian, New Zealand, and Antarctic Birds.* Vol. 1. Melbourne: Oxford University Press.

———. 1993. *Handbook of Australian, New Zealand, and Antarctic Birds.* Vol. 2. Melbourne: Oxford University Press.

Marín, M. 2003. Molt, plumage, body mass, and morphometrics of a population of the White-throated Swift in southern California. *Western Birds* 34:216–224.

Marks, J. S. 1993. Molt of Bristle-thighed Curlews in the northwestern Hawaiian Islands. *Auk* 110:573–587.

Marks, J. S., R. L. Redmond, P. Hendricks, R. B. Clapp, and R. E. Gill, Jr. 1990. Notes on longevity and flightlessness in Bristle-thighed Curlews. *Auk* 107:779–781.

Martínez, M. M. 1983. Nidificación de *Hirundo rustica erythrogaster* (Boddaert) en la Argentina (Aves, Hirundinidae). *Neotropica* 29:83–86.

Mayr, E., and M. Mayr. 1954. The tail molt of small owls. *Auk* 71:172–178.

McGraw, K. J., and G. E. Hill. 2000. Differential effects of endoparasitism on the expression of carotenoid- and melanin-based ornamental coloration. *Proceedings of the Royal Society of London, Series B* 267:1525–1531.

McGraw, K. J., E. A. Mackillop, J. Dale, and M. E. Hauber. 2002. Different colors reveal different information: how nutritional stress affects the expression of melanin- and structurally based ornamental plumage. *Journal of Experimental Biology* 205:3747–3755.

McIlhenny, E. A. 1937. Life history of the Boat-tailed Grackle. *Auk* 54:274–295.

McKnight, S. K., and G. Hepp. 1999. Molt chronology of American Coots in winter. *Condor* 101:893–897.

McNeil, R., M. T. Diaz, and A. Villenueve. 1994. The mystery of shorebird over-summering: a new hypothesis. *Ardea* 82:143–152.

Meijer, T. 1991. The effect of a period of food restriction on gonad size and moult of male and female Starlings *Sturnus vulgaris* under constant photoperiod. *Ibis* 133:80–84.

Mewaldt, L. R., and J. R. King. 1978. Latitudinal variation of postnuptial molt in Pacific coast White-crowned Sparrows. *Auk* 95:168–178.

Michener, H., and J. R. Michener. 1940. The molt of House Finches of the Pasadena region, California. *Condor* 42:140–153.

Michener, J. R. 1953. Molt and variations in plumage pattern of mockingbirds at Pasadena, California. *Condor* 55:75–89.

Middleton, A. L. A. 1986. Seasonal changes in plumage structure and body composition of the American Goldfinch, *Carduelis tristis*. *Canadian Field Naturalist* 100:545–549.

Miller, A. H. 1928. The molts of the Loggerhead Shrike *Lanius ludovicianus* Linnaeus. *University of California Publications in Zoology* 30:393–417.

———. 1931. Systematic revision and natural history of American shrikes (*Lanius*). *University of California Publications in Zoology* 38:11–242.

———. 1933. Postjuvenal molt and the appearance of sexual characters of plumage in *Phainoptila nitens*. *University of California Publications in Zoology* 38:425–446.

Miller, M. R. 1986. Molt chronology of Northern Pintails in California. *Journal of Wildlife Management* 50:57–64.

Minton, C. D. T., and L. Serra. 2001. Biometrics and moult of Grey Plovers, *Pluvialis squatarola*, in Australia. *Emu* 101:13–18.

Monteiro, L. R., and R. W. Furness. 1996. Molt of Cory's Shearwater during the breeding season. *Condor* 98:216–221.

Moum, T., S. Johansen, K. E. Erickstad, and J. F. Piatt. 1994. Phylogeny and evolution of the auks (subfamily Alcinae) based on mitochondrial DNA sequences. *Proceedings of the National Academy of Sciences* 91:7912–7916.

Mulvihill, R. S., K. C. Parkes, R. C. Leberman, and D. S. Wood. 1992. Evidence supporting a dietary basis for orange-tipped rectrices in the Cedar Waxwing. *Journal of Field Ornithology* 63:213–216.

Mulvihill, R. S., and C. C. Rimmer. 1997. Timing an extent of the molts of adult Red-eyed Vireos on their breeding and wintering grounds. *Condor* 99:73–82.

Murphy, M. E. 1999. Energetics and nutrition of moulting. Pp. 527–535 in *Proceedings of the 22nd International Ornithological Congress,* ed. N. J. Adams and R. H. Slotow. Johannesburg: BirdLife South Africa.

Murphy, M. E., and J. R. King. 1984. Dietary sulfur amino acid availability and molt dynamics in White-crowned Sparrows. *Auk* 101:164–167.

———. 1986. Composition and quantity of feather sheaths produced by White-crowned Sparrows during the postnuptial molt. *Auk* 103:822–825.

———. 1991. Nutritional aspects of avian moult. *Acta 20th Congressus Internationalis Ornithologici* 1990:2186–2193.

Murphy, M. E., J. R. King, and J. Lu. 1988. Malnutrition during the postnuptial molt of White-crowned Sparrows: feather growth and quality. *Canadian Journal of Zoology* 66:1403–1413.

Myers, J. P., J. L. Maron, and M. Sallaberry. 1995. Going to extremes: why do Sanderlings migrate to the Neotropics? Pp. 520–535 in *Neotropical Ornithology,* ed. P. A. Buckley, M. S. Foster, E. S. Morton, R. S. Ridgely, and F. G. Buckley. Ornithological Monographs no. 36. Washington, D.C.: American Ornithologists' Union.

Nelson, J. B. 1975. The breeding biology of frigatebirds: a comparative review. *Living Bird* 14:113–155.

Nesbitt, S. A. 1975. Feather staining in Florida Sandhill Cranes. *Florida Field Naturalist* 3:28–30.

Nesbitt, S. A., and S. T. Schwickert. 2005. Wing-molt patterns—a key to aging Sandhill Cranes. *Wildlife Society Bulletin* 33:326–331.

Newton, I. 1973. *Finches.* New York: Taplinger Publishing Company.

Nice, M. M. 1962. Development of behavior in precocial birds. *Transactions of the Linnaean Society of New York* 8:1–211.

Nolan, V., Jr., E. D. Ketterson, C. Ziegenfus, D. P. Cullen, and C. R. Chandler. 1992. Testosterone and avian life histories: effects of experimentally elevated testosterone on prebasic molt and survival in Dark-eyed Juncos. *Condor* 94:364–370.

Nunn, G. B., and S. E. Stanley. 1998. Body size effects and rates of Cytochrome b evolution in tube-nosed seabirds. *Molecular Biology and Evolution.* 15:1360–1371.

O'Briain, M., A. Reed, and S. MacDonald. 1998. Breeding, moulting, and site fidelity of Brant (*Branta bernicla*) on Bathurst and Seymour islands in the Canadian High Arctic. *Arctic* 51:350–360.

O'Hara, P. D., D. B. Lank, and F. S. Delgado. 2002. Is the timing of moult altered by migration? Evidence from a comparison of age and residency classes of Western Sandpipers *Calidris mauri* in Panamá. *Ardea* 90:61–70.

Olson, S. L. 1985. The fossil record of birds. Pp. 70–238 in *Avian Biology,* vol. 8, ed. D. S. Farner, J. R. King, and K. C. Parkes. New York: Academic Press.

Oring, L. W. 1968. Growth, molts, and plumages of the Gadwall. *Auk* 85:355–380.

Osorno, J. L. 1999. Offspring desertion in the Magnificent Frigatebird: are males facing a trade-off between current and future reproduction? *Journal of Avian Biology* 30:335–341.

Owre, O. T. 1967. *Adaptations for Locomotion and Feeding in the Anhinga and the Double-crested Cormorant.* Ornithlogical Monographs no. 6. Washington, D.C.: American Ornithologists' Union.

Palmer, R. S., ed. 1962. *Handbook of North American Birds.* Vol. 1. New Haven, CT: Yale University Press.

———. 1976. *Handbook of North American Birds.* Vol. 2 and 3. New Haven, CT: Yale University Press.

Parkes, K. C. 1967. Prealternate molt in the Summer Tanager. *Wilson Bulletin* 79:456–458.

Parmelee, D. F., H. A. Stephens, and R. H. Schmidt. 1967. The birds of southeastern Victoria Island and adjacent small islands. *National Museum of Canada Bulletin* 222.

Passmore, M. F. 1984. Reproduction by juvenile Common Ground-Doves in south Texas. *Wilson Bulletin* 96:241–248.

Paulson, D. R. 1983. Predator polymorphism and apostatic selection. *Evolution* 27:269–277.

———. 1993. *Shorebirds of the Pacific Northwest.* Seattle: University of Washington Press.

Paulus, S. L. 1984. Molts and plumages of Gadwalls in winter. *Auk* 101:887–889.

Pearson, D. J. 1984. The moult of the Little Stint *Calidris minuta* in the Kenyan Rift Valley. *Ibis* 126:1–13.

Pérez, G. E., and K. A. Hobson. 2006. Isotopic evaluation of interrupted molt in the northern breeding populations of the Loggerhead Shrike. *Condor* 108:877–886.

Petrides, G. A. 1945. First-winter plumages in the Galliformes. *Auk* 62:223–227.

Phillips, A. R. 1974. The first prebasic molt of the Yellow-breasted Chat. *Wilson Bulletin* 86:12–15.

———. 1977. Sex and age determination of Red Crossbills (*Loxia curvirostra*). *Bird-Banding* 48:110–117.

Piersma, T., and J. Jukema. 1993. Red breasts as honest signals of migratory quality in a long-distance migrant, the Bar-tailed Godwit. *Condor* 95:163–177.

Pitelka, F. A. 1958. Timing of molt in Steller Jays of the Queen Charlotte Islands, British Columbia. *Condor* 60:38–49.

Poole, A., and F. Gill, eds. 1992–2002. The Birds of North America. Nos. 1–716. Philadelphia, PA: The Birds of North America Inc.

Potts, G. R. 1971. Moult in the Shag *Phalacrocorax aristotelis*, and the ontogeny of the "*Staffelmauser.*" *Ibis* 113:298–305.

Prevost, Y. 1983. The moult of the Osprey *Pandion haliaetus*. *Ardea* 71:199–209.

Prince, P. A., S. Rodwell, M. Jones, and P. Rothery. 1993. Molt in Black-browed and Grey-headed Albatrosses *Diomedea melanophris* and *D. chrysostoma*. *Ibis* 135:121–131.

Prys-Jones, R. P. 1982. Molt and weight of some land-birds on Dominica, West Indies. *Journal of Field Ornithology* 53:352–362.

Pyle, P. 1995. Incomplete flight-feather molts and age in North American near-passerines. *North American Bird Bander* 20:15–26.

———. 1997a. Molt limits in North American passerines. *North American Bird Bander* 22:49–90.

———. 1997b. *Identification Guide to North American Birds.* P. 1. Bolinas, CA: Slate Creek Press.

———. 1997c. *Flight-feather Molt Patterns and Age in North American Owls.* Monographs in Field Ornithology no. 2. Colorado Springs: American Birding Association.

———. 1998. Eccentric first-year molt patterns in certain tyrannid flycatchers. *Western Birds* 29:29–35.

———. 2005a. Molts and plumages of ducks (Anatinae). *Waterbirds* 28:208–219.

———. 2005b. First-cycle molts in North American Falconiformes. *Journal of Raptor Research* 39:378–385.

———. 2005c. Remigial molt patterns in North American Falconiformes as related to age, sex, breeding status, and life-history strategies. *Condor* 107:823–834.

———. 2006. *Staffelmauser* and other adaptive strategies for wing molt in larger birds. *Western Birds* 37:179–185.

———. 2007. Revision of molt and plumage terminology in ptarmigan (Phasianidae: *Lagopus* spp.) based on evolutionary considerations. *Auk* 124:508–514.

———. 2008. *Identification Guide to North American Birds.* P. 2. Bolinas, CA: Slate Creek Press.

———. In press. Age determination and molt strategies in North American alcids. *Marine Ornithology*.

Pyle, P., and S. N. G. Howell. 1995. Flight-feather molt patterns and age in North American woodpeckers. *Journal of Field Ornithology* 66: 568–581.

———. 2004. Ornamental plume development and "prealternate molts" of herons and egrets. *Wilson Bulletin* 116:287–292.

Pyle, P., S. N. G. Howell, and G. M. Yanega. 1997. Molt, retained flight feathers, and age in North American hummingbirds. Pp. 155–166. In *The Era of Allan Phillips: a Festschrift,* comp. R. W. Dickerman. Albuquerque, NM: Horizon Communications.

Pyle, P., S. L. Jones, and J. M. Ruth. 2008. Molt and aging criteria for four North American grassland passerines. Biological technical publication FWS/BTP-R6011-2008. Washington, D.C.: U.S. Department of the Interior, Fish and Wildlife Service.

Pyle, P., A. McAndrews, Pilar Veléz, R. L. Wilkerston, R. B. Siegel, and D. F. DeSante. 2004. Molt patterns and age and sex determination of selected southeastern Cuban landbirds. *Journal of Field Ornithology* 75:136–145.

Pyle, P., and P. Unitt. 1998. Molt and plumage variation by age and sex in the California and Black-tailed gnatcatchers. *Western Birds* 29:280–289.

Raitt, R. J., Jr. 1961. Plumage development and molts of California Quail. *Condor* 63:294–303.

Rasmussen, P. C. 1988. Stepwise molt of remiges in Blue-eyed and King shags. *Condor* 90:220–227.

Richardson, D. M., and R. M. Kaminski. 1992. Diet restriction, diet quality, and prebasic molt in female Mallards. *Journal of Wildlife Management* 56:531–539.

Robb, M., K. Mullarney, and The Sound Approach. 2008. *Petrels Night and Day.* Poole, Dorset: The Sound Approach.

Robbins, M. B., and B. C. Dale. 1999. Sprague's Pipit (*Anthus spragueii*). No. 439 in The Birds of North America, ed. A. Poole and F. Gill. Philadelphia, PA: Birds of North America Inc.

Rogers, D. I. 2006. Hidden costs: challenges faced by migratory shorebirds living on intertidal flats. PhD diss., Charles Sturt University.

Rogers, D. I., P. Collins, R. E. Jessop, C. D. T. Minton, and C. J. Hassell. 2005. Gull-billed Terns in northwestern Australia: subspecies identification, moults, and behavioural notes. *Emu* 105:145–158.

Rohwer, F. C., and M. G. Anderson. 1988. Female-based philopatry, monogamy, and the timing of pair formation in migratory waterfowl. Pp. 187–221 in *Current Ornithology,* vol. 5, ed. R. F. Johnston. New York: Plenum.

Rohwer, S. 1986. A previously unknown plumage of first-year Indigo Buntings and theories of delayed plumage maturation. *Auk* 103:281–292.

———. 1999. Time constraints and moult-breeding trade-offs in large birds. Pp. 568–581 in *Proceedings of the 22nd International Ornithological Congress,* ed. N. J. Adams and R. H. Slotow. Johannesburg: BirdLife South Africa.

Rohwer, S., and G. S. Butcher. 1988. Winter versus summer explanations of delayed plumage maturation in temperate passerine birds. *American Naturalist* 131:556–572.

Rohwer, S., L. K. Butler, and D. Froehlich. 2005. Ecology and demography of east-west differences in molt scheduling of neotropical migrant passerines. Pp. 87–105 in *Birds of Two Worlds: The Ecology and Evolution of Migration,* ed. R. Greenberg and P. Marra. Baltimore, MD: Johns Hopkins University Press.

Rohwer, S., and A. E. Edwards. 2006. Reply to Howell on primary molt in albatrosses. *Western Birds* 37:245–248.

Rohwer, S., S. D. Fretwell, and D. M. Niles. 1980. Delayed maturation in passerine plumages and the deceptive acquisition of resources. *American Naturalist* 115:400–437.

Rohwer, S., and M. S. Johnson. 1992. Scheduling differences of molt and migration for Baltimore and Bullock's orioles persist in a common environment. *Condor* 94:992–994.

Rohwer, S., W. P. Klein, Jr., and S. Heard. 1983. Delayed plumage maturation and the presumed prealternate molt in American Redstarts. *Wilson Bulletin* 95:199–208.

Rohwer, S., and J. Manning. 1990. Differences in timing and number of molts for Baltimore and Bullock's orioles: implications to hybrid fitness and theories of delayed plumage maturation. *Condor* 92:125–140.

Rohwer, S., A. G. Navarro, and G. Voelker. 2007. Rates versus counts: fall molts of Lucy's Warblers (*Vermivora luciae*). *Auk* 124:806–814.

Rohwer, S. A. 1971. Molt and the annual cycle of the Chuck-Will's-Widow *Caprimulgus carolinensis. Auk* 88:485–519.

Rowley, I. 1988. Moult by the Galah *Cacatua roseicapilla* in the wheatbelt of Western Australia. *Corella* 12:109–112.

Rusch, D. H., S. Destefano, M. C. Reynolds, and D. Lauten. 2000. Ruffed Grouse (*Bonasa umbellus*). No. 515 in The Birds of North America, ed. A. Poole and F. Gill. Philadelphia, PA: Birds of North America Inc.

Salomonsen, F. 1939. Moults and sequence of plumage in the Rock Ptarmigan [*Lagopus mutus* (Montin.)]. *Videnskabelige Meddeleser fra Dansk Naturalhistorisk Forening* no. 103.

———. 1941. Mauser und Gefiederfolge der Eisente (*Clangula hyemalis* (L.)). *Journal für Ornithologie* 89:282–337.

———. 1949. Some notes on the moult of the Long-tailed Duck (*Clangula hyemalis*). *Avicultural Magazine* 55:59–62.

———. 1968. The moult migration. *Wildfowl* 19: 5–24.

Schmutz, J. K., and S. M. Schmutz. 1975. Primary molt in *Circus cyaneus* in relation to nest brood events. *Auk* 92:105–110.

Schreiber, R. W., and N. P. Ashmole. 1970. Sea-bird breeding seasons on Christmas Island, Pacific Ocean. *Ibis* 112:363–394.

Schreiber, R. W., E. A. Schreiber, D. W. Anderson, and D. W. Bradley. 1989. Plumages and molts of Brown Pelicans. *Natural History Museum of Los Angeles County Contributions in Science* 402:1–43.

Sealy, S. G. 1977. Wing molt of the Kittlitz's Murrelet. *Wilson Bulletin* 89:467–469.

Selander, R. K. 1958. Age determination in the Boat-tailed Grackle. *Condor* 60:355–376.

Selander, R. K., and D. R. Giller. 1960. First-year plumages of the Brown-headed Cowbird and Redwing Blackbird. *Condor* 62:202–214.

Serra, L. 2001. Duration of primary moult affects primary quality in Grey Plovers *Pluvialis squatarola*. *Journal of Avian Biology* 32:377–380.

Shannon, P. W. 2000. Plumages and molt patterns in captive Caribbean Flamingos. *Waterbirds* 23 (special publ. 1):160–172.

Shugart, G. W., and S. Rohwer. 1996. Serial descendent primary molt of *Staffelmauser* in Black-crowned Night-Herons. *Condor* 98:222–233.

Sibley, C. G. 1957. The evolutionary and taxonomic significance of sexual dimorphism and hybridization in birds. *Condor* 59:166–191.

Sibley, C. G., and J. E. Alquist. 1990. *Phylogeny and Classification of the Birds of the World*. New Haven, CT: Yale University Press.

Sibley, C. G., and B. L. Monroe, Jr. 1990. *Distribution and Taxonomy of Birds of the World*. New Haven, CT: Yale University Press.

Sibley, D. A. 2000. *The Sibley Guide to Birds*. New York: Knopf.

Sibley, F. C. 1970. Winter wing molt in the Western Grebe. *Condor* 72:373.

Siegfried, W. R. 1971. Plumage and moult of the Cattle Egret. *Ostrich*, supplement 9:153–164.

Skewes, J., C. Minton, and K. Rogers. 2004. Primary moult of the Ruddy Turnstone *Arenaria interpres* in Australia. *Stilt* 45:20–32.

Snow, D. W., and B. K. Snow. 1967. The breeding cycle of the Swallow-tailed Gull *Creagrus furcatus*. *Ibis* 109:14–24.

Snyder, N. F. R., E. V. Johnson, and D. A. Clendenen. 1987. Primary molt in California Condors. *Condor* 89:468–485.

Snyder, N. F. R., and K. Russell. 2002. Carolina Parakeet (*Conuropsis carolinensis*). No. 667 in The Birds of North America, ed. A. Poole and F. Gill. Philadelphia, PA: Birds of North America Inc.

Spear, L. B., and D. G. Ainley 1998. Morphological differences relative to ecological segregation in petrels (family: Procellariidae) of the Southern Ocean and Tropical Pacific. *Auk* 115:1017–1033.

Stangel, P. W. 1985. Incomplete first prebasic molt of Massachusetts House Finches. *Journal of Field Ornithology* 56:1–8.

Steenhof, K., and J. O. McKinley. 2006. Size dimorphism, molt status, and body mass variation of Prairie Falcons nesting in the Snake River Birds of Prey National Conservation Area. *Journal of Raptor Research* 40:71–75.

Stiles, F. G. 1980. The annual cycle of a tropical wet forest hummingbird community. *Ibis* 122:322–343.

———. 1995. Intraspecific and interspecific molt patterns of some tropical hummingbirds. *Auk* 112:118–132.

Stiles, F. G., and A. J. Negret. 1994. The nonbreeding distribution of the Black Swift: a clue from Colombia and unsolved problems. *Condor* 96:1091–1094.

Stiles, F. G., and L. L. Wolf. 1974. A possible circannual molt rhythm in a tropical hummingbird. *American Naturalist* 108:341–354.

Stokkan, K.-A. 1979. Testosterone and daylength-dependent development of comb size and breeding plumage in male Willow Ptarmigan (*Lagopus lagopus lagopus*). *Auk* 96:106–115.

Stonehouse, B. 1962. The Tropic Birds (Genus *Phaethon*) of Ascension Island. *Ibis* 103b:124–161.

Storer, R. W., and J. R. Jehl, Jr. 1985. Moult patterns and moult migration in the Black-necked Grebe *Podiceps nigricollis*. *Ornis Scandinavica* 16:253–260.

Storer, R. W., and G. L. Nuechterlein. 1985. An analysis of plumage and morphological characters of the two color forms of the Western Grebe (*Aechmophorus*). *Auk* 102:102–119.

Stout, B., and F. Cooke. 2003. Timing and location of wing molt in Horned, Red-necked, and Western grebes in North America. *Waterbirds* 26:88–93.

Strauch, J. G., Jr. 1985. The phylogeny of the alcidae. *Auk* 102:520–539.

Stresemann, E. 1963. Variation in the number of primaries. *Condor* 65:449–459.

Stresemann, E., and V. Stresemann. 1966. Die Mauser der Vogel. *Journal für Ornithologie* 107:1–447.

Stresemann, V. 1948. Eclipse plumage and nuptial plumage in the Old Squaw, or Long-tailed Duck (*Clangula hyemalis*). *Avicultural Magazine* 54:188–194.

Stutchbury, B. J., and S. Rohwer. 1990. Molt patterns in the Tree Swallow (*Tachycineta bicolor*). *Canadian Journal of Zoology* 68:1468–1472.

Stutchbury, B. J. M., and E. S. Morton. 2001. *Behavioral Ecology of Tropical Birds*. London: Academic Press.

Sullivan, B. L., and J. Liguori. 2009. Active flight feather molt in migrating North American raptors. *Birding* 41(4): 34–45.
Sullivan, J. O. 1965. "Flightlessness" in the Dipper. *Condor* 67:535–536.
Summers, R. W., L. G. Underhill, and R. P. Prys-Jones. 1995. Why do young waders in southern Africa delay their first return migration to the breeding grounds? *Ardea* 83:351–357.
Sutton, G. M. 1932. Notes on the molts and sequence of plumages in the Old-squaw. *Auk* 49:42–51.
———. 1935. The juvenal plumage and postjuvenal molt in several species of Michigan sparrows. *Cranbrook Institute of Science Bulletin* 3:1–36.
———. 1943. The wing molt of adult loons: a review of the evidence. *Wilson Bulletin* 55:145–150.
Swaddle, J. P., and M. S. Witter. 1997. Food availability and primary feather molt in European Starlings, *Sturnus vulgaris. Canadian Journal of Zoology* 75:948–953.
Tarvin, K. A., and G. E. Woolfenden. 1999. Blue Jay (*Cyanocitta cristata*). No. 469 in The Birds of North America, ed. A. Poole and F. Gill. Philadelphia, PA: Birds of North America Inc.
Taylor, W. K. 1970. Molts of the Verdin *Auriparus flaviceps. Condor* 72:493–496.
Telfair, R. C., II, and M. L. Morrison. 1995. Neotropic Cormorant (*Phalacrocorax brasilianus*). No. 137 in The Birds of North America, ed. A. Poole and F. Gill. Philadelphia, PA: Birds of North America Inc.
Thompson, B. C., and R. D. Slack. 1983. Molt-breeding overlap and timing of pre-basic molt in Texas Least Terns. *Journal of Field Ornithology* 54:187–190.
Thompson, C. W. 1991. The sequence of molts and plumages in Painted Buntings and implications for theories of delayed plumage maturation. *Condor* 93:209–235.
Thompson, C. W., and A. S. Kitaysky. 2004. Polymorphic flight-feather molt sequence in Tufted Puffins (*Fratercula cirrhata*): a rare phenomenon in birds. *Auk* 121:35–45.
Thompson, C. W., and M. Leu. 1994. Determining homology of molts and plumages to address evolutionary questions: a rejoinder regarding Emberizid finches. *Condor* 96:769–782.
Thompson, C. W., M. L. Wilson, E. F. Melvin, and J. D. Pierce. 1998. An unusual sequence of flight-feather molt in Common Murres and its evolutionary implications. *Auk* 115:653–669.
Tickell, W. L. N. 2000. *Albatrosses.* New Haven, CT: Yale University Press.
———. 2003. White plumage. *Waterbirds* 26:1–12.
Tomback, D. F. 1998. Clark's Nutcracker (*Nucifraga columbiana*). No. 331 in The Birds of North America, ed. A. Poole and F. Gill. Philadelphia, PA: Birds of North America Inc.
Tordoff, H. B. 1952. Notes on plumages, molts, and age variation of the Red Crossbill. *Condor* 54:200–203.
Trivelpiece, W. Z., and J. D. Ferraris. 1987. Notes on the behavioural ecology of the Magnificent Frigatebird *Fregata magnificens. Ibis* 129:168–174.
Tyler, S. J. 2004. Family Motacillidae (Pipits and Wagtails). Pp. 686–786 in *Handbook of the Birds of the World,* vol. 9, ed. J. del Hoyo, A. Elliott, and D. A. Christie. Barcelona: Lynx Ediciones.
Unitt, P. 2004. Effect of plumage wear on the identification of female Red-winged and Tricolored blackbirds. *Western Birds* 35:228–230.
Unitt, P., K. Messer, and M. Théry. 1996. Taxonomy of the Marsh Wren in southern California. *Proceedings of the San Diego Society of Natural History* 31.
Valle, C. A., T. de Vries, and C. Hernández. 2006. Plumage and sexual maturation in the Great Frigatebird *Fregata minor* in the Galapagos Islands. *Marine Ornithology* 34:51–59.
Van Rossem, A. J. 1925. Flight feathers as age indicators in *Dendragapus. Ibis* (series 12) 1:417–422.
———. 1936. Birds of the Charleston Mountains, Nevada. *Pacific Coast Avifauna* 24:1–145.
Veit, A. C., and I. L. Jones. 2004. Timing and patterns of growth of Red-tailed Tropicbird *Phaethon rubricauda* tail streamer ornaments. *Ibis* 146:355–359.
Voelker, G. 1997. The molt cycle of the Arctic Tern, with comments on aging criteria. *Journal of Field Ornithology* 68:400–412.
Voelker, G., and S. Rohwer. 1998. Contrasts in scheduling of molt and migration in eastern and western Warbling Vireos. *Auk* 115:142–155.
Voitkevich, A. A. 1966. *The Feathers and Plumage of Birds.* New York: October House.
Walkinshaw, L. H. 1949. The Sandhill Cranes. *Cranbrook Institute of Science Bulletin* 29:1–202.
Watson, A. 1973. Moults of wild Scottish ptarmigan, *Lagopus mutus,* in relation to sex, climate, and status. *Journal of Zoology* (London), series B, 171:207–223.
Watson, G. E. 1962. Molt, age determination, and annual cycle in the Cuban bobwhite. *Wilson Bulletin* 74:28–42.
Webster, J. D. 1942. Notes on the growth and plumages of the Black Oystercatcher. *Condor* 44:205–211.
———. 1958. Systematic notes on the Olive Warbler. *Auk* 75:469–473.
Weller, M. W. 1968. Notes on some Argentine anatids. *Wilson Bulletin* 80:189–212.
———. 1970. Additional notes on the plumages of the Redhead (*Aythya americana*). *Wilson Bulletin* 82:320–323.

Wetmore, A. 1936. The number of contour feathers in Passeriform and related birds. *Auk* 53: 159–169.
Williams, L. E., Jr. 1961. Notes on wing molt in the yearling Wild Turkey. *Journal of Wildlife Management* 25:439–440.
Williams, L. E., Jr., and R. D. McGuire. 1971. On prenuptial molt in the Wild Turkey. *Journal of Wildlife Management* 35:394–395.
Williamson, F. S. L. 1956. The molt and testis cycles of the Anna Hummingbird. *Condor* 58:342–356.
Williamson, K. 1975. The annual post-nuptial moult in the Wheatear. *Bird-Banding* 28:129–135.
Willoughby, E. J. 1966. Wing and tail molt of the Sparrow Hawk. *Auk* 83:201–206.
———. 1971. Biology of larks (aves: Alaudidae) in the central Namib Desert. *Zoologica Africana* 6:133–176.
———. 1986. An unusual sequence of molts and plumages in Cassin's and Bachman's sparrows. *Condor* 88:461–472.
———. 1991. Molt of the genus *Spizella* (Passeriformes, Emberizidae) in relation to ecological factors affecting plumage wear. *Proceedings of the Western Foundation of Vertebrate Zoology* 4:247–286.
———. 2007. Geographic variation in color, measurements, and molt of the Lesser Goldfinch in North America does not support subspecific designation. *Condor* 109:419–436.
Willoughby, E. J., M. Murphy, and H. L. Gorton. 2002. Molt, plumage abrasion, and color change in Lawrence's Goldfinch. *Wilson Bulletin* 114:380–392.
Winkler, H., and D. A. Christie. 2002. Family Picidae (Woodpeckers). Pp. 296–555 in *Handbook of the Birds of the World,* vol. 7, ed. J. del Hoyo, A. Elliott, and J. Sargatal. Barcelona: Lynx Ediciones.
Wiseman, A. J. 1977. Interrelation of variables in postjuvenal molt of Cardinals. *Bird-Banding* 48:206–223.
Wishart, R. A. 1985. Moult chronology of the American Wigeon, *Anas americana,* in relation to reproduction. *Canadian Field Naturalist* 99:172–178.
Wood, H. B. 1950. Growth bars in feathers. *Auk* 67:486–491.
Woodall, P. F. 2001. Family Alcedinidae (Kingfishers). Pp. 130–249 in *Handbook of the Birds of the World,* vol. 6, ed. J. del Hoyo, A. Elliott, and J. Sargatal. Barcelona: Lynx Ediciones.
Woolfenden, G. E. 1967. Selection for a delayed simultaneous wing molt in loons (Gaviidae). *Wilson Bulletin* 79:416–420.
Wyndham, E. 1981. Moult of Budgerigars *Melopsittacus undulatus. Ibis* 123:145–157.
Wyndham, E., J. L. Brereton, and R. J. S. Beeton. 1983. Moult and plumages of Eastern Rosellas *Platycercus eximius. Emu* 83:242–246.
Young, B. E. 1991. Annual molts and interruption of the fall migration for molting in Lazuli Buntings. *Condor* 93:236–250.
Yuri, T., and S. Rohwer. 1997. Molt and migration in the Northern Rough-winged Swallow. *Auk* 114: 249–262.
Zann, R. 1985. Slow continuous wing-moult of Zebra Finches *Poephila guttata* from southeast Australia. *Ibis* 127:184–196.
Zenatello, M., L. Serra, and N. Baccetti. 2002. Trade-offs among body mass and primary moult patterns in migrating Black Terns *Chlidonias niger. Ardea* 90:411–420.
Zusi, R. L. 1996. Family Rynchopidae (Skimmers). Pp. 668–677 in *Handbook of the Birds of the World,* vol. 3, ed. J. del Hoyo, A. Elliott, and J. Sargatal. Barcelona: Lynx Ediciones.

INDEX

Note: Page reference in **boldface** type refer to photographs.